Brian H. Kaye

Science and the Detective

Distribution:

VCH, P.O. Box 10 1161, D-69451 Weinheim, Federal Republic of Germany

Switzerland: VCH, P.O. Box, CH-4020 Basel, Switzerland

United Kingdom and Ireland: VCH, 8 Wellington Court, Cambridge CB1 1HZ, United Kingdom

USA and Canada: VCH, 220 East 23rd Street, New York, NY 10010–4606, USA

Japan: VCH, Eikow Building, 10-9 Hongo 1-chome, Bunkyo-ku, Tokyo 113, Japan

ISBN 3-527-29251-9 (hardcover)
ISBN 3-527-29252-7 (softcover)

Brian H. Kaye

Science and the Detective

Selected Reading in Forensic Science

Weinheim · New York
Basel · Cambridge · Tokyo

Professor Brian H. Kaye
Laurentian University
Ramsey Lake Road
Sudbury, Ontario P3E 2C6
Canada

Published jointly by
VCH Verlagsgesellschaft, Weinheim (Federal Republic of Germany)
VCH Publishers Inc., New York, NY (USA)

Editorial Directors: Dr. Peter Gregory, Dr. Ute Anton
Production Manager: Dipl.-Wirt.-Ing. (FH) Hans-Jochen Schmitt

Cover Illustration: © Bruce Bond's Artwork, 5809 Sugarbush Lane, Greendale, WI 53129, USA.
Cartoons on the cover pages of chapters 1, 7 and 9 by Philip Harms.

Library of Congress Card No. applied for

A catalogue record for this book is available from the British Library

Die Deutsche Bibliothek – CIP-Einheitsaufnahme
Kaye, Brian H.:
Science and the detective : selected reading in forensic science
/ Brian H. Kaye. – Weinheim ; New York ; Basel ; Cambridge ;
Tokyo : VCH, 1995
ISBN 3-527-29251-9 Pb.
ISBN 3-527-29252-7 brosch.

Printed on acid-free and chlorine-free paper

Composition: Hagedornsatz, D-68519 Viernheim
Printing and Bookbinding: Druckpartner Rübelmann, D-69495 Hemsbach

Printed in the Federal Republic of Germany

This book is dedicated to my two grandsons
Adrian and Alessandro Kaye

Biography

Dr. Brian Kaye was born in Hull, Yorkshire, England, in 1932. He obtained his B.Sc., M.Sc., and Ph.D. degrees from London University after studying at the University College of Hull, where he was a George Fredrick Grant Memorial Scholar. After working as a scientific officer at the British Atomic Weapons Research Establishment (Aldermaston) he taught physics at Nottingham Technical College from 1959 to 1963. He then moved to Chicago, where he was a Senior Physicist in the Chemistry Division of the IIT Research Institute (the Research Institute of the Illinois Institute of Technology). There he studied problems as different as why dirt sticks to the fibers of carpet to the design of better propellants for space rockets.

Since 1968 he has been Professor of Physics at Laurentian University in Sudbury, Ontario. He specializes in powder technology, which deals with the manufacture and properties of cosmetics, explosives, powdered metal pigments, drug powders, food powders, and abrasives. He has written a standard text on characterizing powders and authored over 100 scientific papers.

In 1977 his interest in the complex structure of soot involved him in the new subject of fractal geometry, an interest that led to the books "A Random Walk Through Fractal Dimensions," which is now in its second edition, and "Chaos & Complexity. Discovering the Surprising Patterns of Science and Technology," both published by VCH. The philosophical side of science has always interested him and has been complemented by his activities as a methodist local preacher in the Sudbury region of Ontario, Canada. He is just as likely to be found holding a service in a protestant church as he is to be lecturing on fractal geometry and chaos theory at the University.

Foreword

As an instructor of an undergraduate course in forensic anthropology, the wide appeal of forensic science to students has never ceased to amaze me. This book is the product of a distance-education course offered by Dr. Kaye through Laurentian University. His goal has always been to make forensic science accessible to all interested parties, especially to those lacking a science background. In my estimation Dr. Kaye has certainly succeeded. His book is the answer to anyone interested in whodunnits and the science behind the detection of fraud and other crimes.

It has been my pleasure and privilege to examine chapters of this book at various stages of development. The chapters offer a wide range of information on all aspects of forensic investigation. Although the book is geared for people without a science background, I found myself engrossed in even the smallest detail. Dr. Kaye has the uncanny ability to teach the derivation and history of scientific terminology and concepts without the "dryness" of academic papers. From the opening discussion of the term "forensic," to the discussion of drawing conclusions from human remains found in suspicious circumstances, the author presents information that is as interesting to the undergraduate student as it is to members of the legal community unfamiliar with many of the techniques of forensic analysis.

This begs the question, who would benefit from this book? Given the claim that this book is for non-scientists, it would certainly seem that members of the legal community needing a primer on topics such as drug detection, particle analysis, or even ballistics, could get the basics they need, as well as an introduction to the literature. Make no mistake, this book is not a definitive academic volume on forensic science; however, it is an excellent overview for laymen and interested professionals in other areas.

Regardless of your background, I am certain you will be both charmed and informed by this book, as I have been.

Scott I. Fairgrieve
Laurentian University

November, 1994

Preface

Several years ago, Laurentian University made it mandatory for non-science students to take an introductory science course. I felt that if students were required to take such a course, it should be one especially designed to show how modern science impinges on every aspect of their life. Accordingly, I started to offer a course which dealt with news stories breaking in the media, especially in general interest magazines. Teaching such a topical course proved to be a very demanding exercise and gradually we phased out the "hot off the press" aspect of the course and replaced the original curriculum with a series of courses that can function as the required elective for non-science majors and, at the same time, serve as a physics course for a liberal-science degree. (A liberal-science education is one that is rich in conceptual knowledge, providing a basis for understanding science without necessarily becoming a scientist.) A second book with similar aim, called "Golf Balls and Other Interesting Missiles," is in preparation.

As I taught the original general science course I found that many students did not have trouble with the concepts of science, provided they knew the vocabulary. I am firmly convinced, after teaching on and off for forty years, that pouring mathematics onto the heads of students before conceptual mastery has been achieved is very discouraging for the students. Therefore, in this book great care is taken to explain the vocabulary used, and to link new words to well-known words to aid the memory process. Mathematics are absent because the formulae are not needed at the level of exposition chosen. My experience is that many students will tackle more difficult texts once they have learned why it is useful to gain more advanced mastery of the subject. Given their educational objectives, why "Forensic Science?" First of all, I must confess that I am a detective-story addict. Sherlock Holmes and Agatha Christie were my first loves (although I think Agatha cheated in laying false trails). I think I have watched every Perry Mason episode (my favorite forensic expert) as well as Quincy, Matlock, Murder She Wrote, Columbo, and others. When I taught the topical course, students always wanted to know if the science used in last night's detective thriller was correct, and hence the structure for this book. Also, during my professional life I have occasionally participated in forensic work and I have had the privilege of knowing people such as McCrone and Dravnieks, who were always willing to discuss their work. I also worked with some of the pioneers in fingerprint processing and so I was able to combine my love of detective stories with my professional activities as a scientist.

As I worked on the text of the book it was necessary to restrict the topics so that a well-rounded work could be written in the time available. In selecting the topics presented, I did not wander too deeply into the gory details of bloody murder, since other books have covered the fleshy details of such work, and my stomach prefers the physical aspects of forensic science! (My daughter Sharon, who typed Chapter 11, suggests that you do not read it before a meal!)

I also chose topics of interest to arts majors, such as verifying the age of archaeological finds and the detection of art fraud. I included new studies of old cases, such as the Dreyfus case in France, the alleged murder of Napoleon, and the parentage of Anastasia, who claims to be a survivor of the Russian Revolution. I hope you will enjoy the exploration of forensic science presented in this book. I can assure you that I will continue to monitor the forensic sagas of the television screen so that future versions of this book will be as up-to-date as possible.

Brian H. Kaye
Professor of Physics
Laurentian University

November, 1994

Acknowledgments

I wish to acknowledge the encouragement given to me by my students who were eager to learn the art and science of modern detective work. I received encouragement and enthusiastic support to formalize my lecture notes into book form from my editors at VCH – Dr. Ute Anton and Dr. Peter Gregory. The editor in the Centre for Continuing Education at Laurentian University, Debra Beatty-Kelly, made many helpful suggestions as to the content and polished some of my Yorkshire English. As always, Garry Clark, my Research Associate, gave stirling support in diagram preparation, reference collection and he participated in many planning discussions. My two daughters Sharon and Alison helped with the typing and Julie Liimatainen worked on the script. Sonja Humphreys, one of my research students, undertook the difficult task of obtaining diagram copyright clearances. Scott Fairgrieve (the author of the foreword) read the script carefully and has encouraged me to keep working in the field of forensic science. Many thanks to all of the above and to the Director and Assistant Director of Continuing Education at Laurentian University, Marian Croft and Denis Mayer.

Brian H. Kaye
Professor of Physics
Laurentian University

November, 1994

Table of Contents

Wordfinder XVI

Chapter 1: Science and the Fight Against Crime 1

1.1 What is Forensic Science? 3
1.2 Misuse of Forensic Science 4
1.3 No Solicitors Please! 5
1.4 Expert Witnesses and Unique Evidence? 7

Chapter 2: Fingerprints 11

2.1 The History of Fingerprinting 13
2.2 Developing Latent Fingerprints 15
2.3 Storing and Using Fingerprint Images 34

Chapter 3: Footprints and Other Vestiges 45

3.1 Footprints, the Downfall of Idol Priests! 47
3.2 Recording and Developing Footprints at the Scene of the Crime 48
3.3 Tiresome Evidence 56
3.4 First and Last Impressions 66

Chapter 4: Fragmentary Evidence and Tell-Tale Dust 69

4.1 Locard's Principle 71
4.2 Pigmented and Refractive Fragments 72
4.3 Tell-Tale Dust 82
4.4 Hairs and Fibers 86
4.5 Who Killed Napoleon? 94
4.6 Fragmentary Evidence and the Saga of the Great Fiddle 96
4.7 Fragmentary Evidence Points to Modern Fabrication of Supposedly Antique Documents 98

Chapter 5: Bullets, Bombs and Body Armor 103

5.1 Gunshot Residues 105
5.2 Has the Suspect Handled Explosives Recently? 120

5.3 What Can Ballistics Tell Us? 122
5.4 Body Armor 127

Chapter 6: Electronic Ears and Eyes 133

6.1 Is That You Charlie? 135
6.2 Should We eat More Carrots? 143
6.3 Speed Reading 147
6.4 What About Those Watching Electronic Eyes? 150

Chapter 7: Bloodhounds Real and Synthetic 159

7.1 Dogged Detective Work 161
7.2 Why do Onions Make You cry? 162
7.3 Dog Detective Performances 167
7.4 Red Herrings and Work Fatigue 170
7.5 K.9 Synthetic Bloodhounds 171
7.6 Olfactronic Signatures of Individuals, Locations and Compounds 176

Chapter 8: Drunken Drivers, Drugged Individuals and Distracted Horses 181

8.1 Alcohol Abuse 183
8.2 Testing for Alcohol 187
8.3 Detecting and Analyzing Drugs 191
8.4 Tell-Tale Vestiges of Drug Usage in Hair 202
8.5 Were the Witches of Salem High on LSD? 205
8.6 Drugs and Athletic Performance 207
8.7 Drugs and Distracted Horses 210

Chapter 9: Poisoned Arrows and Dangerous Bulgarian Brollies 213

9.1 Poisoned Points 215
9.2 Omelettes, Flypapers and Weed Killers – A Look at how Arsenic has Been Used in the Commission of Crime 221
9.3 Bitter Almonds and Sudden Death 226
9.4 Curiosity (and a Poisonous House Plant) Killed the cat! 227
9.5 Was Van Gogh a Victim of Plant Poisoning? 228
9.6 Lethal Hamburgers and a new Wrinkle on Botulism 231
9.7 Seafood Zombies 233
9.8 Deadly Mushrooms and Killers from the Crypt 235
9.9 Crazy Cats and Mad Hatters 239
9.10 Poisonous Plates and Sweet Wine 244
9.11 The Synergistic Killers 251
9.12 Frye the Witness 260

Chapter 10: Forgery and Fraud 265

10.1 Are You Sincere? 267
10.2 Paper, the Forger's Nemesis? 269
10.3 Ink and Toner 276
10.4 Counterfeit Currency and Checks 287
10.5 False Stamps and Postal Fraud 292
10.6 Jurassic Adventure in the Land of Fraudulent Archaeological Specimens 299
10.7 Artistic Frauds 312
10.8 Fraud in the Marketplace 315
10.9 Checking up on Suspicious Signatures and Documents 318
10.10 Fraud in Science 326

Chapter 11: Bodies, Bones and Blood 333

11.1 Talkative Bodies 335
11.2 What Bones can Tell Us 338
11.3 "Your Brother's Blood is Crying out to Me from the Ground" .. 348

Chapter 12: Alphabet Soup and Genetic Fingerprinting 353

12.1 Coded Information for Building the Body 355
12.2 Constructing the Genetic Fingerprint 360
12.3 Using Genetic Fingerprints 363
12.4 Lies, Damn Lies and Statistics! 370

Author Index 373

Subject Index 375

Word Finder

A
AA 111
absinthe 230
accelerometer 318
acinetobacter calciacatieus 30
acoustic spectrogram 137
adenine 357
agaricus campestris 235
AIDS 366
alchemist 240
alkaline 163
alkaloids 194
alveoli 187
amanita 236
amber 299
amino acids 15
anabolic steroids 207
anaerobic 232
anaerobic bacteria 232
anatase 100
Ångström 113
anode 113
anthropologist 55
anti-radar devices 149
antibody 202
antigen 202
aperture 40
aphrodisiac 223
Apocrypha 47
archaeopteryx 311
argon-ion laser 23
arsenic 94
arsenic trioxide 221
asbestos 251
assassin 200
atom 30
atomic absorption (AA) 111
autopsy 91
autoradiography 30

B
ballistic signature 318
band shift 369
barbecue syndrome 231
barrister 6
beat frequency 148
biological toxin 220
biopsy 92
bit mapping 154
bitter almonds 226
black box 150
blue asbestos 251
boot polish 55
bootlegger 189
borenol 211
Botulism 232
brief 6

C
caffeine 209
calcite 66
calibre 124
camphor 211
cancer 360
cannabis 200
capillary action 279
capillary tubes 279
carbon dating 306
carbon monoxide 238
carrier gas 116
Cartesian geometry 224
castor bean plant 216
cathode 113
chlorophyll 235
chromatogram 116
chromatography 115
chromosomes 356
chronicle 63
cilia 163
cloning 357
codeine 193
coherent light 22
colchicine 227
colloid science 29
comparison microscope 123
coup poudres 234
criminalistics 1
criminology 1
crypt 47
cryptographer 47
crystal glass 79
curare 216
cytology 120
cytosine 357

D
denatured 221
deoxyribonucleic acid (DNA) 357
depression hysteresis 50
desorbed 176
diastema 343
diatoms 338
diffraction of light 36
diffraction patterns 35
diffusion of an odor 168
digitalis 229
DNA probes 362
Doppler effect 147
Doppler gun 147
dropsies 229

E
EDAX 97
EDXRF 75
electrode 18
electrostatic deposition analysis (ESDA) 67
electrostatics 16
eluant 115
eluted 115
energy dispersive x-ray fluorescence (EDXRF) 75
energy dispersive x-ray spectroscopy (EDAX) 97
entomologist, forensic 336
epidemiology 224
eraser 63
egot 205
ergot poisoning 207
ESDA 67
ethyl alcohol 184
etymology 268
Euclidean geometry 152
excipient 200
extender oils 56

F
fax machine 42
fluidized bed 173
fluoram 23
fluorescent 20
fly agaric 236
forensic anthropologist 260
forensic linguistics 324
forensic ondontologist 335
forensic science 1
foxglove 229
fractal dimension 152
fractal geometry 152
fractal transform 154
frequency spectrum 137
Frye procedure 260
fungi 92
fungicide 92
fuzz busters 149

G
gamma rays 93
gas chromatography (GC) 116
gas chromatograph 189
genes 357
genetic fingerprinting (profiling) 355
gold sol 29
Griess test 121
guanine 357

H
half-life of an isotope 306
hamburger disease 231
hard x-rays 75
hardware 13
harmonics 137
Hashish 200
hematologist 349
Henry System 14
high energy photons 93
high explosive 105
HIV 366
humidity 31
hysteresis 50

I
illuminate 21
immune system 202
immunoassay 202
incandescent 21
incoherent light 22
indenture 58
indentured servants 58
infrared radiation 20
inorganic compounds 239
instant verification ink 276
interference fringes 36
interference holography 50
ion 175
iron oxide 77
isotopes 31

J
Jekyll and Hyde 217
jequirity plant 217
Jurassic 299
Jurassic Park 299

K
keratin 91
kipper 170

L
lands 105
laser 21
latent fingerprints 15
lead oxide 77
lexicographer 7
liquid air 82
liquid chromatography (LC) 114
Locard's principle 71
logarithmic scale 217
long lived isotopes 92
low explosive 105
luminescence 21

M
masking agents 209
matrix 39
medulla 91
meniscus 280
Mercurochrome 53
mercury pollution 239
mercury vapor 241
methyl alcohol 184
Ménière's disease 229
microgram 76
micrograms (mg) 116
micrometer 19
micron 20
Minamata 239
mistletoe 217
mitochondria 365
molecular biology 357
monochromatic light 22
morphine 193
multichromatic light 22
mycology 207

N
nanograms (ng) 92, 116
nanometer 19
neural network 154
neutron activation analysis (NAA) 92
Ninhydrin test 28
nitrocellulose 29
nosocomial disease 236

O
opium 192
optical spatial filtering 40
optical variable device 290
optically variable ink (stamp) 297
organic compounds 239
osteons 339
OVD 290

P
palaeography 320
paradoxical 36
parallel processing 42
pathological 325
pathologist 325
PCR 365
peer 260
penetration pressure 128
perylene 63
pheromone 168
phonemes 141
phosphorescence 21
phosphors 21
photocell 37
photon activation analysis (PAA) 92
photon 93
pixel 151
plagiarism 329
plaster of Paris 48
polymerase chain reaction 365
pore patterns 25
pozzolana 97

probability scale 217
probenecid 209
pseudonym 346
pseudoephedrine 209
Pumice powder 66
pumice 66
pyrolysis gas chromatography 74
pyrotechnics 74

Q
quartz 79

R
radar 147
radio immune assay (RIA) 203
radio-carbon dating 306
radioactivity 31
radiochemistry 30
red herring 170
refracted 78
refractive index 78
restriction enzymes 360
revolver 124
ricin 216
rigor mortis 336
rock salt 20
rolled fingerprints 15
round of ammunition 124
roxarsone 222
rubber 63

S
saccharin 234
Scanning Electron Microscope (SEM) 73
Scheeles' green 95
Scotland Yard 8
sebaceous glands 15
serendipity 18
serologist 349
Shirley Institute 48
short life radioactive material 92
silica 79
silicon dioxide 79
sincere 268
sleuth 161
soft x-rays 75
software 13
solicitor 6
sound waves 135
spasm signature 318
spatial filtering 40
spectrum 19
stochastic variable 58
stochastic wear signatures 56
Stradivari 96
striations 58
subdural hematoma 346
synchronous fluorescence 63
synergism 258
syphilis 228

T
TAGA 175
TEM pictures 73
templates, speech 141
teratologist 234
tetrodotoxin 233
The Old Bailey 292
therapeutic dosage window 216
thermoluminescence testing 307
thin layer paper chromatography (TLPC) 115
thymine 357
titanium dioxide 66
toadstool 235
toner 67
toner 282
toxicology 215
toxin 215
toxophiles 215
trace atmospheric gas analyzer 175
transmission electron micrographs (TEM) 73
transmutation 31
tribology 18
truffle 236
trimethoxy arsenic 95
tuning fork 135

U
ultraviolet radiation 20
urea 15

V
Van Gogh 228
venereal disease 228
verotoxigenic coli 231
vestige 55
Vinland map 98
Visuprint System 28
voice lineup 142
volatile 163
Voodoo 234
VTEC 231
Vucetich System 14

W
whistle blowing 327
white arsenic 221
white asbestos 251
whiting 66

Chapter 1

Science and the Fight Against Crime

Chapter 1

Science and the Fight Against Crime

1.1 What is Forensic Science?

Judging by the number of popular television shows in which a master detective fights the war against crime, many people are fascinated by the solving of crimes. Increasingly science is involved in unraveling crimes both in television dramas and in real-life situations. The application of science in the solving of mysteries and crime is known by two different names. One is **forensic science**, the other is **criminalistics.**[1,2] The term forensic comes from the Latin word *forum*, which means "the market-place." This is because justice in Roman society was administered in the market-place. We have continued to use the name forum for a public speaking place, and we have adopted forensics as the name for anything related to the administration of justice. The term criminalistics is obviously derived from the word crime. A crime is technically described as an offense against a person or society. **Criminology** is the study of why criminals have committed crimes and how they should be rehabilitated. In this book, we use the term forensic science to describe the use of science to solve mysteries and fight crime.*

Apart from the general fascination with detective stories and television shows based upon the fight against crime, it is becoming necessary for citizens to know more about forensic science as crime increasingly impinges directly on everyday life. The possibility of serving as a member of a jury and the resultant need to assess scientific evidence are also reasons why one might need a basic knowledge of forensics.

---

* In the United States the American Academy of Forensic Science has a section called criminalistics. According to Paterson, criminalistics includes evidence such as bloodstains, glass, and soil, which are to be identified and interpreted.[3] Other areas of forensic science include forensic pathology, forensic toxicology, forensic anthropology, forensic dentistry (also known as forensic ondontology), questioned document examination, voice analysis, polygraphs, fingerprinting, and ballistics.

1.2 Misuse of Forensic Science

In the late 1980s and early 1990s there were several spectacular cases where earlier convictions had to be reversed by courts of appeal because of the misuse of forensic science. One of the most important cases where forensic evidence was found to be incorrectly presented and interpreted in a court of law was a case in Great Britain known as "The Saga of the Maguire Seven". This case involved the conviction of seven people in connection with the terrorism associated with the crisis in Northern Ireland.[4] In this case the alleged terrorists were convicted of using illegal explosives. Later, evidence appeared to indicate that samples tested by forensic experts had been contaminated by the scientists carrying out the tests. This contamination of the evidence lead to false evidence being submitted indicating that the seven people accused of the crime had handled explosives. (We shall review this case in detail in Chapter 5 of this book.)

In a review article assessing the impact of this type of spectacular miscarriage of justice, Hamer gives some examples of how increased awareness by the defense counsel of the value of forensic evidence presented at a trial has played an important role in the acquittal of accused persons.[4] These cases occurred in Great Britain but could have taken place elsewhere. In the first case mentioned in the article by Hamer, John Priss, a truck driver, was convicted in 1973 of raping and murdering a woman. At his trial, the most important forensic evidence centered on the fact that some group A blood cells were found in semen stains on the woman's pants. Priss was a blood group A person, but the forensic specialist who gave evidence at the trial did not state that the murdered woman was also a blood group A person. This evidence did not come out at the original trial because the forensic specialist did not voluntarily bring the information to the attention of the jury, and was not asked for it by the defense counsel. Priss was freed on appeal because important information had been omitted at the original trial.

In another case, involving armed robbery, fibers found in a stolen getaway car matched those of the suspect's trousers. Forensic experts for the defense asked if any tests had been carried out on trousers of the man who owned the car that had been stolen and used by the robbers. It turned out that the clothing fibres of the owner of the car also matched those found in the car. Therefore, the fiber evidence was rendered meaningless and the accused was acquitted. Caddy, the head of one of the Masters of Science programs in forensic science at a British university, feels that a course in forensic science should be part of every lawyer's training.[2,4] Hamer supports Caddy's claim:

lawyers frequently admit that they mentally switch off for the forensic evidence at a trial and that barristers avoid asking detailed questions of forensic witnesses for fear of asking one question too many. The fear of forensic science amongst lawyers runs from bottom to top of the legal profession.[4]

Currently forensic science tends to be a specialty subject taught at the graduate level, but there is an obvious need for an introduction to acquaint the citizens and the lawyers with the potential usefulness and inherent limitations of scientific evidence presented in court. This book is intended as a textbook for such an introductory course for liberal arts and science students. It does not aim at training people to become forensic scientists, but hopefully, after reading the book, readers will have an appreciation of the potential and the limits of scientific evidence in the fight against crime.

1.3 No Solicitors Please!

When attempting to bring those presenting scientific evidence and those studying such evidence to a common understanding of the significance of evidence in a trial, a major problem is the mutual obscurity of the language used by specialists from different fields. Thus the lawyer may be overwhelmed by scientific technical language, but in the same way the scientist attempting to give evidence may be frustrated by the special procedures and language of the legal profession. Those who become involved with forensic science need to develop an understanding of the specialized vocabulary used by both lawyers and scientists. However, one of the problems encountered when attempting to understand legal and scientific vocabulary in a particular situation or trial is that the vocabulary used both in the law profession and, to some extent, in science, varies from one country to another, as do the structures of the legal system. For example, when describing the role of forensic science in a trial, Hamer makes the following statements:

Britain has an adversarial system of justice, which aims to establish the truth by probing the strength of a prosecution and defense case. Europe has an inquisitorial system of justice. In Europe an investigating judge appoints a forensic scientist to carry out tests independent of the prosecution.[4]

Caddy tells us:

one of the problems of the inquisitorial system is that the judge may not appoint a forensic scientist who would have the right expertise.

He goes on to say that:

> *it is not a perfect system, I would rather have the adversarial system provided it is properly supported.*[4]

The difference between these two basic systems will become apparent as we explore various case histories in the different chapters.

The title of this section draws attention to the problem caused by different terminology even in the English-speaking world. When I came to live in North America I was surprised to see signs on some public buildings that said "no solicitors or hawkers allowed".

In Great Britain, a **solicitor** is a lawyer who advises clients and prepares necessary information (called a **brief**) for another lawyer who specializes in presenting information at a trial. Readers familiar with the television show "Rumpole of the Bailey" will be familiar with this difference between these two types of lawyers. Very often a client is referred to Rumpole by a solicitor who prepares the background documentation for Rumpole to use at the trial. Rumpole is what is known as a **barrister**. In Canada lawyers call themselves barristers and solicitors and, indeed, lawyers here perform both roles in the delivery of service. In the United States, the term solicitor is not used to describe a member of the legal profession, but rather one who tries to obtain money for some good cause or individual. Sometimes such solicitors can be very persistent. For this reason, they are often banned from carrying out their activities in a public building. Canadians also use the word solicitor in this sense. The different uses of the word solicitor in British and North American English arise from the fact that the word solicitor comes from the Latin word *sollicito* meaning "to arouse [emotions]." In the legal system of the United Kingdom, a solicitor came to mean the person who asked for information or asked for mercy at a trial, but in North American English, it came to mean one who asks for support for a cause or an individual. Both usages are correct; however, the British legal expert visiting North America would be very surprised to see the sign "no solicitors or hawkers."

Rather than keep interrupting the discussion in the textbook with lengthy explanations of the differences among attorneys, solicitors, barristers, and so on, important new words are printed in boldface type and defined when they first appear in the text. These words and the location of the definitions are then listed in the Word Finder located at the front of the book.

1.4 Expert Witnesses and Unique Evidence?

When either the prosecution or the defense wishes to introduce an expert witness during the course of a trial, an essential step is to establish the credentials of the witness and have the person accepted as an expert by both the prosecution and the defense. Establishing the necessary credentials of an expert can sometimes become almost trivial. For instance, a few years ago a colleague of mine was asked to help investigate the possible nuisance being created by noise at a gravel quarry. Until that time, my colleague had never been involved in such investigations, but the Department of Physics at Laurentian University had a meter for measuring noise levels and he knew how to operate the equipment. He therefore went out to the site of the quarry and took measurements with the meter. At the trial the fact that my colleague knew how to read the meter and was a Professor of physics was sufficient in the minds of both the prosecution and the defense for my colleague to be accepted as an expert witness. Afterwards, on the strength of his initial involvement with the noise problem at the gravel quarry, he became involved in several trials on industrial-noise nuisance.

This straightforward acceptance of his expert status contrasts with my experience when I acted as a witness for the U. S. Internal Revenue Service in a tax case. The problem involved a study of the pulverization process, which the defendants claimed was part of a mining process and not a process subsequent to the mining of sand. (The sand was being pulverized to make abrasives for cleaning powders.) The key issue hinged upon the meaning of the word pulverization as it was defined in 1930, when the legislation was written by the American Congress. The trial was fairly lengthy and complex. A friend of mine happened to overhear some discussions in a cafeteria among lawyers and experts for the defense about the evidence that I was known to be preparing for the prosecution. My friend heard that my expertise was going to be challenge on the grounds that although I was a powder expert, I was a physicist who therefore could not be considered as an expert in pulverization, which is an engineering problem. What the defense lawyers didn't know was that for several years I had been involved in writing entries for technical dictionaries. The expert who writes entries for dictionaries is called a **lexicographer**. Since the key issue in the trial was the meaning of the word pulverization, by presenting me not as a physicist but as a lexicographer, the prosecuting lawyer was able to outflank the forthcoming objection to my expertise by the defense lawyers. Unprepared for such a description of the expert, the defending lawyers had to let the expert status go unchallenged, and in evidence we were able to pursue the meaning of the word pulverization both from a lexicographic and a scientific perspective. Originally, the prosecuting lawyers had approached me to act as an expert witness on the strength of the fact that I had written the article on techniques for measuring the size of the powders in a standard reference book – Perry's Handbook of Chemical Engineering.[5] In their view this fact by itself would establish my expert witness credentials in court. The possible tactics of the opposing lawyers caused a shift in the way in which my credentials as an expert witness were presented and accepted in court.

When expert witnesses present evidence in a courtroom situation, they must not only be able to demonstrate the significance of the evidence they are presenting, but also the uniqueness of that information.[6,7] Very often police collect evidence that they know is proof that a certain person has been involved in a crime, but which they are unable to use in court because the defense counsel will be able to ask questions about the uniqueness of the evidence presented. This is demonstrated by an incident in my own scientific career, which nearly involved me as an expert witness in a murder crime. In 1963, I was contacted by an inspector of Scotland Yard to discuss the characterization of clay found on the body of a murder victim. The inspector had read an account in some technical literature of an instrument I had developed for measuring the size distribution of tiny particles of clay.[8] These are used in the coating of paper to give it both a glossy appearance and the physical properties required for high quality color pictures, like those found in magazines such as Time and Discover. **Scotland Yard** is the location of the Metropolitan Police Headquarters in London. Specialists of Scotland Yard are often called in by other police forces in Great Britain to help in the solving of a crime.[9]

The inspector was convinced that the victim's body had been moved from one location to another and that clay from the area where the body was found did not match the clay on the trousers. I was asked by Scotland Yard to look into the sizing of the clay fine particles found on the victim and those in the sample of clay from the suspected murder site. After several discussions, it was decided that, indeed, the instrument I had developed could be used for such a study. However if the evidence on the size distribution of the fragments of clay was presented in court, a good defense counsel would immediately challenge the specialist concerning the uniqueness of the clay found at the two different locations being studied. Therefore, establishing uniqueness would involve measuring the size distribution of clay material from hundreds if not thousands of different locations in Great Britain. It was decided not to pursue this line of investigation even though the inspector himself was convinced that size analysis of the clay would establish the difference between the two locations. A few years after my discussions with Scotland Yard, I was pleased to discover that scientists at the British Home Office Central Research Office (the British organization concerned with the developments of forensic science technology) were attempting to set up a reference bank of size distribution of soils so that evidence of the kind I had discussed with the inspector from Scotland Yard could be used in court.[10,11]

In the discussions of forensic studies contained in this book, we attempt to study the uniqueness of the evidence and the problems associated with any ambiguity in the interpretations of this evidence.

In a subject as vast as forensic science it is not possible to discuss all of the different subjects in detail; therefore, to some extent the actual topics covered in this book reflect my own experiences and interests in the applications of the physical sciences to problems in forensic science. For this reason, perhaps the discussion of the biological aspects of forensic science is a little skimpy for some readers' preferences. Those readers are referred to existing texts on forensic medicine and pathology in forensic investigations.[12–14]

As the technology of forensic science has developed, some scientists have re-examined famous historic problems of criminal investigations and looked for possible new explanations of historic events. For instance, optical-computer specialists have re-examined the handwriting specimens from the Dreyfus case (see Chapter 10) while other specialists have analyzed the famous persecution of the Witches of Salem near Boston and feel these outbursts could actually have been triggered by L. S. D. poisoning from a fungus growing on rye plants (see Chapter 8). Some historic forensic investigations will be revisited as we undertake our travels through the technology of forensic science.

References

1. G. Davies (Ed.), *Forensic Science. ACS Symposium Series 13*, American Chemical Society, Washington DC, 1975.
2. Educational opportunities in forensic science are discussed in Reference 1 and in J. L. Paterson (Ed.), *Forensic Science: Scientific Investigation in Criminal Justice*, AMS Press, New York, 1975.
3. J. L. Paterson, "The team Approach to Forensic Science," in W. J. Curran, A. L. McGarry, C. S. Petty (Eds.), *Modern Medicine. Psychiatry and Forensic Science*, F. A. Davies Co., Philadelphia, PA, 1980, 991.
4. M. Hamer, "Forensic Science Goes on Trial," *New Scientist*, 9 November 1991, 30.
5. R. H. Perry, C. H. Chilton (Eds.), *Chemical Engineering Handbook*, Fifth Edition, McGraw-Hill, New York, 1973.
6. P. J. Neufeld, N. Colman, "When Science Takes the Witness Stand," *Scientific American 262* (1990), 46.
7. For a discussion of the way evidence can be presented to a jury see the latter part of K. Gold, "If All Fails, Read the Instructions," *New Scientist*, 13 June 1992, 38.
8. The instrument that was developed for measuring the size distribution of clay fine particles is discussed in B. H. Kaye, *Direct Characterization of Fine particles*, Wiley, New York, 1981.
9. J. Broad, *Science and Criminal Detection*, Macmillan, London, 1988.
10. R. J. Dudley, "The Particle Size Analysis of Soils and Its Use in Forensic Science – The Determination of Particle Size Distribution in the Silt and Sand Fractions," *Journal of Forensic Science Society 16* (1977), 219.

11. R. J. Dudley, K. W. Smalldon, "A comparison of Distributional Shapes with Particular Reference to a Problem in Forensic Science," *International Statistical Review 46* (1978), 53.
12. C. J. Polsten, *The Scientific Aspects of Forensic Medicine*, Contemporary Science Paperbacks Number 40, Oliver and Boyd, Edinburgh, 1969.
13. C. Wilson, *Written in Blood: A History of Forensic Detection*, Thorsons, London, 1988.
14. C. Joyce, E. Stover, *Witnesses from the Grave*, Bloomsbury, London, 1966.

Chapter 2

Fingerprints

Chapter 2

Fingerprints

2.1 The History of Fingerprinting

In many good detective stories and in real life, fingerprints left behind at the scene of a crime constitute important evidence.[1] The availability of intelligent computers to search out and recognize stored fingerprints is revolutionizing some of the techniques used to fight crime. However, work still needs to be done on the development of hidden fingerprints left at the scene of the crime. This fact is illustrated by the job advertisement shown in Figure 2.1. The term **hardware** used in this advertisement refers to instruments, as distinct from **software**, which is computer jargon for the program and instructions built into a computer. Thus, the physical instrument used to transfer the image to the computer is referred to as hardware and the instructions for searching out and recognizing the fingerprint from stored records constitutes the software of the system. In this chapter, we will take a comprehensive look at the modern techniques being developed for making hidden fingerprints visible and for transferring developed images to a computer for the recognition processing.[2,3]

We know that thousands of years ago the Chinese recognized that the impression of a fingerprint on a document was a unique signature. In the western world, one of the first scientists to recognize the importance of fingerprint patterns was Malpighi, Professor of anatomy at the University of Bologna in the 1680s. The English engraver and author Thomas Beswick (1753–1828) made wood engravings of the patterns of his own fingerprints and used them as trademarks in the books that he published. In 1823, Dr. J. E. Purkinje, Professor of anatomy at the University of Breslau, published a treatise commenting on the diversified, rich patterns present in fingerprints. He classified them into general pattern types.

The first discussion of how fingerprints might be used to identify individuals was published in the English scientific journal *Nature* in 1880, by Dr. Henry Faulds of the Tsukiji Hospital in Tokyo, Japan.[4] This missionary doctor observed the wide diversity amongst individual fingerprints and the fact that these patterns remained unchanged throughout life. This latter fact was an important step in the recognition that fingerprints can be used to identify an individual.

POLICE SCIENTIFIC DEVELOPMENT BRANCH
FINGERPRINT RESEARCH GROUP

SCIENTIFIC OFFICER

to £19,378

Home Office

A vacancy has arisen for a specialist chemist to join a small team involved with the development and introduction of a range of reagents, processes and hardware for the detection of fingerprints in Police fingerprint laboratories and at scenes of crime.

The work involves some fundamental research and taking concepts through from laboratory formulation, or prototype, testing, to evaluation and implementation in police forces throughout the U.K.

The post is located at the Police Scientific Development Branch Laboratories outside St Albans, Hertfordshire.

Qualifications
Applicants should have a first or second class honours degree in chemistry. Recent graduates will be considered, but knowledge or experience of surface chemistry, organic synthesis and hardware development would be an advantage.

Salary will be in the range £13,449 to £19,378.

The Home Office is an equal opportunities employer and welcomes applications from suitably qualified people regardless of race, sex, disability or marital status.

Figure 2.1. Developing techniques for making the structure of a fingerprint visible to the investigating officer is still an important aspect of research in forensic science technology. (Used by permission of the Police Scientific Development Branch, Home Office, U. K.)

Sir William James Herschel was probably the first police officer to use fingerprints for identifying people. Sir William was a British official in Bengal, India. After he read the original publication by Dr. Faulds, he wrote a letter to the publisher of the journal *Nature*, noting that for the last twenty years he had used a system of fingerprinting to identify government pensioners as well as prisoners.[5] The first known extensive collection of fingerprint records dates back to Sir Francis Galton. In the late nineteenth century, Galton made an extensive study of the way in which various biological features are inherited. As a result of his work, the British government set up a committee to consider the advisability of adopting fingerprinting as a method of identifying criminals. One of the original members of the committee was Sir Edward Richard Henry. Later, he became the director of Scotland Yard. The committee studying the implementation of fingerprinting criminals finished their report in 1901, and the system of pattern classification adopted as a result of this study is often known as the "Henry System." In Spanish-speaking countries, fingerprint patterns are classified based on a system developed by the Argentinean scientist Juan Vucetich. Scientists in English-speaking countries claim that the **Henry System** is simpler and more comprehensive than the **Vucetich System**.

In the United States, New York state authorities began to fingerprint prisoners in 1903. On November 2, 1904, the warden of the U. S. penitentiary of Levenworth, Kansas, was authorized to take fingerprints of federal prisoners. By the early 1970s, it was estimated that the U. S. security authorities had over 200 million fingerprints on file.

In the early days of the use of fingerprinting to combat crime, people were naturally suspicious that there could be several individuals with the same fingerprints. However, as the use of fingerprinting developed, it was demonstrated experimentally that amongst the millions of fingerprints studied in the world, no two are exactly the same. Even identical twins have different fingerprints. Biologists have estimated that, if we take into account the entire world population, we would have to go through several generations before a repeat finger pattern could be discovered. In Figure 2.2(a), a typical fingerprint record card used to record fingerprint impressions by a police force in the United States is shown.[6] These prints have been made by rolling the thumb (or other finger) on a black ink pad and then pressing the finger onto a white card. Such fingerprints are known as **rolled fingerprints.** When fingerprints are inadvertently left at the scene of a crime by the criminal, they are not usually visible to the investigating officer. Until they are made visible, they are described as **latent fingerprints**. In Latin, the word *latent* means "hidden." A very important stage in any police investigation is to treat latent fingerprints so that they become visible. In the next section, we will look at the various techniques for developing latent fingerprints.

2.2 Developing Latent Fingerprints

If we were to look at a fresh fingerprint made on a glass surface through a microscope, we would see that the fingerprint on the surface is made up of minute droplets of various sizes. The droplets range in size from 1 to 20 micrometers. The chemical content per square millimeter of an average fingerprint smear is shown in Table 2.1. The meaning of the various chemical terms given in this table are also summarized in the table.[3] The freshly formed fingerprint contains approximately one millionth of a gram of material and consists mainly of sweat (99 % water, sodium and potassium chloride.) There are also trace amounts of **amino acids** and **urea** plus fatty substances transferred from other parts of the body to the fingers. (There are no **sebaceous** glands, which give out fatty substances, on the hands.) Figure 2.2 (b) shows a few examples of developed fingerprints. To make latent fingerprints visible, very fine powder is brushed lightly over the surface believed to be carrying the fingerprints. On many surfaces the grains of the powder stick to the liquid droplets forming the fingerprint. On some surfa-

Constituent	Quantity per mm^2
Chlorides	> 10 µg
Amino acids	10 - 100 µg
Urea	~ 1 µg
Ammonia	< 0.5 µg
Sebum	5 - 100 µg

µ is a symbol meaning one millionth (10^{-6}).
µg is the symbol for one millionth of a gram also called a microgram.

Amino acids are the chemical substances, all containing nitrogen, which the body uses to make proteins.

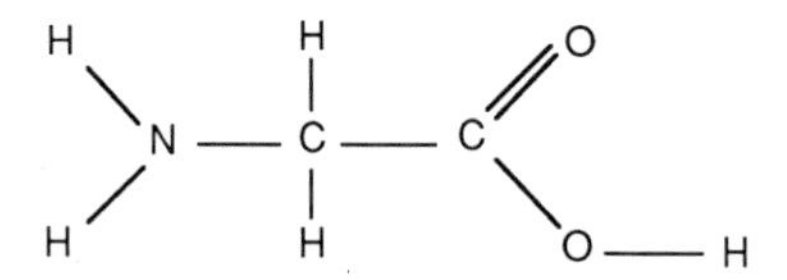

Urea is the main product of protein metabolism. It is excreted by the body in urine. (Metabolism - the chemical and physical changes that take place within a functioning body and enable it to grow and/or continue operating.)

Ammonia is a chemical made from hydrogen and nitrogen and has the formula

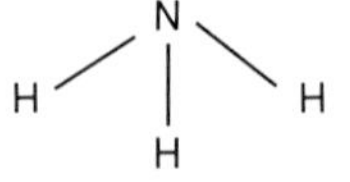

Sebum is the oily substance secreted by the sebaceous glands and reaching the surface of the skin through small ducts that lead to hair roots. Sebum provides a thin film of fat over the skin which slows the evaporation of water and has an antibacterial effect.

In the chemical formulae above
H : Hydrogen N : Nitrogen C : Carbon O : Oxygen

Table 2.1. The chemical constituents of a human fingerprint.

ces, however, the droplets of the fingerprint are absorbed into the surface carrying the prints, and the powder does not stick to the ridges of the fingerprint.

In some situations the adherence of the powder grains to the structure of the fingerprint can be enhanced using what is known as electrostatic forces. **Electrostatics** is the study of the behavior of material carrying electric charges on its surface, which are not free to move about the surface. For the purpose of understanding how electrostatic technology can aid in the development of a latent fingerprint, it is sufficient to recall the high school introduction to electrostatics in which a rubber rod is rubbed with a piece of fur and then used to attract pieces of paper. (The same type of electrostatic charge is involved in the party trick of rubbing a balloon against your hair and then sticking the balloon to the wall.)

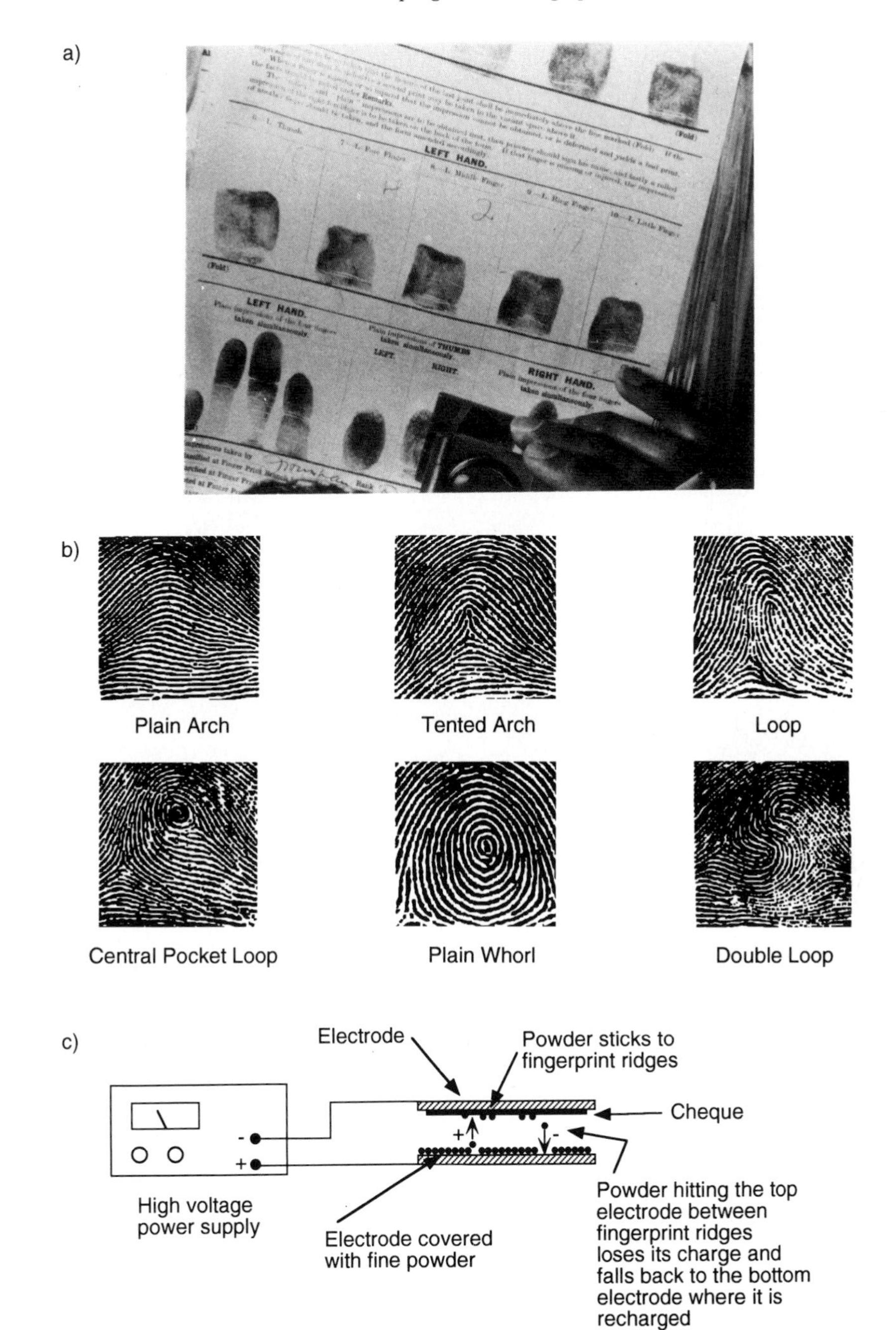

Figure 2.2. Fingerprints can often be detected at the scene of a crime and are used to track down and convict criminals. a) Typical fingerprint card. b) Various structures found in fingerprints. c) A fingerprint can be made visible with electrostatically charged powder.[7]

You can carry out an electrostatic forces experiment for yourself by combing your hair vigorously with a plastic comb. The friction between the hair and the plastic surface of the comb results in the transfer of electric charges between the two materials, and in this charged state the comb will pick up small pieces of paper from the surface of a desk.

The generation of electrostatic charge by rubbing is described as **tribology**. This word is taken from the Greek word meaning "to rub." Tribology, the study of what happens when two surfaces are rubbed together, is an important branch of modern physics and engineering. We will have several occasions in this book to discuss the transfer of electric charge, and thus fragments of material, from one surface to another when surfaces are rubbed together.

To develop a fingerprint with the help of electrostatics, the system shown in Figure 2.2(c), can be used.[7] The item that one wishes to examine for possible fingerprints, such as a forged check, is placed upon a flat metal plate connected to the terminal of a source of high voltage. (Scientists measure electrical potential in volts.) For the system shown in Figure 2.2(c), 10,000 volts is sufficient. The flat plate connected to the source of high voltage is described as an **electrode**. (This word simply means a roadway for electricity. It is derived from two Greek words: *electra*, the word for amber, the first material known to exhibit electrostatic behavior; and the Greek word for road, *hodos*.) The lower metal electrode of the system of Figure 2.2(c) connected to the positive terminal of the power supply is covered with a fine powder. These grains of powder become positively charged, and, as a consequence, they are driven towards the negative electrode to which the specimen under investigation is attached. As a result of the high voltage driving the grains of powder towards the electrode which carries the specimen, the grains hit the object with considerable speed. On the surface of the specimen, which contains chemicals from the ridges of the fingerprint pattern, the arriving dust fine particles stick strongly to the pattern of the fingerprint. Those fine particles hitting the gaps between the ridges lose their original electric charge to the electrode. They then become oppositely charged and are quickly repelled back to the base electrode where the process begins again. After a short time, the developing fingerprint can be clearly seen outlined by the fine particles stuck to the ridges of the print.

Some people would say that the next method of developing fingerprints, laser luminescence, was discovered accidentally. I would prefer to say that the development of this technique is an example of serendipity. The dictionary describes **serendipity** as "the gift for finding valuable or agreeable things not sought after." The word was created by Horace Walpole in a letter to his friend in 1754. He coined the word from the details of a story where someone discovered an unexpectedly beautiful princesses in a place called Serendip (an old name for what is now known as Sri Lanka).

The development of laser luminescence came about because of a serendipitous encounter in a Toronto suburb between an Ontario police officer and a scientist of Xerox Canada Limited. The scientist from Xerox was complaining to his neighbor that in the equipment he was using to develop xerographic copies, fingerprints on some of the internal parts of the machine would fluoresce when illuminated by laser light, and that this fluorescence interfered with the process of making copies of various documents. Immediately the police officer realized that what was a problem for the scientist from Xerox was an opportunity for those who need to see hidden fingerprints. Out of this chance conversation grew a new technique for developing latent fingerprints called laser luminescence.[8,9]

To understand how laser luminescence makes fingerprints visible we will need to develop some concepts of applied optics. Light energy is described by the scientist as electromagnetic radiation. We do not need to understand electromagnetism, except to know that the energy in electromagnetic radiation travels in pulses and that the distance between pulses in the radiant energy is described as the wavelength of the radiation. The fact that white light is a mixture of many different colors was discovered by the famous British physicists Sir Isaac Newton (1642–1727). In 1672, he described the way in which a glass prism can spread out white light into what he called a spectrum of light. This is shown in Figure 2.3. He coined the word **spectrum**, from the Latin word for image or apparition, to describe how the prism had made the various colors in white light appear on a screen.

Red light in the visible spectrum has a wavelength of the order of 700 nanometers. The unit **nanometer** (nm) is one billionth of a meter (10^{-9} m). To gain a physical appreciation for something as small as 700 nanometers, it should be noted that a course black hair has a width of 100,000 nanometers. Another unit of measurement that we will have occasion to use in our exploration of forensic science is the micrometer (μm). The **micrometer** (or

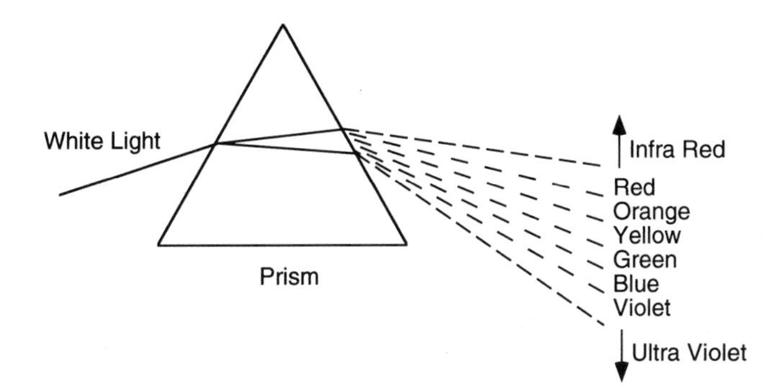

Figure 2.3. A prism will refract (bend) a light beam to display a spectrum of color. Each wavelength of light is bent by a different degree by the glass of the prism thus spreading out the colors and forming the spectrum.

micron) is one millionth of a meter (10^{-6} m). Using this unit of measurement we say that the wavelength of red light is 0.7 micrometers. Normally, the human eye cannot see an object smaller than 30 micrometers without the use of an optical instrument such as a microscope. The skin is sensitive to grit larger than 30 micrometers.

Newton distinguished seven clear colors in the spectrum: red, orange, yellow, green, blue, indigo, and violet. We say that ordinary glass is transparent to the colors of the visible spectrum because a glass prism will spread out white light into these colors, forming a continuous spectrum of visible light. If, however, we were to investigate the energy arriving at the earth with a prism made out of rock salt, we would find invisible electromagnetic energy below the red end of the visible spectrum. (Ordinary table salt, sodium chloride, can occur in very large crystals known as **rock salt**. Lenses and prisms can be made from such crystals.) Infra means "below." Therefore scientists call radiant energy longer in wavelength than red light **infrared radiation**.

If we used a prism made out of quartz, we would find quite a lot of energy beyond the violet end of the spectrum. Scientists call this type of radiation **ultraviolet radiation**. Ultra means "above." (It is the ultraviolet radiation from the sun which causes the skin to tan and can cause skin cancer after repeated, prolonged exposure.) When we take a journey in a car, the ordinary glass windows of the car block the ultraviolet light so that we do not get a suntan while riding in the vehicle. Ultraviolet lamps intended for tanning without going out into the sun have a quartz window which lets ultraviolet radiation through. In popular speech, ultraviolet light is sometimes called "black light" because lamps popularly used to cause certain materials to "glow in the dark" have a very dark, nearly black coating on them and give off very little "visible light." Theater groups such as The Famous People Players make their props and puppets visible by using special paints, which give off visible light when exposed to the black light illuminating the stage. The people operating the puppets remain unseen by the audience because they wear black suits and hoods, which absorb whatever visible light which may be present, rendering them "invisible." In forensic science, ultraviolet light is widely used to inspect suspected forgeries (see Chapter 10).

If we direct radiation at a substance, in practice some of the radiation is absorbed. This energy absorbed will stimulate the molecules of the material and they will then give out a different type of radiation as they go back to their so-called ground state. The radiation emitted is classified as either fluorescent or phosphorescent, depending on whether the emission of the secondary radiation stops immediately or continues after the primary "stimulating" radiation is switched off. If the emission of the secondary radiation stops immediately after the stimulating radiation is switched off, the material is described as being **fluorescent**. This name comes from the fact that one of the first chemical to give out light when stimulated with radiation was a substance known as a fluoride (a compound of the gas fluorine).

If a substance goes on emitting secondary light after the stimulating source is removed, the effect is called **phosphorescence**. If phosphorescence continues for a considerable time after the radiating source is extinguished, the effect can be easily distinguished from fluorescence. In other cases, the phosphorescence may only last for a fraction of a second, and when this happens it is hardly worth trying to distinguish between the two types of behavior. The screen of a television set is covered with powders known as **phosphors**, which give out light when the electron beam scanning behind the screen hits the phosphor-coated surface. The phosphors on the screen are designed to glow for about the same period of time as a complete scan of the electron beam. When a television set has been turned off in a dark room, one can see the glowing of the screen for some time. Whiteners added to washing powders are also phosphors: They absorb ultraviolet light and give out light in the blue part of the visible spectrum. If one switches off the room light when someone is dressed in a white shirt that has been washed in washing powder containing whiteners, the shirt appears to float across the room as a bluish-white ghost. The word phosphor comes from two Greek words meaning "carrier of light." Phosphorus, the chemical, was given its name because it glows in the dark. The word *lumen* in Latin means "light." When we **illuminate** a room we throw light upon it. In recent times, because of the confusion in the exact meaning of the words phosphorescence and fluorescence, scientists have started to use the word **luminescence**, which means "giver of light," without indicating for how long light is given out when the stimulation stops. Most diamonds placed under ultraviolet lamps fluoresce vividly while an imitation stone will not fluoresce at all. This test can be used in the detection of imitation jewelry. Many substances around us in everyday life fluoresce when illuminated with ultraviolet light. Traces of these substances, which are picked up by the fingers cause fingerprints to glow when illuminated with the appropriate form of laser light.

To explain what we mean by the "appropriate form" of laser light, we need to understand the difference between light from a laser light source and an ordinary incandescent light bulb. The ordinary light bulb, which has a wire filament, gives out light because the central wire is heated to a high temperature by an electric current. It is said to be **incandescent**.

A laser light gives out a special type of light by a process similar to that by which an ultraviolet whitener gives out blue light when it absorbs radiant ultraviolet light. The **laser** was invented in the early 1960s. Its name comes from the initial letters of the phrase, "**l**ight **a**mplification by **s**timulated **e**mission of **r**adiation." When a laser is operated, radiant energy is fed into the center of the laser and then, when certain conditions are reached, many of the molecules in the laser material suddenly emit radiation at the same time. Furthermore, the pulses of the energy coming from the laser light source are

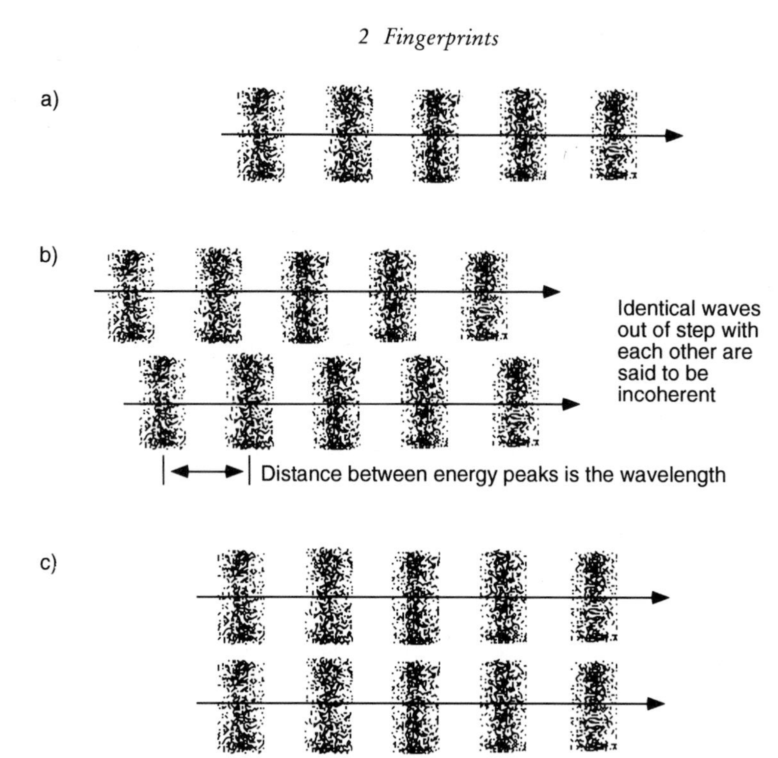

Figure 2.4. Some characteristic properties of light. a) Light is made up of pulsing electromagnetic energy. b) Light of one wavelength is called monochromatic. Identical waves out of step with each other are said to be incoherent. c) Light waves in step with each other are said to be coherent.

synchronized with each other. As explained in Figure 2.4, when the light emitted from an energy source has only one wavelength, the light is said to be **monochromatic**. (This takes its name from the Greek words meaning "one color.") Light from an incandescent bulb has light of many wavelengths and is said to be **multichromatic**. The original experiment was carried out by Newton, who spread out white light from the sun into its constituent colors and showed that sun light is multichromatic. The sun is in fact an incandescent source of light. Its surface temperature is of the order of 6,000 degrees centigrade. In ordinary monochromatic light, the pulses of light energy are not in step with each other and the light is said to be **incoherent**. (See Figure 2.4(b)) In laser light, because of the way in which the bursts of energy are released, all the different pulses of light are in step with each other and the light coming out of the laser is said to be **coherent light** (See Figure 2.4(c)). In technical terms, laser light differs from light from an incandescent source because it is monochromatic, coherent, and very powerful. Also laser light does not spread out. We will find that in some applications in forensic science, it is the *power* of the laser that we utilize. In other situations, it is the *monochromatic* nature of the light and in yet others it is the *coherence* of the light-beam energy.[10,11]

The scientist from Xerox, whom we discussed earlier, discovered that the very powerful light from a laser was stimulating the emission of light from the fingerprint by a process called fluorescence. The fluorescent compounds in the fingerprint had been picked up by fingers in the course of everyday life. For example in Canada, mail is sorted by robots that read a pattern of fluorescent bars representing the postal code printed on the envelopes. Anyone handling used envelopes picks up this fluorescent material on their fingers.

In subsequent co-operative research between the Ontario provincial police and scientists from Xerox, equipment was developed, which used the light from an argon-ion laser to make the fingerprints left on a surface fluoresce. Different types of lasers often take their name from the gas or substance filling the center of the laser system. In this case, we know that the gas argon was being used in the laser. Another widely-used relatively inexpensive laser is a helium-neon laser, which uses these two gases instead of argon. In Figure 2.5, the basic equipment and the optical phenomena involved in the development of a latent fingerprint by laser luminescence are shown. The lines on the graph of Figure 2.5(b) show the wavelengths of light that can be absorbed by both the fluorescent substances in the fingerprint and the fluorescent radiation, which is given out by the stimulated fluorescent material. The **argon-ion laser** used emits significant energy at seven different wavelengths, as indicated by the vertical lines in the diagram. The camera used to photograph the fluorescent fingerprints is fitted with an optical filter that cuts out the light given out directly by the laser. Thus only the fluorescent energy is recorded on the photographic film of the camera.

The laser excitation of fluorescence in a fingerprint can be used to develop latent prints on many surfaces, which cannot be studied using the powder and brush technique. For instance, the laser excitation technique can develop fingerprints on plastic and rubber surfaces, painted walls, cloth, wood, and metals. In tests carried out by Carey and MacClement at the National Research Council in Ottawa, it was found that fingerprint florescence comes from daily contact with such items as motor oil, paints, and inks, rather than from any natural substance given out by the skin. These scientists also found that the laser excitation of a fingerprint pattern could be enhanced if the surface to be examined was sprayed with a chemical called **fluoram**. This is because a fingerprint always contains traces of stable amino acids, not visible under the laser alone (see Table 2.1). (Fluoram is a proprietary name for a commercially secret chemical. The name comes from the two words fluorescent and amino.) This compound forms new compounds with the amino acids, which then fluoresce. Using this enhancement technique, clear fingerprints could be found using laser luminescence on book pages and documents up to ten years after the fingerprint was created.

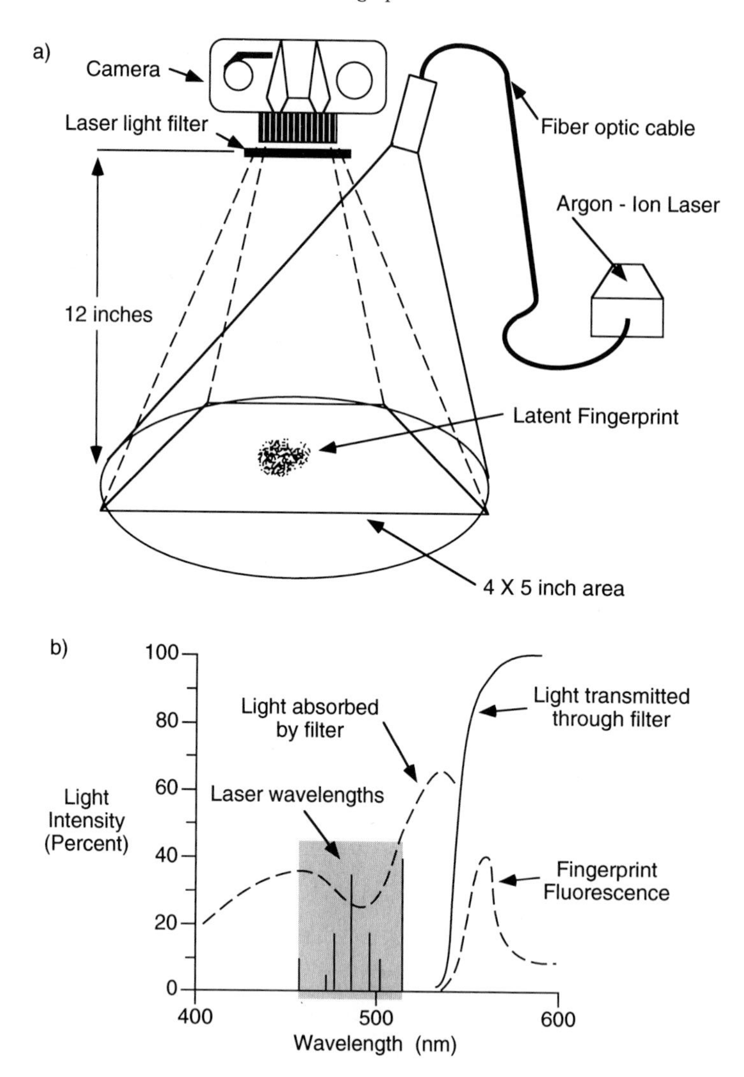

Figure 2.5. a) The camera recording the presence of a luminescent fingerprint is fitted with a filter that prevents the laser light itself from reaching the film in the camera but lets the longer wavelength light, produced by fluorescence from the fingerprint, pass. b) The wavelengths of light that are absorbed by both the fluorescent substance in the fingerprint and the fluorescent radiation.

Scientists studying the use of laser luminescence say that under the microscope the fluorescent fingerprints show a beautifully detailed structure of pores amid the usual whorls and ridges of the fingerprint. These pores, natural outlets for sweat, do not show up using conventional techniques. Like the fingerprint patterns themselves, they are uniquely associated with an individual's skin. Because many fingerprints collected by the police are incomplete, and therefore not useful for identification, **pore patterns** can provide a powerful additional identification information feature.

In Figure 2.6, successful development of several latent fingerprints by the laser luminescence technique on different materials is illustrated. In Figure 2.7, another example of the power of the laser luminescence technique to develop latent fingerprints is shown. On a plastic shopping bag, fingerprints are invisible. However, when illuminated with laser light with a wavelength of 514 nanometers, the fingerprint stands out clearly and the printing on the bag nearly disappears. The white background of the bag appears black, because the plastic of the bag absorbs the light of the wavelength given out by this particular laser.[12–15]

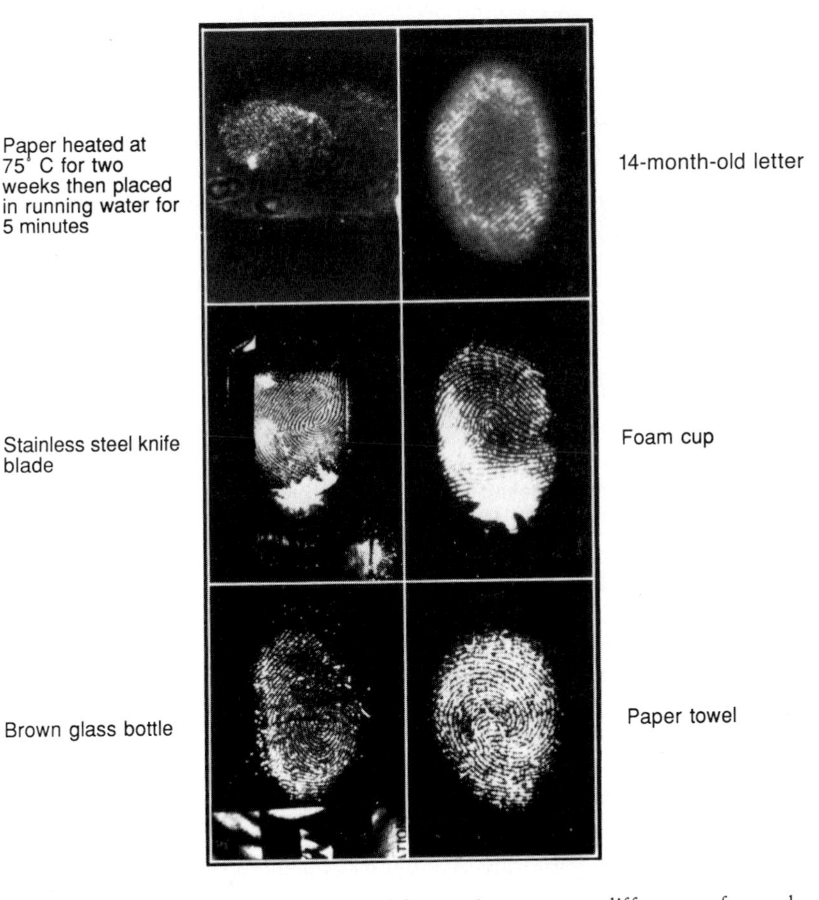

Figure 2.6. Laser luminescence can help to find fingerprints on many different surfaces where normal fingerprint detection procedures would not work. (Reprinted with permission of ASTM.)

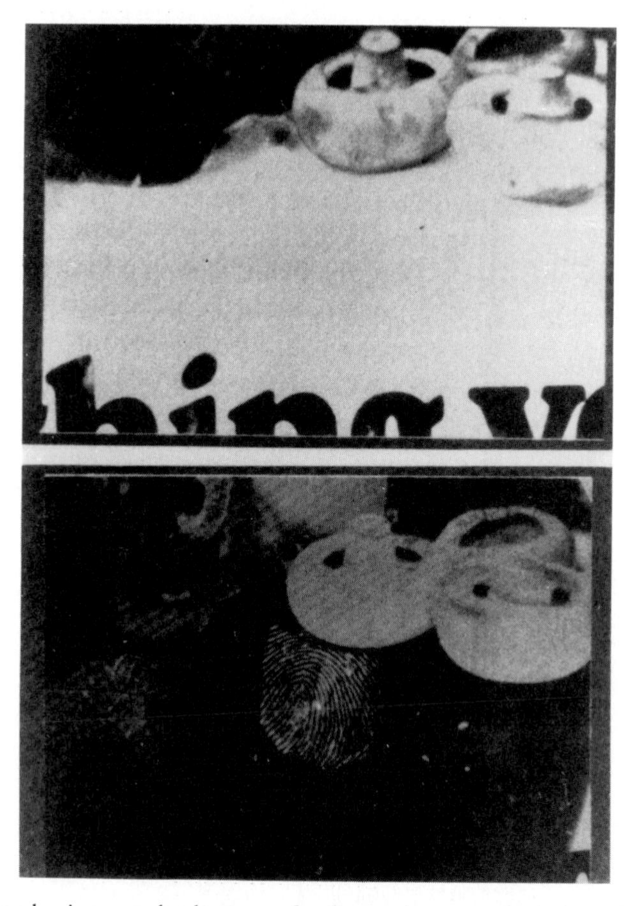

Figure 2.7. Laser luminescent development of a fingerprint on a plastic bag. (London, England Metropolitan Police Forensic Laboratory, described by Broad in *Science and Criminal Detection*, Macmillan, London, 1988.)

Another method for developing latent fingerprints, the discovery of which involved an element of serendipity, is illustrated in Figure 2.8(a). This technique was developed by a scientist famous for his experiments in optics, Dr. Tolansky. While developing a plating technique for making a metal coated mirror, he found that fingerprints developed on glass surfaces, which had not been cleaned properly before coating with metal vapor. In this technique, the surface to be coated is placed above a piece of metal foil, which is melted by an electric heater and evaporates onto the glass surface to form the mirror. The air is pumped out of the equipment to create a vacuum. This means that when the metal foil is heated, the evaporating metal atoms are not slowed down by the air molecules between the heater and the specimen to be coated. This technique was developed for forensic science by the

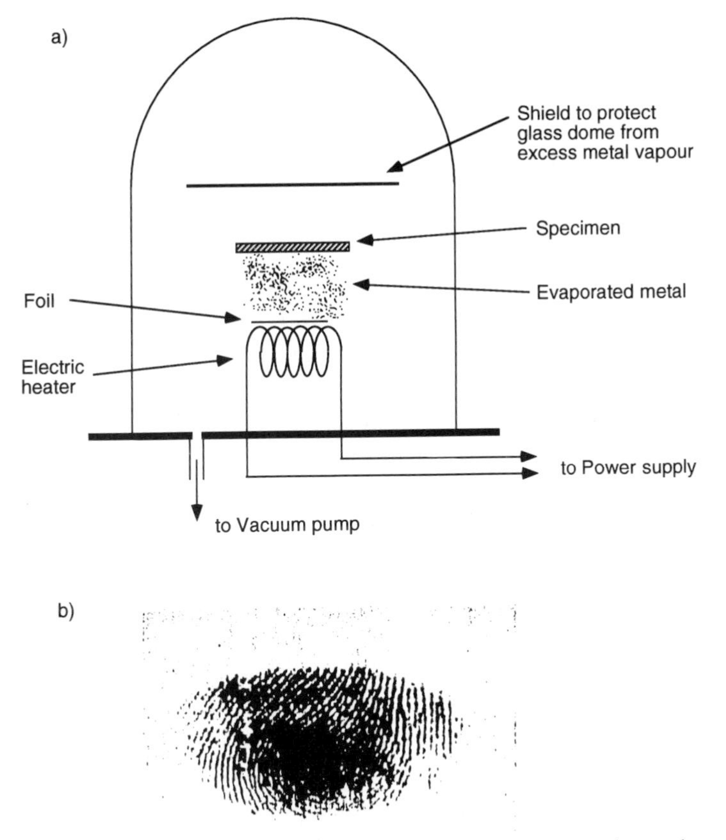

Figure 2.8. The technique used to put the reflective coating on mirrors can be used to develop latent fingerprints on surfaces. a) Sketch of the system used to develop fingerprints by metal evaporation. b) A fingerprint developed by first coating the specimen with gold, which sticks to the material around the fingerprint, then finishing with cadmium, to fill in the print and provide contrast.

French scientist Ceccaldi in 1968. The print quality developed by this method is very good when first a thin layer of gold is evaporated onto the material to be investigated. Then, once the fingerprint starts to develop, one switches to evaporating the metal cadmium, generating a picture that contains sufficient detail for legal identification and use in a court setting.[2,3] A fingerprint developed in this way is shown in Figure 2.8(b). This fingerprint was found on a piece of cotton fabric. Compared to the normal fingerprint, the structure is reversed showing that the material from the ridges on the skin prevent the deposition of the metallic vapor and that the plating material collects in the valleys between the ridges of the fingerprint.

Chemical techniques have been used for many years to develop latent fingerprints. When the fingerprint is on a piece of paper, one can use a spray of silver nitrate solution. The silver nitrate reacts with the chlorides in the fingerprint debris to create a visible deposit, which can then be photographed and used to search for matching fingerprints. Biochemists routinely test solutions for the presence of amino acids using the chemical indantrione hydrate. In short form this test is known as the **ninhydrin** test. If a solution of this material is sprayed onto a surface suspected of carrying a fingerprint, the pattern of the ridges becomes visible as the chemical reacts with the amino acids. Chemical development of fingerprints using silver nitrate or ninhydrin is not suitable for fabrics or rough textured surfaces such as wood. One way of treating surfaces of wood and fabric that may be carrying latent fingerprints is to expose them to iodine vapor, which reacts with chemicals in the fingerprint. Iodine vapor development of a fingerprint is not a useful technique if the fingerprint is more than twenty-four hours old because the water in the fingerprint evaporates. The reaction does not easily occur with a dry latent fingerprint. This problem was tackled by two scientists, Dr. Sasson and Dr. Anati at the Casali Institute of Applied Chemistry at the Hebrew University in Israel. This work was carried out in collaboration with Israeli forensic expert Dr. Joseph Almog. They built a device that simultaneously ejects steam and iodine vapor onto the surface being studied. The debris from the dried out fingerprint absorbs the moisture, which then reacts with the iodine. Using this device, they were able to develop fingerprints which were two months old.

A variation of the iodine vapor technique, developed in Toronto, can be used to take fingerprints from a murder victim up to 105 hours after the crime has been committed. Using this technique, iodine crystals and other chemicals are mixed in a 35 cm plastic tube. Heat from the hand of the operator holding the tube is sufficient to cause the iodine to vaporize. Then, by blowing into the tube, the vapor is spread over a small area of the victim's body. This vapor reacts with the moisture in the fingerprints left on the body.[16] A pliable silver plate, which is approximately 0.03 cm thick, is pressed against the area where the print appears. The pattern is transferred to the metal foil for permanent storage. The use of this flexible silver plate allows the technique to be used on curved areas of the body such as the throat, arms, chest, or thighs, areas often touched during a homicide. Using this technique, prints can also be developed and lifted from wood, plastic, foam, paper, and leather. Experiments on living persons show that fingerprints must be taken within one and a half hours after the prints are made because chemicals coming through the skin of the victim wash the prints away.

A commercially available system for chemically developing a fingerprint, which was also developed in Ontario, is the **Visuprint System**. This was developed empirically by an Ontario Provincial Police officer, Bourdon. Over a period of years, he tested many different chemical vapors as developers for latent fingerprints and discovered that the chemical used to make super glues, cyanoacrylate esters, can also be used for developing latent fingerprints. Compounds in the fingerprint appeared to create a plastic pattern from the fumes of the chemical.[17,18] The plastic replica of the fingerprint developed in this way adheres very firmly to the surface, and it is not easy to brush the pattern away after development. To develop a fingerprint, the object suspected of carrying latent prints is placed in a chamber full of circulating fumes of the chemicals. These develop the prints in a period of approximately thirty minutes. Successful development of prints using the Visuprint system include the following examples:

- 1. Prints that defied traditional fingerprinting methods were obtained from a handgun after four years of storage in police archives.
- 2. A whiskey bottle found immersed in the bilge water of a stolen motorboat yielded prints that led to a conviction.
- 3. After traditional methods failed to develop prints on the interior of a car, the car was sealed and the developer chemical pumped into the car. Using this approach, a multitude of prints were found, which assisted police in their investigations.
- 4. Soot-covered articles have been successfully treated to reveal fingerprints in arson cases.

A scientist at Los Alamos National Laboratory, in the United States, George Saunders, used the techniques of **colloid science** to develop another method for making latent fingerprints visible.[19,20] The name colloid scientist was coined many years ago to describe the specialist who studied very tiny pieces of material dispersed in another carrier material. Many of the substances studied by these specialists were glue-like, and the term colloid comes directly from a Greek word meaning "glue." Very often tiny pieces of materials suspended in a colloidal solution carry electric charges, which stabilize the dispersion state of the finely divided material. A colloidal **gold sol** solution contains tiny pieces of gold of the same order of magnitude as the wavelength of light (800 nanometers). In suspension, these tiny pieces of gold carry negative electric charges. If a check or similar piece of paper, which is believed to carry a fingerprint, is placed in the colloidal gold solution, the tiny pieces of gold are attracted to the proteins in the fingerprint. If the technique works, one can sometimes see a faint image of the fingerprint outlined in gold fine particles, but to make the print more visible the piece of material being investigated is rinsed thoroughly and then placed in a silver solution. Silver is then deposited onto the surface. The silver enhances the image by filling in around the gold highlighting of the fingerprints features. The developed print is then transferred to a thin sheet of **nitrocellulose**, a cotton-like polymer. After further chemical treatment, the print shows up on the material clearly. This technique was first used in March of 1988, when Saunders was able to develop a palm print on a blank check that had been stolen from a military facility in the United States. The silver deposits created by this technique can be seen easily on dark surfaces. Because of this fact, secret service agents in the United States were able to use the method to detect prints on black adhesive tape wrapped around a bomb. The FBI has also applied the technique for developing prints on photo negatives and, in general, the technique can work well on plastic, glass, and metal surfaces. However, unfortunately, the technique cannot be used with American paper currency because the paper used to make this money contains a high percentage of flax and cotton fibers which have considerable amounts of protein associated with them. (This aspect of paper manufacture will be discussed in more detail in Chapter 10 dealing with counterfeit money.)

One of the most difficult surfaces on which to develop fingerprints is the surface of a valuable oil painting. Obviously the development technique must not harm the painting. Harper and his colleagues at the Police Forensic Science Laboratory of the Metropolitan Police in London, England have developed a technique in which bacteria that flourish on the proteins in the fingerprint to generate a visible image.[21] To find a suitable bacteria for growing on the remnants of the fingerprints left on a surface, metropolitan police screened nearly 600 types of bacteria found upon the skin. Out of the bacteria screened for the job they discovered that the bacteria **acinetobacter calciacatieus** works best. Harper notes that this particular variety of bacteria was obtained by chance when he swabbed the forehead of a visitor to the laboratory![21] To develop the fingerprint, a colony of the bacteria are pasted, in a nutrient gel, over the site of the suspected fingerprint. As they multiply and form colonies on the print ridges, they make the fingerprint clear to the observer. Cultures of the acinetobacter calciacatieus bacteria are now kept in freeze dried form at police laboratories ready for use in specialty theft cases. When needed, they are thawed and made into fingerprint developing gel. The technique takes about 24 hours to produce an identifiable print and is still being optimized to obtain clearer prints. On an oil painting treated in this way, one wipes off the harmless gel after photographing the print and the oil painting is left undamaged.

Yet another technique for developing latent fingerprints enables the fingerprint to take a photograph of itself, a technique based on what is known as **autoradiography**. To be able to understand the basis of this method, it is necessary to develop some concepts of the subject known as **radiochemistry**. Scientists have shown that the world is made up of just under 100 naturally occurring elements. For the purposes of this discussion, we can regard elements as being made from a concoction of what is known as protons, neutrons, and electrons. The smallest piece of an element that can exist is known as an **atom**. The number of electrons in an atom determines the chemical nature of an element. The center of the atom is made up of a number of small dense objects known as protons and neutrons. The number of protons matches the number of electrons in the element. The protons carry a positive charge, so that the overall assembly of protons and electrons has no electric charge. The neutrons only add weight to the center of the atom, known as the nucleus. As scientists' knowledge of the elements of the universe increased, they arranged them in a table, as shown in Table 2.2. Eventually, scientists found that the atoms of some elements could have more neutrons than other atoms. This difference caused the different atoms of the elements to have slightly different physical properties but the same chemical activity and properties.

For example, scientists found out that one type of carbon atom could have six protons and eight neutrons, whereas another type of carbon atom could have six protons and six neutrons. Scientists started to represent varied types of atoms by writing the total number of protons and neutrons as a number at the top left hand corner of the symbol representing the element. For example, they write the two forms of carbon as ^{12}C and ^{14}C. The different types of atoms of "the same" element were given the name **isotopes** from the Greek root words *iso* meaning "the same" and *topos* meaning "place." The name indicates that isotopes occupy the same place in the table of the elements even though they have slightly different physical properties. Further studies of the structure of atoms revealed that some types of atoms are unstable and change their nature by giving out high-energy radiation known as **radioactivity**. Thus, it was found that carbon 14 can give out energy and change into a nitrogen atom. Not all the carbon 14 atoms change into nitrogen atoms at the same time, and, as we shall discover in later chapters, we can tell how old a carbon-containing object is by finding how many of its ^{14}C atoms have changed their nature. The change from an atom of one element to an atom of another element is called **transmutation**, a word coined from the Latin root words *trans*, meaning "across," and *mutate*, meaning "to change."

In the first step taken to create an autoradiograph of a latent fingerprint on something such as a piece of fabric, the fabric is enclosed in a container in which the humidity of the air is kept below 50%. (The **humidity** of the air is a measure of the amount of moisture in the air.) Radioactive gas is introduced into this container. The radioactive atoms interact with the fingerprint deposit to become part of the fingerprint debris. Next, the fabric is removed from the container and pressed against a photographic film. The energy of the decaying radioactive material, which has become part of the fingerprint, develops a picture on the film. Two radioactive gases, which have been found to be useful for this purpose, are radioactive iodine and radioactive sulfur dioxide. In Figure 2.9 shows a palm print, as it night appear if developed on cotton fabric using gaseous, radioactive sulphur dioxide. This autoradiographic technique cannot be used if the fabric has been exposed to a high humidity environment for more than a few hours before it is treated with the radioactive gas.

Anyone who has watched a crime movie knows that smart crooks wear gloves when carrying out their activities. If, however, they are caught, they had better have destroyed the gloves used, because scientists have now dis-

	2A 2 IIA	3A 3 IIIB	4A 4 IVB	5A 5 VB	6A 6 VIB	7A 7 VIIB	8 8 VIII	8 9 VIII	8 10 VIII	1B 11 IB	2B 12 IIB	3B 13 IIIA	4B 14 IVA	5B 15 VA	6B 16 VIA	7B 17 VIIA	0 18 VIIIA
1.0079 $_{1}$**H**																	4.0026 $_{2}$**He**
6.941 $_{3}$**Li**	9.0122 $_{4}$**Be**											10.811 $_{5}$**B**	12.011 $_{6}$**C**	14.007 $_{7}$**N**	15.9994 $_{8}$**O**	18.998 $_{9}$**F**	20.180 $_{10}$**Ne**
22.990 $_{11}$**Na**	24.305 $_{12}$**Mg**											26.982 $_{13}$**Al**	28.086 $_{14}$**Si**	30.974 $_{15}$**P**	32.066 $_{16}$**S**	35.453 $_{17}$**Cl**	39.948 $_{18}$**Ar**
39.098 $_{19}$**K**	40.078 $_{20}$**Ca**	44.956 $_{21}$**Sc**	47.88 $_{22}$**Ti**	50.942 $_{23}$**V**	51.996 $_{24}$**Cr**	54.938 $_{25}$**Mn**	55.847 $_{26}$**Fe**	58.933 $_{27}$**Co**	58.69 $_{28}$**Ni**	63.546 $_{29}$**Cu**	65.39 $_{30}$**Zn**	69.723 $_{31}$**Ga**	72.61 $_{32}$**Ge**	74.922 $_{33}$**As**	78.96 $_{34}$**Se**	79.904 $_{35}$**Br**	83.80 $_{36}$**Kr**
85.468 $_{37}$**Rb**	87.62 $_{38}$**Sr**	88.906 $_{39}$**Y**	91.224 $_{40}$**Zr**	92.906 $_{41}$**Nb**	95.94 $_{42}$**Mo**	98.906 $_{43}$**Tc** *	101.07 $_{44}$**Ru**	102.91 $_{45}$**Rh**	106.42 $_{46}$**Pd**	107.87 $_{47}$**Ag**	112.41 $_{48}$**Cd**	114.82 $_{49}$**In**	118.71 $_{50}$**Sn**	121.75 $_{51}$**Sb**	127.60 $_{52}$**Te**	126.90 $_{53}$**I**	131.29 $_{54}$**Xe**
132.91 $_{55}$**Cs**	137.33 $_{56}$**Ba**		178.49 $_{72}$**Hf**	180.95 $_{73}$**Ta**	183.85 $_{74}$**W**	186.21 $_{75}$**Re**	190.2 $_{76}$**Os**	192.22 $_{77}$**Ir**	195.08 $_{78}$**Pt**	196.97 $_{79}$**Au**	200.59 $_{80}$**Hg**	204.38 $_{81}$**Tl**	207.2 $_{82}$**Pb**	208.98 $_{83}$**Bi**	208.98 $_{84}$**Po** *	209.99 $_{85}$**At** *	222.02 $_{86}$**Rn** *
223.02 $_{87}$**Fr** *	226.03 $_{88}$**Ra** *																

138.91 $_{57}$**La**	140.12 $_{58}$**Ce**	140.91 $_{59}$**Pr**	144.24 $_{60}$**Nd**	146.92 $_{61}$**Pm** *	150.36 $_{62}$**Sm**	151.97 $_{63}$**Eu**	157.25 $_{64}$**Gd**	158.93 $_{65}$**Tb**	162.50 $_{66}$**Dy**	164.93 $_{67}$**Ho**	167.26 $_{68}$**Er**	168.93 $_{69}$**Tm**	173.04 $_{70}$**Yb**	174.97 $_{71}$**Lu**
227.03 $_{89}$**Ac** *	232.04 $_{90}$**Th** *	231.04 $_{91}$**Pa** *	238.03 $_{92}$**U** *	237.05 $_{93}$**Np** *	244.06 $_{94}$**Pu** *	243.06 $_{95}$**Am** *	247.07 $_{96}$**Cm** *	247.07 $_{97}$**Bk** *	251.08 $_{98}$**Cf** *	252.08 $_{99}$**Es** *	257.10 $_{100}$**Fm** *	258.10 $_{101}$**Md** *	259.10 $_{102}$**No** *	260.11 $_{103}$**Lr** *

Table 2.2. The atoms that make up the universe have been arranged by scientists into what is known as the Periodic Table of the Elements.

Figure 2.9. A palm print such as the one shown above can be developed on a material such as cotton cloth using autoradiographic techniques which employ radioactive sulphur dioxide.

covered that one can develop glove prints from the scene of the crime and compare them with any pair of gloves found in possession of the suspect. Thus, in Figure 2.10, two scientists are shown matching a glove print from the scene of the crime and from gloves discovered in the possession of the suspect.[22] A few years ago Dr. Dravnieks suggested that modern technology has reached such a level of sophistication that it is possible to take a smudged fingerprint and evaporate the volatile organic constituents into an instrument known as a gas chromatograph for characterization (See discussion of synthetic bloodhounds in Chapter 7.) Dr. Dravnieks points out that by using this analysis of the odor of a fingerprint, one should be able to determine the recent occupational history and environmental contacts of a criminal leaving fingerprints at the scene of a crime. Thus, in the near future we can expect scientists to extract useful information even from smudged fingerprints and material left on the inside of a glove.

Figure 2.10. Just as two fingerprints are the same, glove prints are also unique evidence. Above, a print left by a glove is compared with the actual surface of a glove. (The lines on the diagram draw attention to identical aspects of the two patterns.[22])

2.3 Storing and Using Fingerprint Images

The first task faced by the scientist when given a set of prints is to describe the print in a way that can be used by a computer for storing and retrieving information. The description system chosen must also be useful in presenting evidence that two fingerprints, one of a suspect and one retrieved by the search system, are indeed a matching pair. As seen in Figure 2.2 (b) some of the different patterns, along with the words used to describe the different features to be found in fingerprints are shown. Various recording systems for describing the different patterns in a fingerprint are in use and the interested reader will find details of the description systems in several reference publications listed at the end of this chapter.[1,2]

When a police expert presents evidence with regard to fingerprints, current legal practice dictates that the expert must be able to detail at least sixteen matching features when comparing a fingerprint from the scene of a crime and that taken from a suspect or from a fingerprint on file in the police records. This fact is illustrated in Figure 2.11. There is nothing magi-

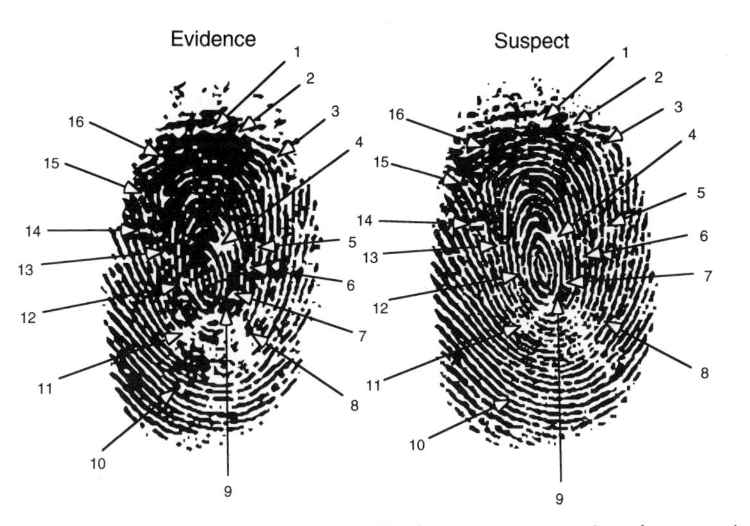

Figure 2.11. When matching fingerprints, the judicial system now requires that 16 points on the fingerprints agree. a) Fingerprint that might be found on a surface at the scene of a crime. b) Fingerprint taken from an individual in detention or already on record.

cal about the number 16; this is the number that has evolved during expert witness presentations in the courts over the last several decades.

Today, using modern optical techniques, we can build robots to search for matched fingerprints much more efficiently than humans. However, because juries would find it difficult to understand how robots can conclude that two fingerprints are the same it will be some time before we see robots on the witness stand. Nonetheless, due to the potentially amazing ability of robots to study, store, and recognize the structure of fingerprints, we will explore the physics behind the methods that robots use to match fingerprints almost instantaneously. At the moment such systems are used to provide checks on who is allowed into a security area. To enter the area, one must place one's finger on a small window where it is inspected by the optical and electrical devices inside the device. The fingerprint pressed onto the window is then matched with the fingerprints stored in a memory to see who is allowed into that particular area. The same type of device would enable a police officer arriving at the scene of a crime to place a suspect's finger on the window of such a device to receive information (in a very short time) on the possibility that this person is a wanted criminal. The robotic fingerprint inspector is an example of what scientists could develop for society if society were really serious about improving the ability of security forces to fight crime.

The first task that robotic inspectors could do in the crime laboratory is to extract features of the patterns, such as those in Figure 2.2(b), by using lasers to generate **diffraction patterns**. To understand what is meant by this statement we must explore the wave nature of light and gain a basic understanding of diffraction and interference of light. It is very difficult to explain the physical nature of light. We have already discussed the basic features of light energy given out by incandescent light sources and by lasers. For many purposes light behaves as if it were made up of waves; if we pass monochromatic, coherent light through a narrow slit, we obtain what is known as a diffraction pattern, as illustrated in Figure 2.12(a). The word **diffraction** is described in the dictionary as: "the spreading of light and other rays passing through a narrow opening or by the edge of a dense body." The word is coined from Latin root words meaning "to break apart." Thus, in a way the narrow slit diffracting light breaks the original beam of light into a pattern known as the diffraction pattern. The diffraction of light is one of the those properties of things around us which we find very difficult to understand. We just have to grasp the observed facts firmly as one of the things that happens in the universe. Scientists quarrelled for many years over the true nature of light. Even now we can only say that when light is generated and when it is absorbed it behaves as if it were single packets of energy whereas when it is being transmitted it moves like a wave. The fact that light is sometimes described using the idea of packets of energy (the smallest packet is called a photon) and that at other times it is a wave, is said to be **paradoxical**. A paradox is a statement, the two halves of which appear to contradict each other. (For good textbook discussions of the nature of light see references 23 and 24.) When studying diffraction patterns it has been shown that the narrower the slit, the wider the diffraction band. In a Chapter 10 the use of diffraction patterns to discover fraud in the sale of high priced specialty wool will be explored. In that chapter we will see that the narrower the wool fiber, the wider the diffraction pattern.

When diffracted light from two slits illuminated by the same source of light is allowed to overlap, the light not only spreads out, but energy from the light beams interacts to form patterns known as **interference fringes**, as illustrated in Figure 2.12(b). Such parallel line systems are known as diffraction gratings. Most modern spectrometers no longer use glass prisms to spread out the light but use diffraction gratings made of many parallel lines drawn on a transparent surface. The same phenomenon can be observed in reflected light. If one holds a long-playing record up to sunlight the "rainbow" reflected from the surface is generated by the grooves of the phonograph record, which act like a diffraction grating. In the same way the surface of a compact disc will act as a diffraction grating to spread out the colors of the light falling onto the surface of the grating.

When looking at the picture of the diffraction grating shown in Figure 2.12(c) it can be seen why a scientist looking at a fingerprint, such as those shown in our various diagrams, immediately suspects that the fingerprint will act as a diffraction grating when one shines laser light through a reduced negative of the fingerprint. One reduces the fingerprint photogra-

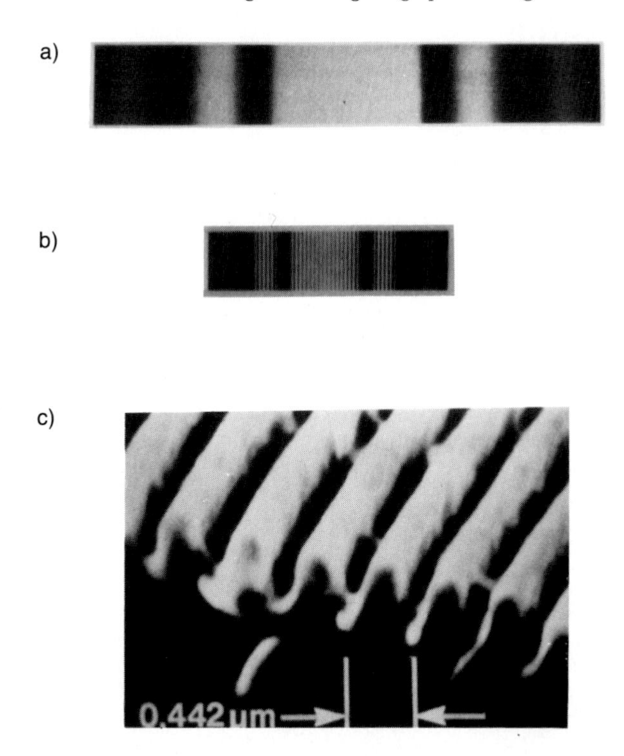

Figure 2.12. When light passes through a narrow slit, the energy "spreads out" in a process known as light diffraction. a) Diffraction of monochromatic light from a single, narrow slit. b) When monochromatic light is diffracted through two parallel narrow slits the light interacts to form a pattern known as interference fringes. c) Magnified picture of the surface of a diffraction grating, which is made up of a multitude of thin parallel slits. (Photo courtesy of Holotek.)

phically making the spacing between the lines small enough compared to the wavelength of light to make an efficient diffraction grating. As computers started to be developed and laser light became widely available in the mid 1960s, many scientists quickly recognized that one could use this type of system to look at the different features present in a fingerprint. In Figure 2.13(b) some of the pictures reported in the pioneering work carried out by Penn and Duffy are shown.[25] One can take these patterns, put them into a computer and teach the computer how to recognize the fingerprints.

One of the devices, which can be used to transfer a diffraction pattern into the memory of a computer, is shown schematically in Figure 2.14. It is a mosaic of photocells, which can look at the various parts of the diffraction pattern. Each of the small **photocell** devices, which can turn light energy into electrical energy, has an address so that the computer knows which of the squares of the mosaic are being lit up by the diffraction pattern (for cla-

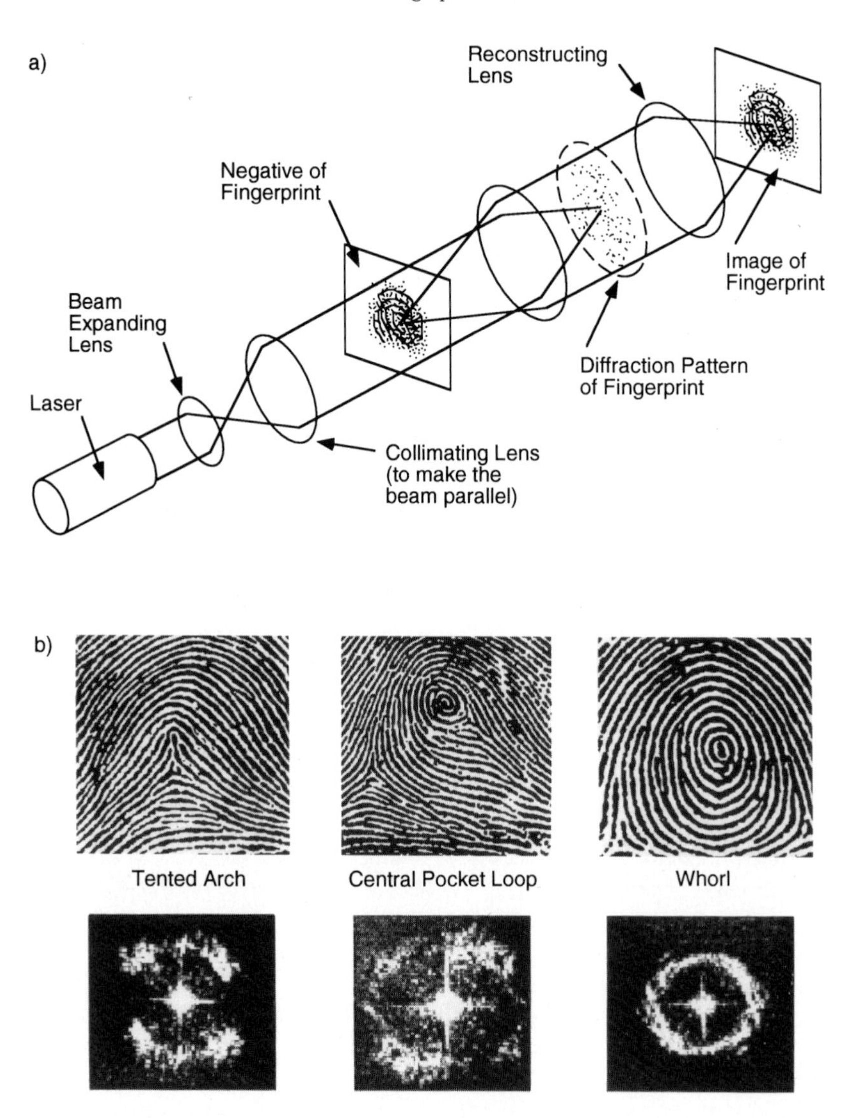

Figure 2.13. To a physicist an enlarged picture of a fingerprint looks similar to a diffraction grating. Not surprisingly illuminating a photographic negative of a fingerprint with laser light generates a diffraction pattern in which the energy distribution represents various features of the fingerprint.[20] a) Sketch of the equipment used by Penn and Duffy to characterize the structure of a fingerprint. b) Typical diffraction patterns of different fingerprints generated by Penn and Duffy.[25]

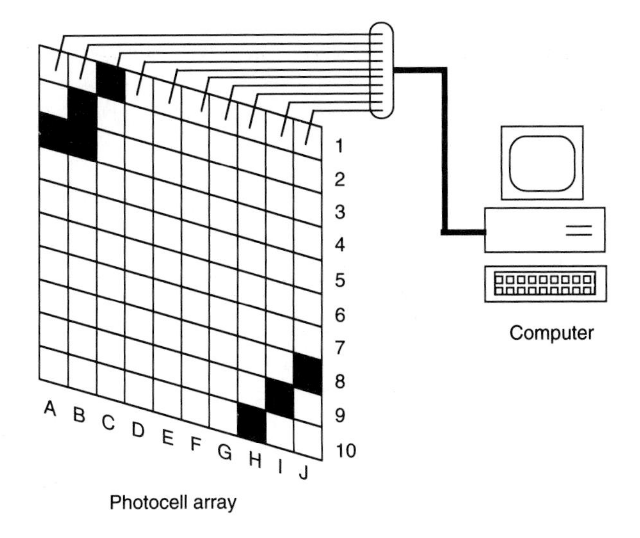

Figure 2.14. One way to transport information from the diffraction pattern into a computer is to place a mosaic array of photocells at the position where its diffraction pattern is produced.

rity only the connections to the first line of photocells is shown in Figure 2.14). Thus in our diagram photocells at the address C1, B2, B3, and A3 are part of the diffraction pattern. The term for an array of photocells such as that shown in Figure 2.14 is a **matrix** of photo cells. One could train a computer to recognize fingerprints by showing different fingerprints to the matrix of photocells and recording the addresses of the photocells that are lit with each fingerprint. Thus, if a certain feature of a fingerprint had been shown to the computer, generating a subsidiary pattern of lit elements of the matrix, this pattern of squares would be recorded in the computer memory. Transforming the diffraction patterns into a long sequence of lit, unlit photocells is described by the scientist as a linear processing system. This name illustrates that all the data from the matrix forms a line of information to be processed by the computer. The various elements of the matrix are known as pixels (short for picture element). In linear processing the pixels queue up to report to the computer whether or not they are receiving light energy.

Much faster methods than this linear information system are available for recognizing the structure of a diffracted pattern of a fingerprint. This technique is known as optical computing. To explain the basis of a possible optical computing technique for fingerprint recognition, we will discuss a simpler problem related to fingerprint recognition – that of the way in which robots have been trained to look for holes in a very fine mesh of sieve being used to prepare diamond powder for polishing lenses. In Figure 2.15(a) the diffraction pattern of an undamaged sieving surface is shown. This pattern is identical to the one that can be seen through the texture of a woven cloth umbrella if one looks at a distant street light. If one uses a very fine sieve, of the type shown in this diagram to separate diamond powder, one must be able to find even one damaged **aperture** (the technical term for the opening in a sieve surface). If a damaged hole goes undetected, larger than expected diamond dust grains can pass through the sieve into the polishing powder being made by the sieving process. Even one larger particle will scratch the lenses being polished with the contaminated diamond polishing powder. However, looking for one large hole amongst millions in the sieving surface is a very difficult problem.

Scientists overcame this problem by using a technique known as **optical spatial filtering**. In the first step one places an undamaged piece of sieve surface in the equipment (which is similar to that used by Penn and Duffy) and generates a diffraction pattern. One then photographs the diffraction pattern to create its negative. Consider now what would happen if one were to place the negative of the diffraction pattern of Figure 2.15(a) at the point in the system where the diffraction pattern is formed, as shown in Figure 2.15 (b). If an undamaged sieve is in the examination position, then the diffraction pattern created by the sieve being examined will exactly match the negative, which is placed at what is known as the diffraction plane. This means that light cannot travel beyond the diffraction plane, since the negative blocks all of the light diffracted by the sieve surface being examined. Thus, there will be no energy to create an image of the sieve and the reconstruction screen will be blank. The operation of putting a negative of the ideal diffraction pattern in the position to stop light from the undamaged sieve is described as **spatial filtering**. If one placed a damaged sieve in the examination plane, then the light diffracted by the damaged apertures would bypass the spatial filter and would create an image at the reconstruction screen. Figure 2.15 (c) shows the picture of a damaged sieve mesh and its filtered reconstructed image, using the system of Figure 2.15(b). It can be seen that all the information on the undamaged apertures has disappeared from the screen and only differences are reproduced. This enables the scientists interested in studying the sieve to focus their attention on the damaged apertures.[26] In the same way, if one photographs the fingerprint of a suspect and placed it in the diffraction plane, then one could run all of the stored fingerprint images through the inspection position and only when the suspect fingerprint matched one in the examination plane, would there be zero illumination on the reconstruction screen. If one now replaced the reconstruction screen with a photocell one could run the store of fingerprints through the examination plane and when a match was made there would be zero signal at the photocell. At this point, the electronic robot could stop the inspection procedure and print out the reference number of the matching fingerprint. As early as the mid 1970s machines had been developed which could inspect 1.5 million fingerprints in two minutes, using this type of fingerprint search system. From a scientific point of view, this type of spa-

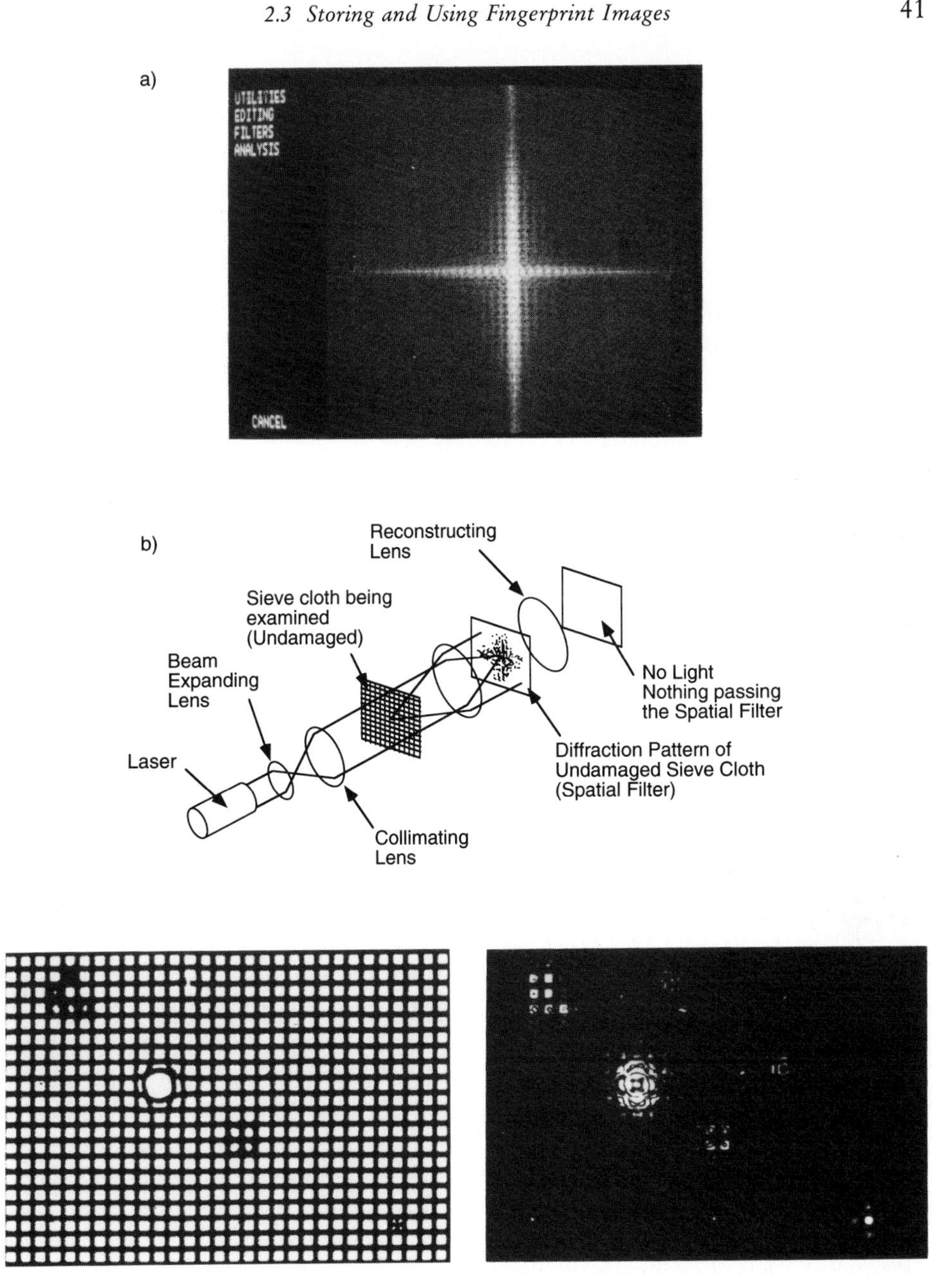

Figure 2.15. In a process known as "optical spatial filtering," the diffraction pattern of a specimen is compared with a known diffraction pattern, so that only the differences between the specimen and the reference pattern become visible.[26] a) Diffraction pattern of an undamaged sieve cloth. b) Sketch of the equipment used for optical spatial filtering. c) The reconstructed image of a damaged sieve surface highlights the damaged areas of the specimen being examined. left: Microscopic photograph of the damaged sieve surface right: Spatially filtered diffraction pattern of the damaged sieve surface.

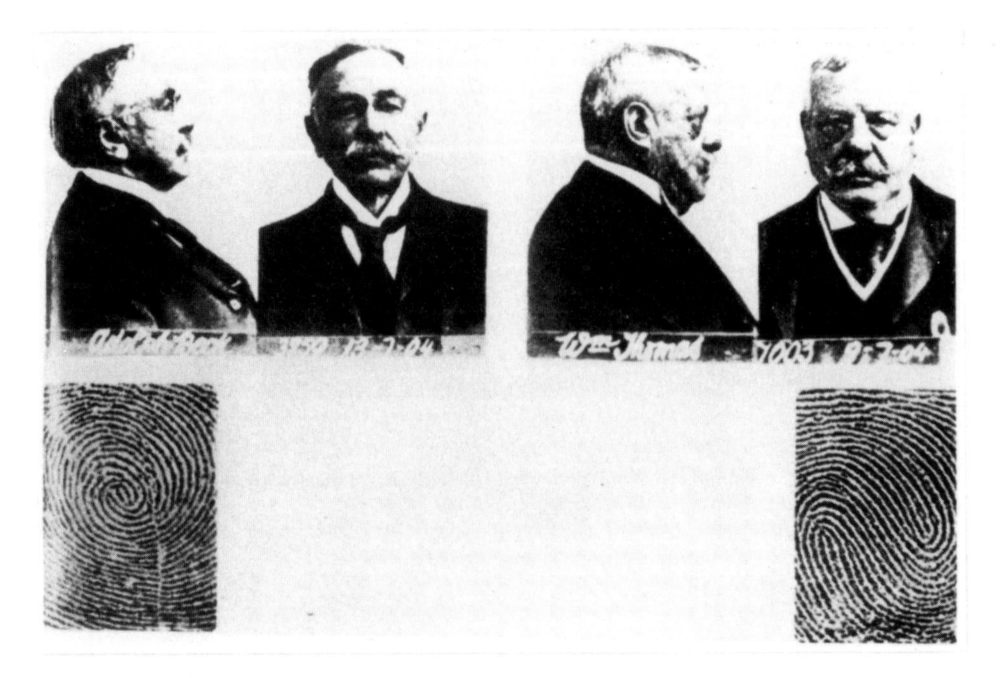

Figure 2.16. Fingerprints can establish the innocence of a suspect as well as convict a criminal. The person on the left was mis-identified by a witness and might well have been convicted if a fingerprint found at the scene of the crime had not proven that he was not the perpetrator.

tial filtering is known as **parallel processing**. It is mathematically a much more powerful technique than the linear processing discussed with respect to the data from a mosaic of photocells. The transformation of an image into a long line of information, using a matrix of photocells is basically the way in which a **fax machine** takes an image and transforms the information of the image into a sequence of data points, which can be sent over a telephone line. It is obvious that spatial filtering, diffraction pattern searching, and other modern computer techniques greatly enhance the ability of crime-combating organizations to use the information to be found in fingerprints left at the scene of a crime.

Fingerprint recognition not only enables convictions to be achieved in court, but they are also useful in establishing innocence. Figure 2.16 shows two people who were confused by witnesses to a crime. However, the person who was mistaken for the criminal had different fingerprints to those left at the scene of the crime.[27] The reader will find an interesting review of some crimes that were solved using fingerprint evidence in the book by Block.[28] With the growing array of sophisticated optical reproduction systems one must also be on guard against the possibility of planting a fingerprint at the scene of the crime. It is therefore important that the fingerprint collected at the scene of a crime is carefully documented. In a famous case that occurred in the early 1970s, a corrupt policeman planted a copied fingerprint from a file record of a previously convicted suspect at the scene of a crime and presented it as evidence in court. The suspect was wrongly convicted of the crime and spent two and a half years in prison before an expert was able to establish that the fingerprint was a planted xerographic copy.

References

1. For a useful and comprehensive overview of the history of the use of fingerprints in crime detection see, L. McGinty, "Fingerprints The Next Data In The Bank," *New Scientist*, 31 October 1974, 320.
2. D. F. Shaw, "Physics In The Prevention and Detection of Crime," *Contemporary Physics 17* (1976), 307.
3. D. F. Shaw, "Crime Counter Measures," *Physics Technology 9* (1987), 192.
4. H. Faulds, "On the Skin Furrows of the Hand," *Nature 22* (1880), 605.
5. W. J. Herschel, "Skin Furrows of the Hand," *Nature 23* (1881) 76.
6. See article on fingerprints in *Funk and Wagnall's Encyclopedia, Vol 10*, Funk and Wagnall Inc., Mahwah, NJ, 1965, 3490.
7. The use of electrostatic forces to enhance the powder development of a fingerprint is described in A. Rose-Innes, "Static Electricity – An Ancient Enigma," *New Scientist*, 6 May 1982, 346.
8. W. F. Frizzel, "Laser Assisted Latent Fingerprint Detection," *Electro-Optical Systems Design*, November 1979, 19.
9. "Fluorescence Favours the Forensic Policeman," *New Scientist*, 13 April 1978, 88.
10. J. P. Wheeler, "Lasers: Surprise Tools for Forensic Scientists," *Industrial Research*, 15 November 1977, 7.
11. R. P. Brennan, *Dictionary of Scientific Literacy*, Wiley, New York, 1992.
12. C. B. Daish, *Light*, Second Edition, The English University Press Ltd., London, 1971.
13. G. R. Noakes, *A Textbook of Light*, Second Edition, Macmillan, London, 1962.
14. J. Orear, *Physics*, Macmillan, New York, 1979.
15. L. Hogben, M. Cartwright, *The Vocabulary of Science*, Heinemann, London, 1969.
16. "New Fingerprinting Method Developed," *Research and Development 21* (1979).
17. Details on the Commercially Available Visuprint System Available from Payton Scientific Inc. of Scarborough, Ontario.
18. S. A. Hains, "Plastic Prints," *Science Dimension 4* (1982), 16.
19. "Goldfinger," *Discover*, March 1992, 16.
20. J. Beard, "Gold Fingerprints Lift the Lid of Illicit Deals," *New Scientist*, 11 November 1989, 33.
21. I. Mason, "Bacteria Fools the Light Fingered Thief," *New Scientist*, 4 June 1987, 40.
22. "To Catch a Criminal – Try Gloveprints," *Popular Mechanics*, July 1976.
23. P. A. Tyler, *Physics for Scientists and Engineers*, Third Edition, Worth, New York, 1991.
24. D. Halliday, R. Resnick, *Physics*, Wiley, New York, 1981.
25. W. A. Penn, D. E. Duffy, "Extraction of Fingerprint Features Using Coherent Optical Techniques," Paper presented at the Spring 1969 meeting of the Optical Society of America, San Diego, California, 1969. The work was carried out at the Electronics Laboratory of the General Electric Company, Syracuse, New York.

26. The scientific procedures for finding damaged apertures in a sieve by spatial filters is discussed in detail in B. H. Kaye, *Direct Characterization of Fineparticles*, Wiley, New York, 1981.
27. F. Smyth, *Cause of Death; The Story of Forensic Science*, Van Nostrand Reinhold Co., New York, 1980.
28. E. E. Block, *Science versus Crime*, Cragmont Publications, San Francisco, 1979.

Chapter 3

Footprints and Other Vestiges

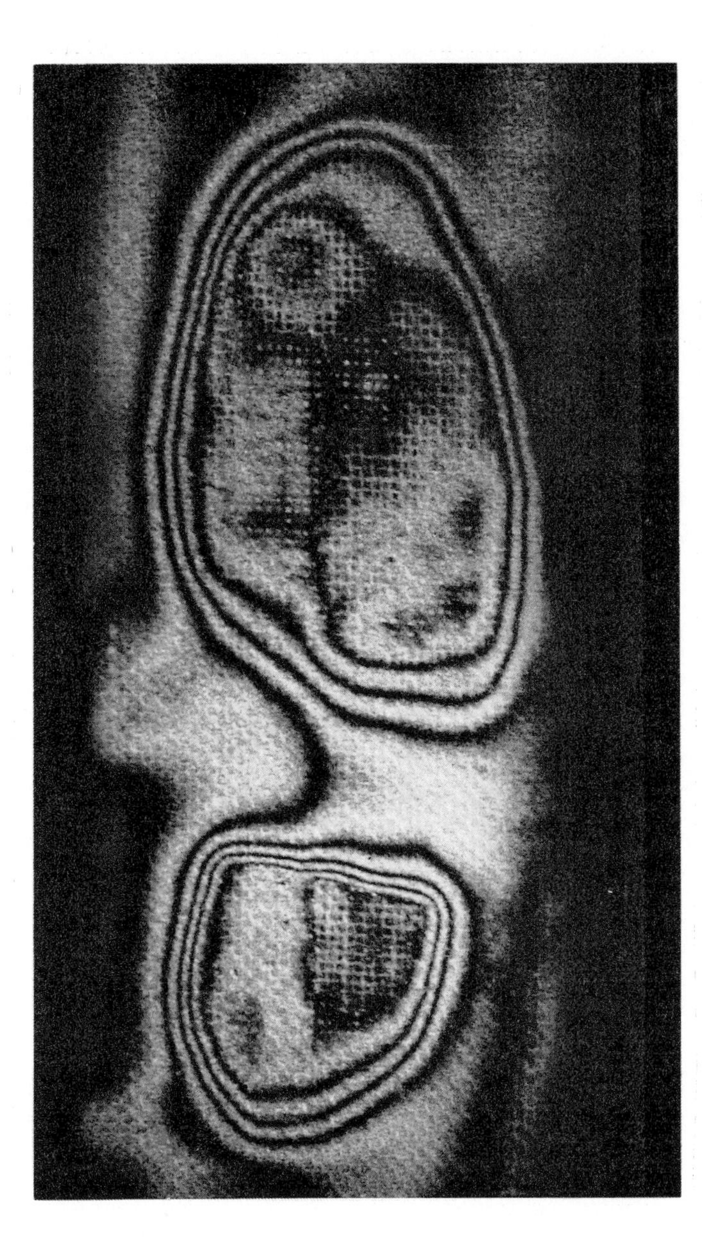

Chapter 3

Footprints and Other Vestiges

3.1 Footprints, the Downfall of Idol Priests!

No, the title does not contain a misprint! We are not going to discuss idle priests. We are going to review how the use of footprints to expose fraud forms part of one of the oldest detective stories in history. The story is told in a book known as the Apocrypha. This comprises a group of religious writings which the Protestant church declared as not being divinely inspired and, therefore, should not form part of the collection of sacred writings known as the Bible. The Roman Catholic church, however, treats this as standard bible literature. If you were to pick up a comprehensive Protestant Bible, you may find a set of books between the Old and the New Testament, known as the Apocrypha. My copy of the Apocrypha is a separate book altogether. The word **apocrypha** originally meant "hidden or unknown." The second part of the word is related to the word **crypt**, which originally meant "a hiding place." A **cryptographer** is someone who sends secret messages. In such messages, the meaning is hidden from the reader. In modern English, the word apocryphal can also mean a story which is not true.[1,2]

In the Apocrypha there is a story entitled "Bel and the Dragon." This story is taken from the end of the Book of Daniel (as presented in the Old Testament). The story is concerned with the relationship between the King of Persia and the priests of an idol, or false god, known as Bel. The King put out choice foods every day as an offering to the idol. The food disappeared after the temple closed for the night. The priests claimed that Bel was a great god who personally accepted with pleasure the offerings of the King. To prove that the priests of Bel were stealing the food for themselves, Daniel arranged for his servant to scatter ashes over the floor of the temple before the King sealed it for the night. When the priests with their wives and children came during the night to eat and drink the offerings left for the idol, they left their footprints in the dust all over the floor. When the temple was opened in the morning, Daniel brought the King to see the footprints in the dust. The King could see for himself that these prints disappeared at the point of the wall where there was a secret entry. These footprints proved to be the downfall of the fraudulent priests of Bel.

3.2 Recording and Developing Footprints at the Scene of the Crime

Many years ago when I was a Boy Scout, our troop leader told us that we were going to make replicas of animal spoors. I hadn't the slightest idea what a spoor was. If I had consulted a dictionary, I would have discovered the following definition: "a track, especially of a hunted animal, derived from the dutch word spoor (a track)." In other words, a spoor is a footprint of an animal. Baden Powell, the founder of the Scouting movement, developed his ideas for Boy Scouts when he was serving in the British army: he fought in the Boer war at the beginning of the 20th century against the descendents of Dutch settlers in what is today called South Africa. The original language of the Dutch settlers of South Africa had changed over the years to become the modern language known as Afrikaans. In Afrikaans spoor means "track" (note that in modern Dutch "spoorweg" is the word for railway). Boy Scouts who find an animal spoor make a replica by mixing plaster of Paris with water and pouring it into the footprint. **Plaster of Paris** is made from Gypsum, which has the formula $CaSO_4 \cdot 2(H_2O)$ where Ca = calcium, S = sulfur, O = oxygen, H = hydrogen. Gypsum comes from a Greek word meaning "chalky." (Chalk is chemically different from Gypsum; it has the formula $CaCO_3$, where C is the symbol for carbon.) The part of the gypsum formula written $2(H_2O)$ is called water of hydration. One can drive out some of the water of hydration by heating gypsum. Plaster of Paris takes its name from the fact that it was first discovered near Paris, France.[1,2] When water is added to plaster of Paris, a paste is formed, which quickly sets to a near rock-hard consistency. An important characteristic of plaster of Paris is that it expands as it sets, thus making a very good copy of the foot print. In forensic science, specialists use the same technique that I learned as a Boy Scout to make casts of footprints and tire prints left on soft surfaces around the scene of a crime.

Modern day science can make copies of footprints which are much less obvious to the eye than those left in soft mud. In Figure 3.1 the pattern of a footprint left on a carpet is made visible using small plastic beads.[4] On many carpets, the act of walking across the surface creates an electrostatic charge on the surface of the carpet in the same way that a plastic comb can become electrostatically charged by combing one's hair. Like small pieces of paper are attracted to an electrostatically charged comb, small plastic beads sprinkled on a carpet are attracted to the parts of the carpet that are charged by the movement of the shoe over it.

This technique for developing footprints on a carpet was developed by Dr. Kurt Greenwood, head of the textiles products division of the Shirley Institute. (The **Shirley Institute** is a large research center located outside Man-

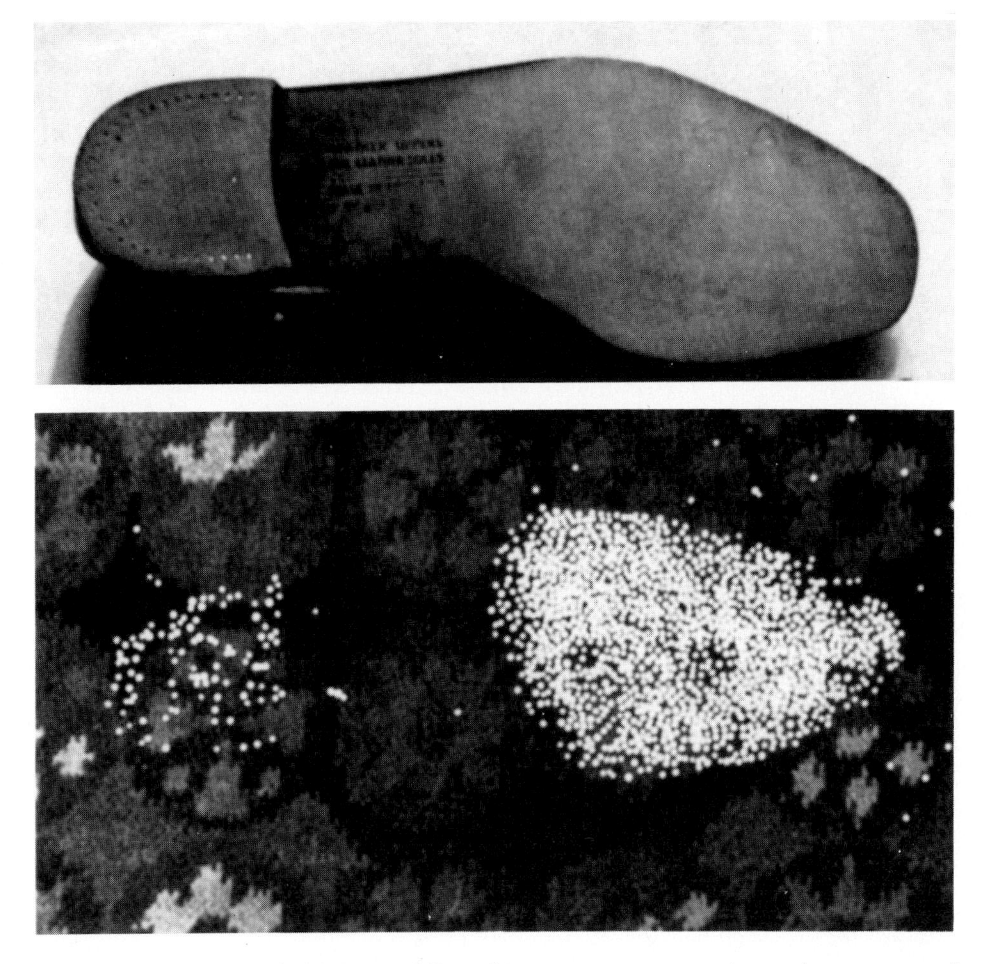

Figure 3.1. Triboelectric charging created by walking across a carpet can leave a latent pattern of footprints on the carpet. These can be developed (made visible) by sprinkling small, light, plastic beads on the carpet. a) Photo of the bottom of a shoe. b) Pattern made visible on a carpet by plastic beads clinging to the electrostatic charges created by the shoe. (Reproduced with permission from *New Scientist*.[4])

chester, England, specializing in the science of textiles.) Dr. Greenwood discovered the technique when he was investigating the problem of how to stop carpets from creating an electrostatic charge and, subsequently, shocking people who walked over them. In many modern carpets tiny metal fibers are woven in with the fibers of the nylon, or other material, used to make the carpet. These tiny metal threads create a conducting path to dissipate the charge built up by the friction generated from walking on the carpet. However, as pointed out by Illner, there are many square miles of carpets in many thousands of homes and offices which do not have this means of discharging electricity.[3,4]

Dr. Greenwood has co-operated with Scotland Yard and has been able to adjust the size and constitution of the beads to the point where the developed footprint will exhibit the whorl and line patterns of new shoes and the wear marks of old shoes. This technique works best on nylon and acrylic carpets, but vinyl carpets will pick up the pattern of charge in a dry atmosphere.[3,4]

Another, perhaps more universal, technique for looking at footprints left at the scene of a crime is based on the fact that many carpets exhibit a property known as **depression hysteresis**. *Hysteresis* is a Greek word meaning "to be late" or "to lag behind." This phenomenon is illustrated in Figure 3.2(a). When, after the pile of a carpet has been depressed by letting something stand on it, the pressure caused by this object is removed, the pile of the carpet does not spring back immediately but takes its time to creep back up to its normal level. If an object such as a table or chair is placed on a carpet for a considerable amount of time, the impressions left by that object are visible for weeks after the object is removed. In a more subtle manner, smaller depressions created by walking over a carpet leave traces not visible to the human eye, which can be made visible using a special technique known as **interference holography**.

In Figure 3.2(b) the equipment used by Dr. Bradford in his development of the technique for studying the footprints left on a carpet is shown. Dr. Bradford first made a hologram of the surface of the carpet. He then waited fifteen minutes before taking a second hologram, using the same photogra-

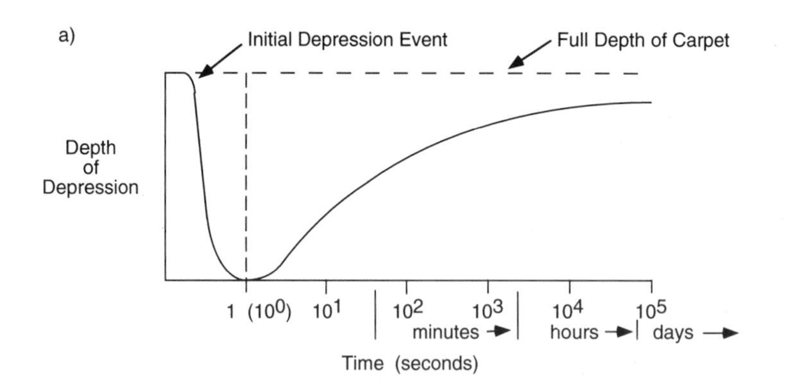

Figure 3.2. Footprints left as depressions on a carpet can be made visible by a technique known as interference holography.[4,5] a) After the pile of the carpet is depressed by a foot or a body, the fibres do not spring back immediately, but recover slowly – a phenomenon called hysteresis. b) Typical optical arrangement used to prepare a hologram. c) Holographic interference pattern created by capturing two images fifteen minutes apart on the same photographic plate, as footprints left on a carpet slowly recover. (Reproduced from *Science Spectrum*.)

b)

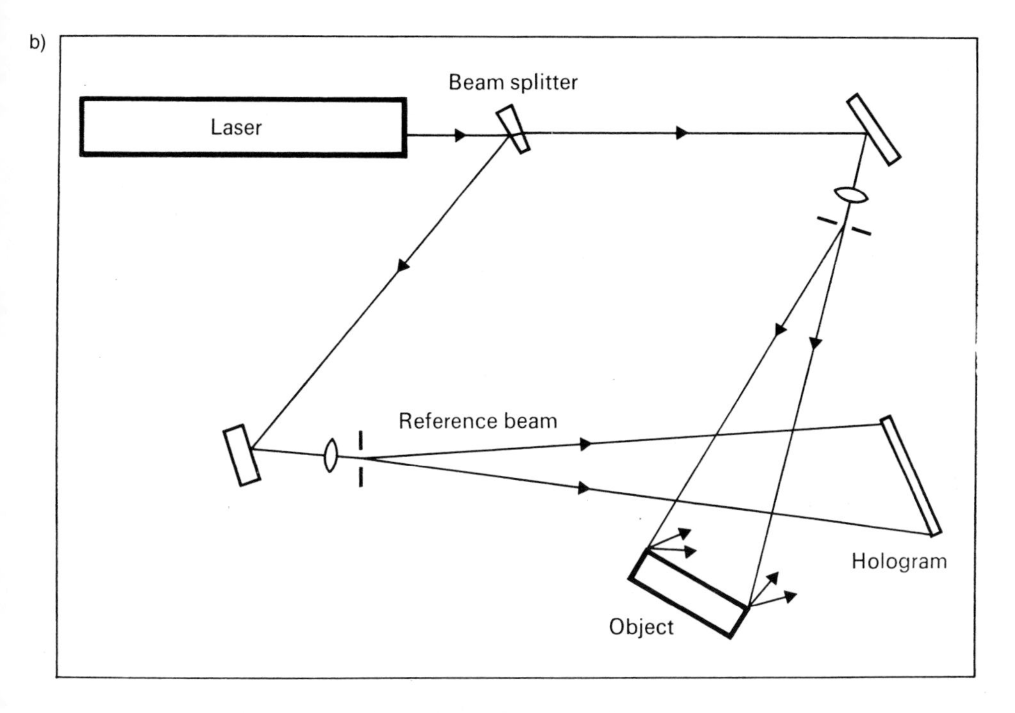

c)

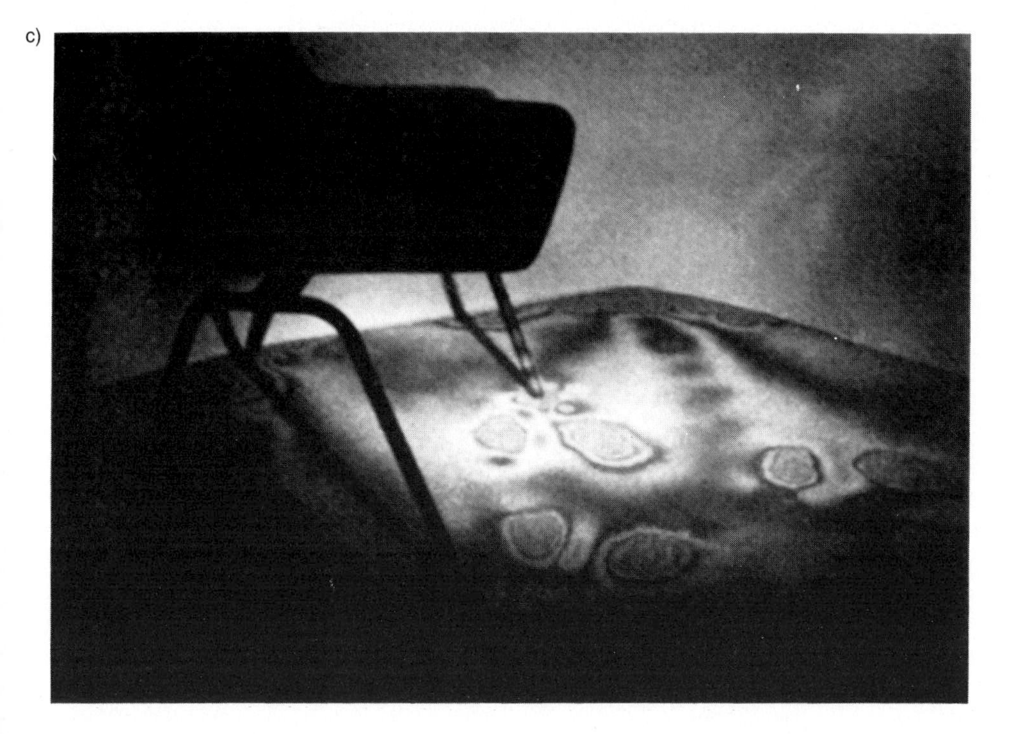

phic plate. In the period of 15 minutes, the surface of the carpet recovered a very small amount (a distance measurable in wavelengths of light rather than in millimeters of depression). When the hologram was illuminated with light to create an image, the two holograms recorded in the same photographic film behaved like the two slits of our interference pattern of Figure 2.12. In the same way that fringes appeared in the diffraction pattern of the two slits, interference fringes appeared in the reconstituted image created by the holograms. The pattern of footprints made visible by the interference fringes is shown in Figure 3.2(c).[5,6]

In Figure 3.3 an enlarged holographic image of a single footprint on a carpet shows the interference fringes generated by Dr. Bradford's technique. The pattern of lines on the footprint looks like contours of a hillside on a map and it shows how the sole of the shoe has worn from the way in which the person walked. This wear pattern is probably unique to an individual's

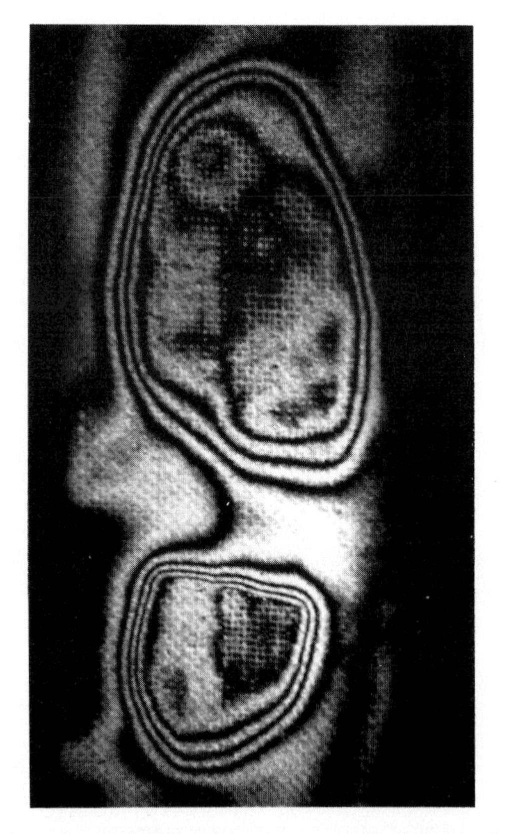

Figure 3.3. Enlarged view of a single footprint on a carpet developed by interference holography. Details of the wear pattern on the bottom of the shoe can be seen. (Reproduced from *Science Spectrum.*)

pattern of walking and could help to identify the person that made the print. Footprints left on tile surfaces can also be developed using Dr. Bradford's technique.

Because some substances absorb infrared radiation more readily than visible light, one can sometimes make a footprint visible by illuminating a surface with infrared light, as shown in Figure 3.4.[5]

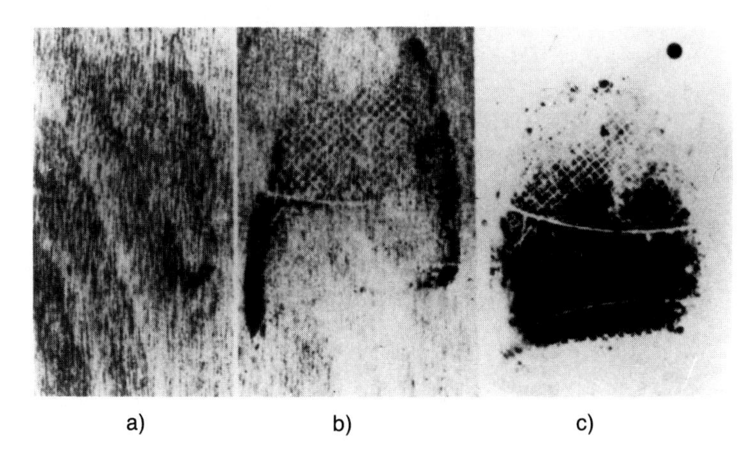

Figure 3.4. A latent footprint can sometimes be made visible using infrared light.[5] a) A wooden surface photographed with white light. b) The surface illuminated with infrared light. c) A print made with the suspect's shoe.

Bloody footprints left on a surface are almost impossible to scrub out. Their remnants can be developed using laser luminescence.[7,8] The results obtained with this technique are illustrated in Figure 3.5. Describing how the pictures of Figure 3.5 were obtained, Laing and co-workers report that they used:

> *Three types of blood...types B and O from humans and another type from swine. No anti-clotting agent was added to the blood. Latent bloody footprints were prepared on red painted board, red chemical fiber carpet, and black cotton cloth as soon as fresh blood was obtained. Some of the footprints on the red painted boards were rubbed strenuously and repeatedly with a brush and oil soap suds in cold water for about two minutes, two days after preparation.*[8]

To develop these invisible footprints, Laing and his co-workers sprayed the surfaces using a solution of a chemical known as **mercurochrome**. They report that in all cases the barely visible footprints became visible by means of fluorescence if the surface was illuminated with light from an argon-ion laser.

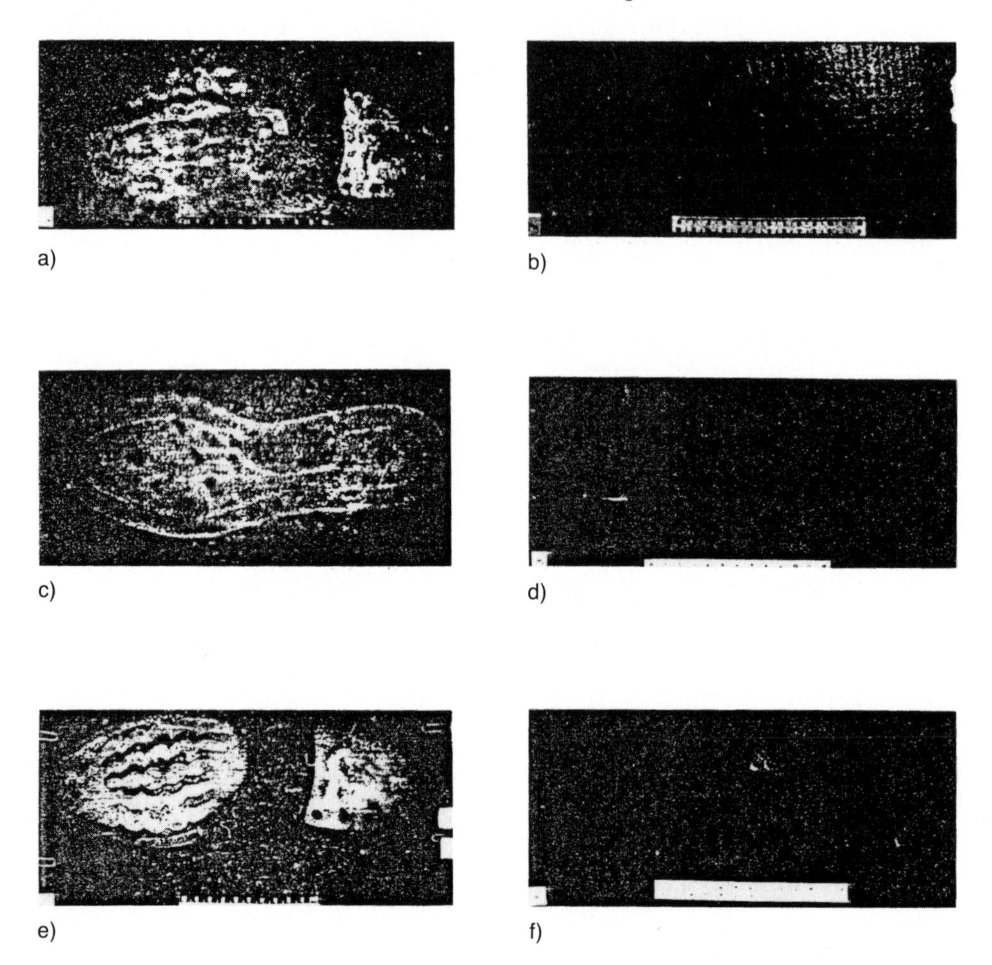

Figure 3.5. Latent bloody footprints can be developed on various surfaces after being sprayed with a chemical "Mercurochrome" and then viewed with light from an argon-ion laser. a) Latent footprint on a red painted board viewed with an argon-ion laser. b) Footprint of (a) viewed in white light. c) Latent footprint on a red carpet made with synthetic fibers viewed with an argon-ion laser. d) Footprint of (c) viewed in white light. e) Latent footprint on black cotton viewed with an argon-ion laser. f) Latent footprint of (e) viewed in white light. (Photos courtesy of Optical Engineering (SPIE Publications)).[8]

When a prosecutor attempts to use footprint evidence in a court case, he must be able to establish the uniqueness of the link between a footprint and the suspect being charged with the crime. Thus, in a court case, which took place in Winnipeg in 1983, the prosecution was conducting a case against a Mr. Nielson, 30, and his former police partner, Mr. Stolar, 35. These two people were charged with first-degree murder in the death, two years earlier, of Mr. Nielson's brother-in-law, Mr. Clear. The prosecution had presented evidence from an anthropologist living in the United States (an **anthropologist** is a specialist who studies the way in which humans have evolved, including their social behavior and biology; the term comes from the Greek word *anthropos* meaning "man") and a forensic scientist from Scotland Yard. Both of these men testified that it was highly likely that a bloody footprint found near the scene of the crime was left by Mr. Nielson. This evidence was challenged by Professor Peter Cavanagh, a Professor of Bio-mechanics at a university in the United States, who acted as a consultant to many shoe manufacturers. Professor Cavanagh paid 200 Pennsylvania State University students $ 5.00 each for their footprints, and an analysis of the data showed that 11 of 200 students who had been foot printed had footprints similar to the bloody one found at the scene of the crime. Furthermore, when he visited Mr. Nielson in jail, he found that the toes of the accused man could not fit into a shoe that would leave a bloody print resembling the one photographed. He concluded, "there was not enough extension in the accused's toes; I couldn't pull them over far enough." The prosecution had claimed that only one in one hundred people have feet as broad as that of the bloody footprint, whereas Dr. Cavanagh's data placed the evidence in the category of one in twenty, a far less convincing probability.[9]

The third word used in the title of this chapter, "vestige," needs some explanation. It comes from the Latin word *vestigium* which means "footprint" In modern English a **vestige** is "a trace of something, especially one that serves as a clue." Vestiges of a footprint may be as damning or useful in evidence as the whole footprint. In October of 1992, a defending lawyer in a court case in Scotland was able to demonstrate that marks on clothing of an accused had been made by the boots of police officers. **Boot polish** is a complex mixture of dyes, wax, fats and oils. Until recently such a mixture was difficult to analyze. A witness for the defence, Michael Cole, a chemist at the University of Strathclyde, was able to demonstrate by chemical analysis that grey scuff marks on the jeans and t-shirt of the accused were most probably caused by the policemen's boots. Stoddart, the accused person, was acquitted not only of charges of causing a breach of the peace, but also of head butting a policeman in the police car and molesting two officers in the police station. The Judge ruled that the evidence showed that the police had used excessive force when arresting the suspect and that Stoddart's subsequent actions constituted self-defence.[10]

Early in the 1990s, the forensic science service in Great Britain commissioned a study by Smith Systems Engineering, a consultancy group based in Guildford, Surrey, into the police use of information on fingerprints and footprints. Geoff Wyss, technical director of the study said:

> *For less serious crimes it is not worth collecting fingerprints. Even if the criminal has not worn gloves it is not worth the effort to eliminate fingerprints of innocent people in small-scale crimes.*[11]

Footprints, however, are often left in unusual places such as on a window ledge, in a flower bed, or on a piece of furniture. A *New Scientist* review of the study of Wyss gives the following figures.[11] In Great Britain each year there are on average 25,000 burglaries reported to the police. Footprints are left at the scene of up to 75 % of these crimes. Local forces sometimes collect information, but this is very ad hoc. Wyss states that a national data base of footwear would comprise around 2,000 patterns that could easily be searched by relatively unskilled officers on an ordinary computer. A relatively small number of features identify a footprint as surely as a fingerprint. Officers could encode the salient features of each print and not only discover the shoe manufacturer, but also link the footprint with those left at the scene of other crimes. Wyss says that image processing could also help forensic scientists to produce evidence from an incomplete print or from prints that have been masked by a highly patterned floor.

Lloyd reports that in some cases analysis of paint chips left behind in hit-and-run automobile accidents can be studied using the technique of fluorescence spectroscopy. This technique is described in the next section when we examine at the way in which one can analyze vestigial evidence left on the road by tire marks.

3.3 Tiresome Evidence

As already discussed, if a vehicle involved in a crime moves over soft terrain, the tire tread can leave tracks and these can be studied by means of plaster casts. These plaster casts can be compared to prints made from actual tires using white powder and black paper. A British forensic scientist, Dr. J. B. F. Lloyd, has shown that latent tire prints left on some surfaces can be made visible using ultraviolet light, as shown in Figure 3.6. Under the stimulus of the ultraviolet light, constituents of rubber tire debris in the latent tire print fluoresce.[12,13] When rubber is made into a car tire it is compounded with other materials known as **extender oils**, which are added to make the rubber more flexible. Extender oil is one of the main materials that fluoresce when left behind on a surface – rubber itself does not leave florescent traces under ultraviolet light. The amount of extender oil in a rubber tire varies from one manufacturer to another, and depends on the age of a tire. Good quality tire prints can be retrieved from little used concrete or paved surfaces. However, bitumen or tar present in the surface being investigated fluoresces strongly and wipes out any pattern created by a tire. A tire print, which became highly visible in ultraviolet light is shown in Figure 3.6(a).

When comparing patterns left at the scene of a crime with those of a suspect's vehicle, one makes use of what can be described as **stochastic wear**

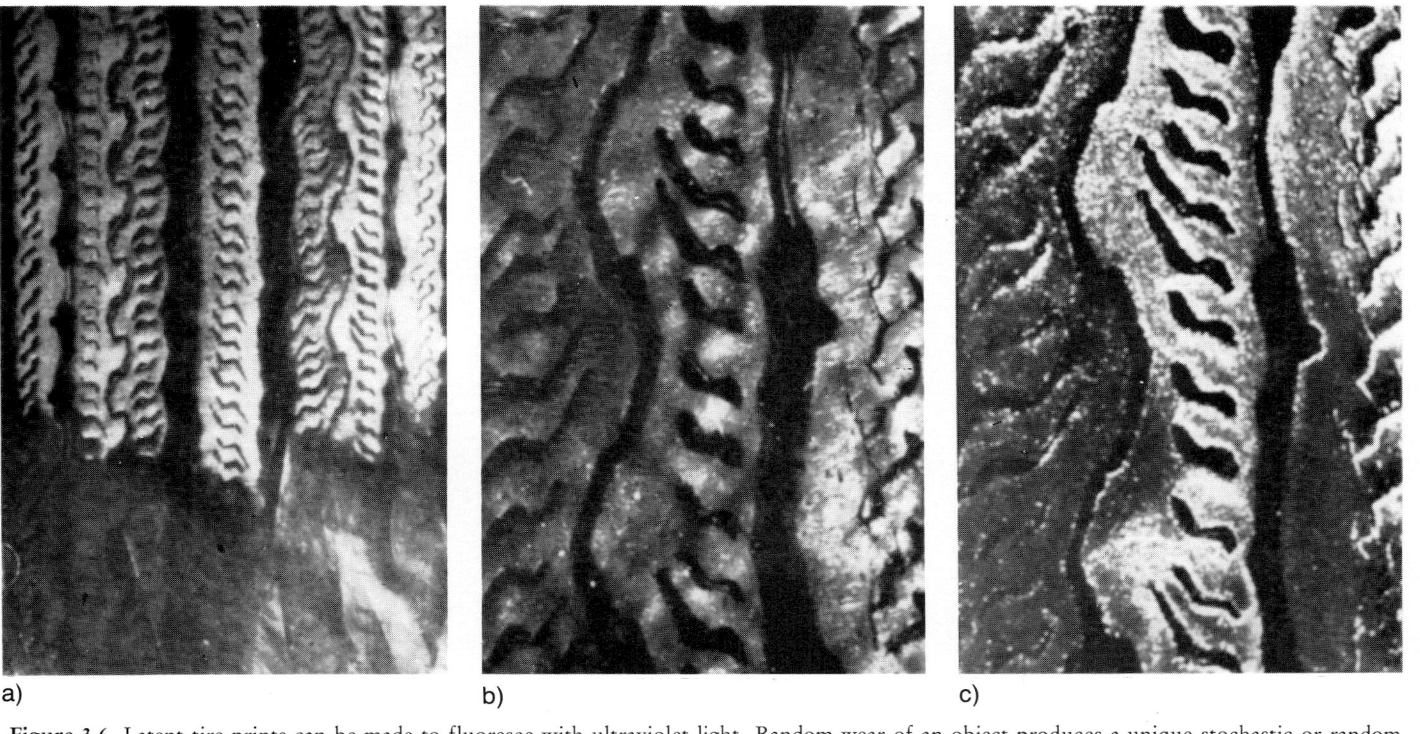

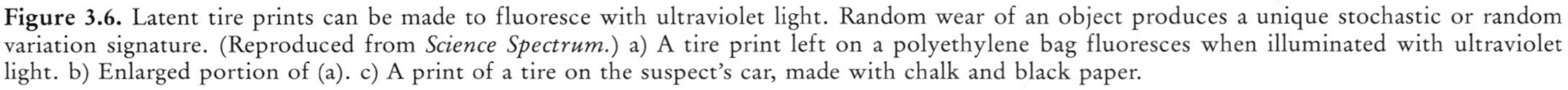

Figure 3.6. Latent tire prints can be made to fluoresce with ultraviolet light. Random wear of an object produces a unique stochastic or random variation signature. (Reproduced from *Science Spectrum.*) a) A tire print left on a polyethylene bag fluoresces when illuminated with ultraviolet light. b) Enlarged portion of (a). c) A print of a tire on the suspect's car, made with chalk and black paper.

signatures. In physics, a **stochastic variable** is one that behaves in an unpredictable and random manner. Many effects contribute to the wear of a tire. These range from forces on the road, the way in which a driver drives, to the terrain over which the tire is driven. Because of the stochastic (random) interaction of these forces, it is virtually impossible for the wear patterns on different tires to be identical. Thus, in comparing the enlarged photographs of Figure 3.6(b) and (c), we can see that nicks in the tire surface are unique to the wear pattern of that tire. The fact that random wear patterns are unique to a particular object is exploited when comparing cut marks made by a knife or marks left on an object with a tool such as a pair of pliers. In Figure 3.7 (a) and (b) stochastic wear signatures of a knife and a pair of pliers are shown. In the same way, the two pieces of a fractured object have a pattern unique to these two parts. Thus, in Figure 3.7 (c) the two surfaces of a broken wire are shown.[14] (The pictures of Figure 3.7 (c) were made with a scanning electron microscope. The operation of this instrument is described in the next chapter.) The pattern of random lines constituting a stochastic wear signature, such as that of Figure 3.7 (b), are called **striations**. The word striation comes from the Latin word for the furrow left by a plow in a field. With a little imagination, the set of lines in some stochastic wear signature looks like the picture of a ploughed field.

It is interesting to note that the unique patterns left by tearing a piece of paper, or fracturing an object, were utilized in medieval times to protect apprentices or servants who could not read signed contracts. The contract for services and training was torn in two. The servant or apprentice was given one half of the paper, and the master retained the other half. This was done because it was virtually impossible to create a copy (constituting a changed contract) because of the unique pattern of the torn edge of the document. The torn edge of the document looked as if it had been nibbled by a sharp-toothed mouse. Hence, the document was called a document of **indenture**: from the Latin words for "bitten into with teeth," the same Latin word has given us the word dentures, for false teeth. In other words, the **indentured servant** or apprentice had a unique, rough, stochastic signature, in the form of their half of the contract, to help protect them against fraud.

The uniqueness of the structure of broken surfaces was used in ancient times to establish the credentials of messengers used in war time. Thus, a piece of clay would be broken in two and the messenger would have one half; the general, who was to receive a message later on in the campaign, would keep the other half. If the messenger appeared with commands from the other general, the messenger had first to show that his piece of pottery matched the broken edge held by the general. Putting the two broken halves together created a validation symbol. (The word symbol comes from the Greek word *syn* meaning "together" and *ballein* meaning "to throw;" the symbol of validation was created by "throwing the two pieces together." Ballein has also given us the term ballistic missile, for a missile that is

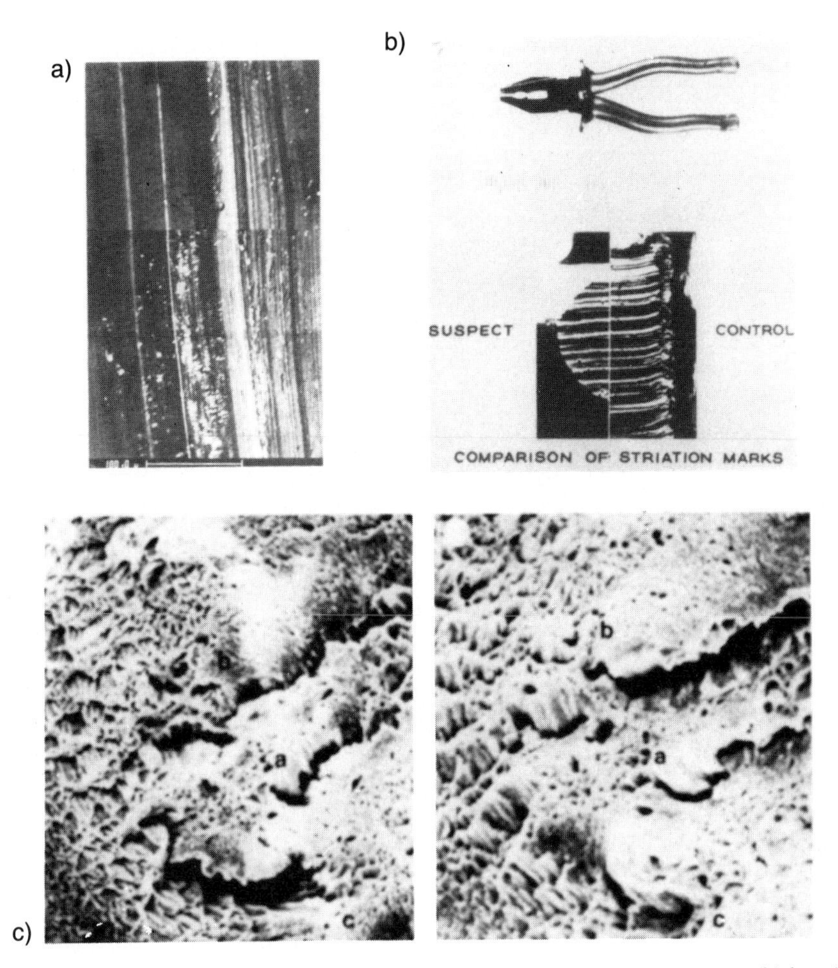

Figure 3.7. Random wear and breakage patterns can leave "stochastic signatures," which uniquely identify objects associated with a crime.[14] a) Scratches left by a knife at the scene of a crime. b) Scratches left by the jaws of a pair of pliers at the scene of a crime compared with those left by the suspect's pliers. c) Comparison of a broken surface with its matching surface on the other piece of the broken object.

thrown at a target.) In modern-day court procedures the matching of two pieces of broken material can establish their unique correlation. Thus, if a fractured piece of the marker light of a car results in one piece being left on the victim of a hit-and-run accident, then matching this piece with the broken edge of the remainder of the light would constitute unique evidence that the vehicle was involved in the crime. Again, in Figure 3.8 the unique matching of the fractured edge of a knife tip, found in the body of a murder victim, with the broken end of a knife, in the possession of a suspect, establishes that this knife was used when the murder was committed.

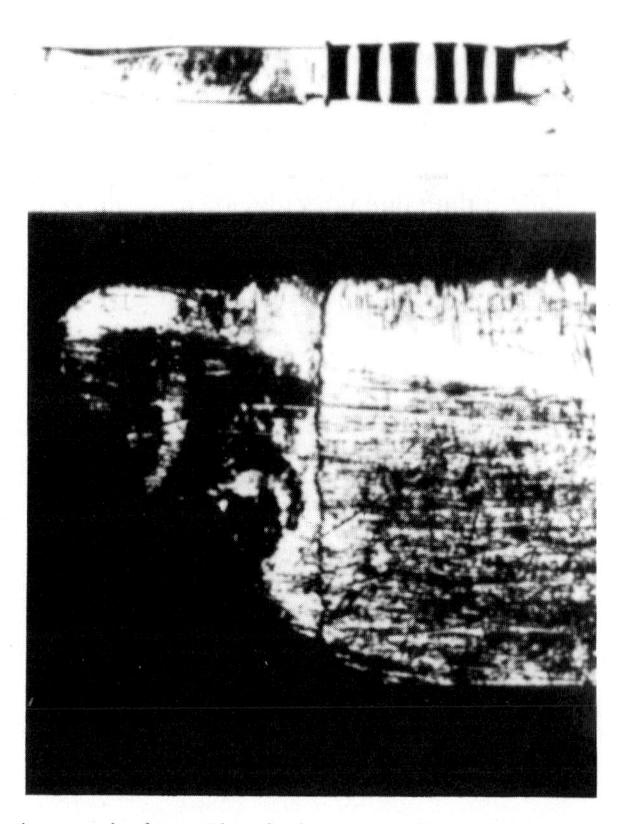

Figure 3.8. The unique match of two sides of a fracture provides evidence that the tip of the knife found at the scene of the crime and the main blade of the suspect weapon belong together.[14]

Ultraviolet light can sometimes be used to make latent tire prints visible, as shown in Figure 3.6. In this case compounds left behind in the tire print debris fluoresced. In other instances infrared illumination can make the tire pattern visible. Figure 3.9(a) shows the back of a jacket belonging to a victim of a hit-and-run automobile accident, as viewed with white light. When the same article of clothing is viewed in infrared light, some of the tire marks from the vehicle become visible, as shown in Figure 3.9(b). In this case the latent print becomes visible because the material of the jacket reflects the infrared light, whereas the dirt and debris from the tire print absorbs it.[15]

Sometimes a print on a surface can be made visible using a combination of chemical and physical techniques. Thus, Lloyd reports that if a weak print is sprayed with lacquer consisting of a solution of poly(vinyl acetate) in ethyl acetate, the florescent material hidden within the particles left in the tire

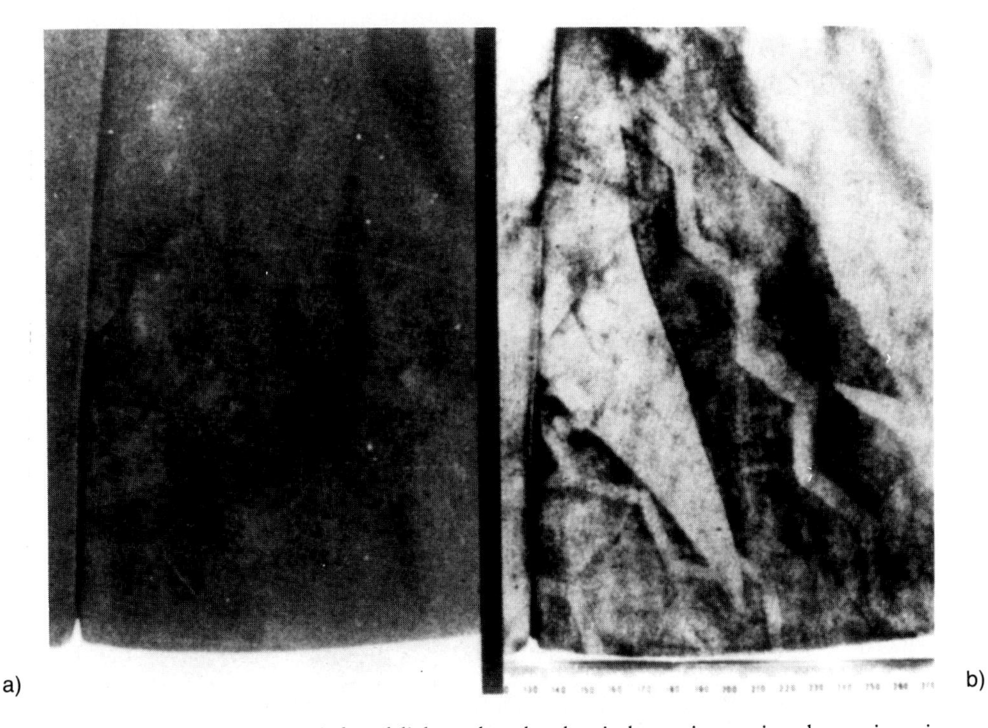

Figure 3.9. In some situations infrared light makes the chemical constituents in a latent tire print visible. a) Jacket of a hit and run victim viewed in white light. b) The jacket of (a) photographed in infrared light.[15]

print will be brought to the surface of the rubber debris by the ethyl acetate. The particles come to rest on the surface of the debris as the lacquer evaporates.[12,13] In Figure 3.10 the appearance of a lacquered tire print, when made to fluoresce with ultraviolet light of wavelength 366 nanometers and viewed through a yellow filter, is shown. This tire print was left on a concrete surface, and the difference between Figure 3.10(a) and Figure 3.10(b) shows the improvement made by the lacquer coating technique. The treatment of tire prints with a polyvinyl-based lacquer prevents prints from fading. This technique can also be used to preserve a print which otherwise would dissipate over a few days, because of the effects of both weather and humidity on the surface bearing the print. Lloyd tells us that the fluorescence of lacquered prints may be further intensified by treating them with liquid nitrogen. The intense cold produced by this technique (−186 °C) changes the physical properties of the chemicals in the print, and causes them to phosphoresce and fluoresce when illuminated by stimulating ultraviolet light. If an iodobenzene lacquer is used, the area of the tire print becomes intensely phosphorescent. Dr. Lloyd tells us that in some circumstances rubber-soled shoes

Figure 3.10. Special lacquer can be used to increase the visibility of a tire print when viewed with ultraviolet light. This lacquering of the print can also preserve the print.[12,13] a) Untreated portion of a tire print viewed in ultraviolet light. b) Treated portion of the same tire print viewed in ultraviolet light.

and other plastic materials can leave vestiges on surfaces, which can then be made to fluoresce using the same technique. On the first occasion that fluorescent tire prints were observed during a murder investigation, experts were able to obtain a complete set of prints from the vehicle thought to have carried the victim's body. From these prints the track width, wheel base, and make of the vehicle were determined. The two makes of tires involved were identified and analysis of the fluorescent light enabled a match to be made between the tires on the vehicle and the prints found at the scene of the crime. In the same way, an oil drip was matched to the contents of the vehicle's oil pan. These two matching exercises left no doubt that the expert had identified the vehicle involved in the crime.

The techniques for analyzing fluorescent light given out by the debris in a tire print have been developed by Dr. Lloyd to the point where one can identify the nature of the rubber debris left in, for example, skid marks, rubber debris scraped from an area where the vehicle accelerated or decelerated, and from debris left on a victim's body. The main identification technique used by Dr. Lloyd is known as **synchronous fluorescence**. The physical principles of this method are illustrated in Figure 3.11. The chemical studied in Figure 3.11 (a) is a solution of a compound **perylene**, (which is often found in rubber tire debris), in a solvent known as cyclohexane. The broken curve shows how the florescence at a fixed wavelength changes as the wavelength of the stimulating light is varied. The solid curve superimposed on the dotted line shows the range of wavelengths of fluorescence when a compound is irradiated at a fixed wavelength. When the wavelength of the stimulating light and that of the observed florescent light are varied by the same amount at the same time, the output is a simplified unique fluorescent signature, as shown in the lower part of Figure 3.11 (a). (Synchronism means at the same time; the term comes from the Greek words *syn*, "together," and word *chronos*, "time." The newspaper as a **chronicle** is supposed to be a timely report of news.) In Figure 3.11(b) Dr. Lloyd reported data obtained by studying scrapings taken from a skid mark. He dissolved the bits in an appropriate fluid and studied them by synchronously varying the wavelengths of the stimulating light and the wavelengths at which he observed fluorescence. He carried out the experiment three times using various wavelength differences between stimulating light and observed fluorescence. The data of Figure 3.11 (b) established that a suspect's vehicle had left the skid marks on the road surface. The same type of investigation can be used whenever one suspects that a vestige of a rubber complex has been left at the scene of a crime. For example, this technique has been used to study the debris from a rubber-based **eraser**. (In British terminology an eraser is known as a **rubber**.) In a short article in which a scientist gave recipes for making pencil and ink erasers, the following comments on the various types of erasers found in an average office were made:

A good pencil eraser removes a pencil mark without damaging or defacing the paper. To do so it must lift the fine particles of graphite from the surface without causing the paper to wear. It then must convey the fragments of print away in rolls of crumbly rubber. To achieve the necessary characteristics, the rubber needs to be relatively soft and for this reason rubber-based erasers for use with pencils have a higher proportion of the chemical known in the trade as "factice," which aids crumbling of the rubber. Formulation 1 (of Table 3.1) contains 25 % of volume of rubber and this gives the eraser sufficient strength to rub out hard pencil marks. The second formulation contains only 13 % of rubber since it is intended for use with large areas of soft pencil marks. Ink marks penetrate into the paper surface and so some abrasion of the paper along with the printed letter becomes necessary to remove material on the surface. For this reason an eraser intended for use with ink has three times the amount of abrasive powder in its constitution.[16]

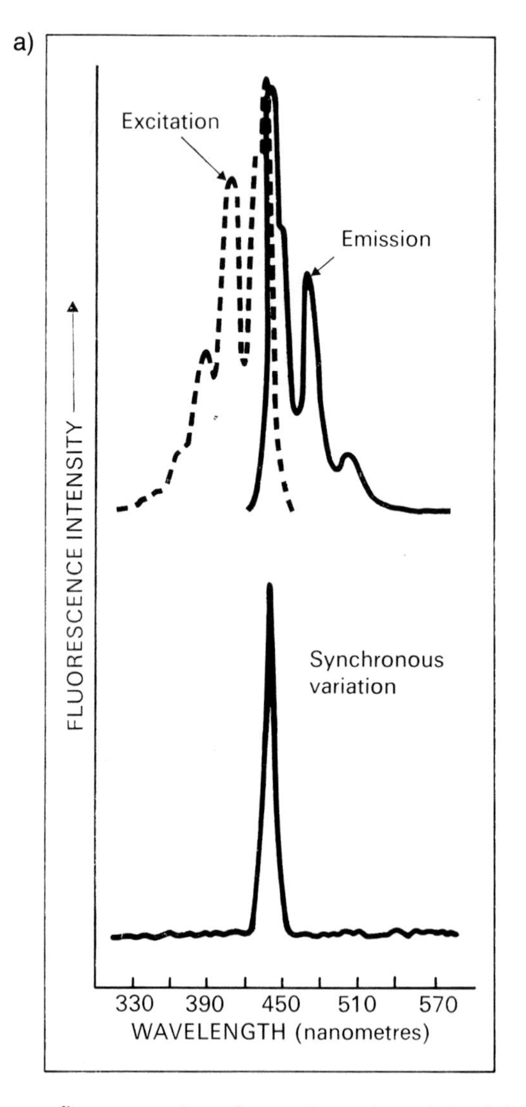

Figure 3.11. Synchronous fluorescence is used to study vestiges of tires left in skid marks. (Reproduced from Science Spectrum.) a) Excitation and emission spectra of perylene, a compound often found in tire prints. b) The correlation between rubber debris left in a skid mark (solid line) and the tires of a suspect's vehicle (dashed line) can be established using synchronous fluorescence studies.

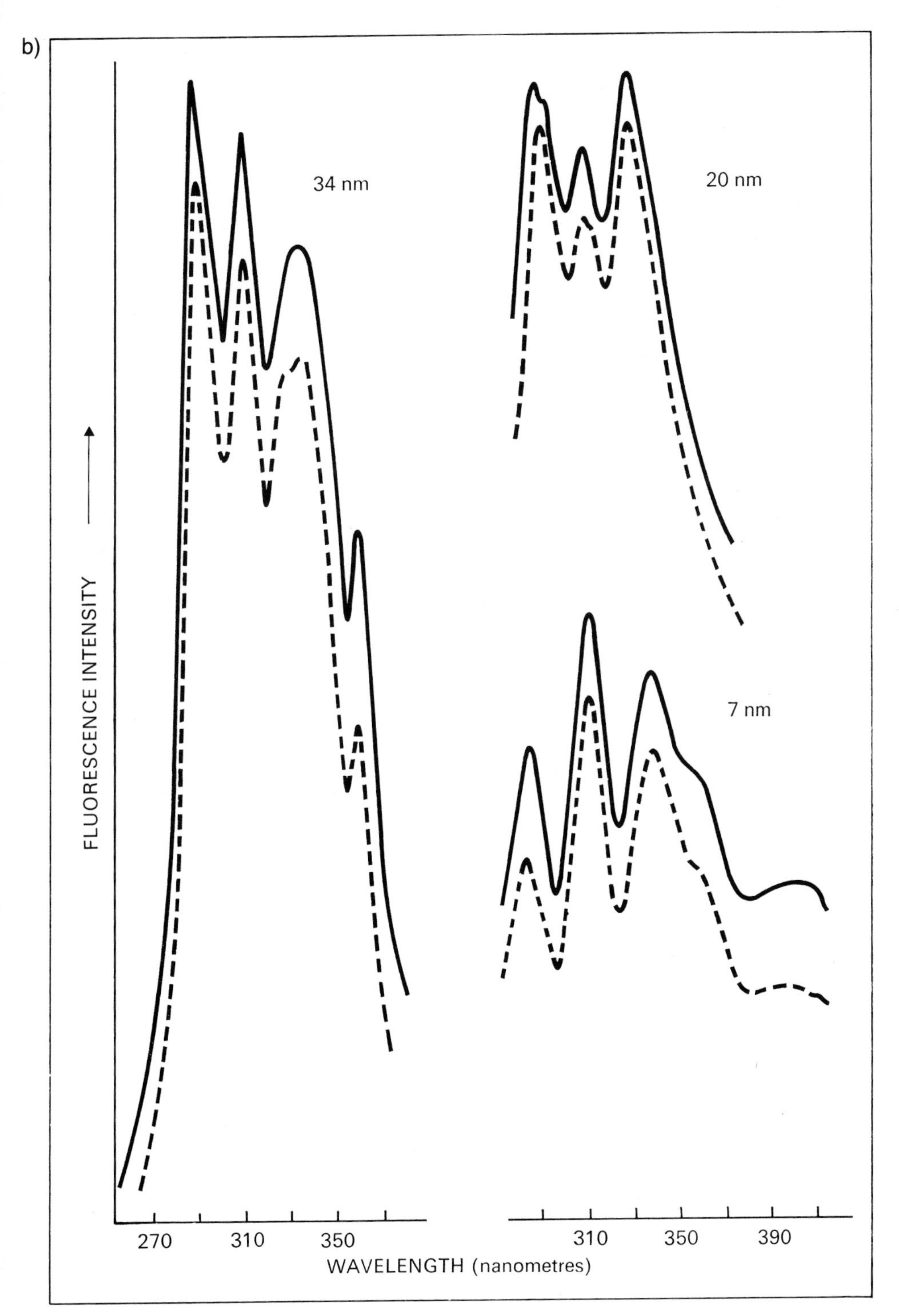

b)
34 nm
20 nm
7 nm
FLUORESCENCE INTENSITY
270
310
350
310
350
390
WAVELENGTH (nanometres)

Table 3.1. Eraser Formulations.[16]

Constituent	Pencil Type 1	Eraser Type Pencil Type 2 (Parts by Weight)	Ink
Natural rubber	100	100	100
White factice	100	400	50
Magnesium oxide	30	30	30
Lithopone	100	–	20
Barytes	80	–	–
Whiting	150	300	–
Pumice	–	50	150
Titanium dioxide	10	–	5
Paraffinic/naphthenic oil	30	100	–
Zinc oxide	10	10	10
Stearic acid	2	2	2
Sulphur	8	7	6

Pumice powder used in erasers is made from the rock **pumice**, which is found around volcanoes. When found in lump form, it is a highly porous rock which floats in water. It is made up of sand-like chemicals and can be broken down to make a very fine polishing powder used in domestic cleansers. **Whiting** is the trade name for a finely divided calcium carbonate occurring in crystalline form called **calcite**. This crystalline form of calcium carbonate is a mild abrasive which has been used in toothpaste. **Titanium dioxide** is added to the mixture to add whiteness, and other ingredients add color. Chemical analysis and fluorescence studies of debris from an eraser can establish a unique match between the eraser and the debris, because of the varied structures of eraser constituents made by different manufacturers.

3.4 First and Last Impressions

Most of the readers will have seen, at one time or another, a movie in which the detective seeking information at the scene of a crime picks up a pad of paper and notices that there are impressions on the surface of the top sheet. These impressions were obviously left behind by writing on the page that was on top, which was then torn off. A simple way of making such impressions visible is to take a soft pencil and place it at an angle to the surface of the paper and to move the pencil back and forth lightly. This creates pencil markings, which do not go into the indentations and, as a consequence, the message which had been written on the top page of the pad becomes visible on the surface of the next piece of paper. The depth of the impression created obviously depends on the type of paper as well as the type of instrument used to write the message.

The introduction of ball-point pens has made this type of retrieval of messages easier, because people tend to press more firmly when using a ball-point pen than they would, or could, if they were using other types of writing instruments. If the top message is written with a felt pen or an ordinary fountain pen, there is much less chance that a significant impression is left on the lower page, as any one who has tried to fill out triplicate forms has discovered for himself. In the late 1970s workers at the London College of Printing developed a technique for making pen impressions left on a piece of paper visible to the naked eye. In this technique a thin plastic film is placed over the paper supposedly bearing impressions and then pulled down tightly onto its surface by sucking air through the paper (most paper is porous). The two surfaces which are now in close contact are charged electrostatically in a manner similar to the method discussed for developing latent fingerprints on a check in Chapter 2. The electrostatic pattern created on the plastic film is then developed by sprinkling dry powdered ink, used in xerographic copying machines (called **toner**), over the plastic surface. The dry ink clings preferentially to the indentations on the page, thus making the writing visible. Heating of the electrostatically developed pattern with an infrared lamp fuses the toner to the surface. This becomes a permanent record of the information present in the indentations on the paper. The scientists who developed this technique are not exactly sure of how the technique works, but it can provide a very clear piece of handwriting as demonstrated by the example shown in Figure 3.12.[15] The plastic film can be peeled off the surface of the document, leaving it unharmed for further storage and/or use.

The electrostatically based technique for developing visible messages from indentations on the paper surface is known as **ESDA** from the phrase **e**lectrostatic **d**eposition **a**nalysis. This technique was used to provide evidence, which led to the acquittal of a person jailed for participating in the murder of a policeman during riots in a North London (England) housing estate in 1985. The individual, Winston Silcott, was acquitted on the basis that the confession, which the police claimed was made by Winston Silcott at the time of his arrest, appeared to be tampered with. Thus, the information that was developed from the indentations on a sheet of note paper that was underneath the one on which his confession was written disagrees with the written confession, indicating that changes may have been made to the top sheet at a later date. In Chapter 5, where we examine the famous terrorist case in Great Britain known as the Birmingham Six, we will find that the same type of evidence was used to show that written confessions had been tampered with.

Figure 3.12. Handwriting impressions left on a note pad can be made readable using a plastic sheet and black toner powder.

References

1. J. Ayto, *A Dictionary of Word Origins*, Arcade Publishing, New York, 1990.
2. *The Merriam Webster New Book of Word Histories*, Merriam Webster, Springfield, MA, 1991.
3. O. Illner, "Tracing the Hidden Footprint," *Science Spectrum 100* (1972), 10.
4. "Keep Trails of Criminal Feet" *New Scientist*, 12 October 1972, 86.
5. W. R. Bradford, "Light on the Invisible Footprint," *Science Spectrum 190* (1976), 2.
6. F. Smyth, *Cause of Death*, Van Nostrand Reinhold, New York, 1980.
7. S. G. Cheng, "The Use of Frequency-Doubled Nd:YAG Laser for Enhancement of Weak Bloody Fingerprints," *J. Forensic Science 33* (1988), 1022.
8. E. J. Liang, S. P. Shen, D. C. Zheng, Y. Zhu, B. Yuan, H. O. Song, "Fluorescence Enhancement of a Latent Bloody Footprint Sprayed with Mercurochrome Reagent," *Optical Engineering 31* (1992), 232.
9. B. Gory "Professor Challenges Footprint Evidence," (news story), *Toronto Globe and Mail*, July 5 1983, A14.
10. W. Brown, "Boot Test Shows Police Force Rubs Off," *New Scientist*, 17 October 1992, 7.
11. "In Burglar Shoes," *New Scientist*, 8 May 1993, 18.
12. J. B. F. Lloyd, "Luminescence Traces Tyres," *Science Spectrum 145* (1976), 2.
13. J. B. F. Lloyd, "Make the Traces Glow," *Industrial Research*, 15 November 1977, 29.
14. O. Johari, I. Corvin, "Scanning Electron Microscopy and the Law: A Review," *Canadian Research 9* (1976), 24.
15. J. Broad, *Science and Criminal Detection*, Macmillan, London, 1988.
16. "Erasers – Pencil and Ink Eraser Formulations" *Natural Rubber Technology 14* (1983),41.

Chapter 4

Fragmentary Evidence and Tell-Tale Dust

Chapter 4

Fragmentary Evidence and Tell-Tale Dust

4.1 Locard's Principle

When two objects, such as a car and a hit-and-run victim, come into contact, tiny pieces of material are inevitably transferred from one object to the other. This exchange of fragmentary material between two bodies in contact is known in forensic science as **Locard's principle**, named after Dr. Edmond Locard. He was a director of the technical police laboratory in Lyon, France, in the early part of the 20th century. The formal statement of Locard's principle is:

> *There is no such thing as a clean contact between two objects. When two bodies or objects come in contact they mutually contaminate each other with minute fragments of material.*[1–3]

Dr. Locard tells us in his writings that he was inspired in his development of forensic science technology by the Sherlock Holmes stories written by Sir Arthur Conan Doyle. Palenik quotes Locard in 1925 as stating:

> *A police expert or examining magistrate would not find it a waste of time to read Doyle's novels. I must confess that if at the Laboratory of Lyon we are interested in any unusual way in this problem of dust it is because of having absorbed the ideas found in the writing of Gross and in Conan Doyle and also because certain investigations in which we became involved happened, so to speak, to force the issue.*

In one of his articles Locard states:

> *the microscopic debris that covers all our clothing and bodies is the mute witness sure and faithful of all our movements and all encounters.*

Perhaps one of the Sherlock Holmes stories that inspired Locard is the story entitled "The Adventures of Shoscombe Old Place." This story begins with the following account:

> *Sherlock Holmes had been bending for a long time over a low-power microscope. He straightened himself up and looked around at me in triumph 'It's glue Watson,' he said, 'unquestionably it is glue. Have a look at these scat-*

tered objects in the field of view, those threads are from a tweed coat. The irregular grey masses are dust. They are epithelia scales on the left. Those brown blobs in the center are undoubtedly glue.'[1]

Holmes then tells Watson that the dust was from a cap found beside a dead policeman. The structure of the dust confirmed the fact that the suspect in the case was a picture frame manufacturer who habitually handled glue.

Hans Gross, the individual referred to by Locard, was an Austrian police official who supposedly was the first to actually use the study of dust to solve a crime. Gross states:

The clothes of a suspect were placed in a strong paper bag and beaten with a stick. The dust collected was examined microscopically. It was found to consist principally of sawdust and glue confirming the occupation of the owner of the clothing as a cabinet maker.[2]

In this chapter, we shall look at the various ways in which forensic scientists can make fragments found at the scene of crime tell their tale of violence and yield incriminating evidence.

4.2 Pigmented and Refractive Fragments

The power of fragmentary material to guide the police in their work is illustrated by a story, which appeared in the Toronto Globe and Mail in December, 1978. The story relates how paint fragments from a car found in a Sudbury wrecking yard led to the arrest of a Toronto man involved in a 1975 hit-and-run accident. This accident left a boy, Howard Sobell, with serious head injuries. The boy was hit by a car while riding his bicycle and the car did not stop. Police routinely check auto-wrecking yards to see if they can find evidence of cars that may have been involved in crimes at an earlier date. In this particular case they found a 1969 Chrysler Coronet, which had been scrapped in the Sudbury yard. (Sudbury is a mining town 250 miles North of Toronto.) Police were able to take paint samples from the car, indicating that this car was involved in the 1975 accident in Toronto. As a consequence of this paint chip evidence a person was charged with criminal negligence, causing bodily harm, and failing to remain at the scene of an accident.

Matching paint specimens found on victims of hit-and-run accidents or at the scene of the crime with those left on the clothing of suspects can sometimes be achieved by visual inspection, usually when the object from which the paint fragments were removed had been painted several times. For this type of flake, the number of layers of paint, the sequence, and the thickness can be matched against each other. In general, one must look at the structure of the paint fragment to be able to establish a relationship between two fragments.[4,5]

Paint is a complex mixture of pigment and other powders suspended in a supporting liquid. The process of paint drying generates a structure in which the pigment and other powders are held in position. In Figure 4.1 a transmission electron microscope picture of a relatively simple paint film is shown. **Transmission electron micrographs** are made by passing a stream of electrons through a material to be photographed. The pattern of absorption created by the various components of the material generates a pattern on a photographic film, which is then developed like a normal photograph. Such pictures are often referred to briefly as **TEM picture** (or images) to distinguish them from pictures taken using a **scanning electron microscope**. Pictures taken with the latter instrument are known as SEM photographs. In SEM pictures the electrons used to create the picture do not pass through the specimen. So-called secondary electrons are collected when a primary-electron beam hits the surface being examined. These secondary electrons are thrown out from the surface because of the energy of impact of the primary electrons. As the beam of electrons scans the surface, the number of secondary electrons is measured, and converted to a signal, which is used to create a picture on a television screen. This image can then be photographed.

In Figure 4.2 two SEM pictures taken by Johari and Corvin are shown. The first picture shows the filament of a car headlight, which was fragmented by a collision when the light was off. The second shows a filament that fractured on impact when the headlight was on and the filament was hot.

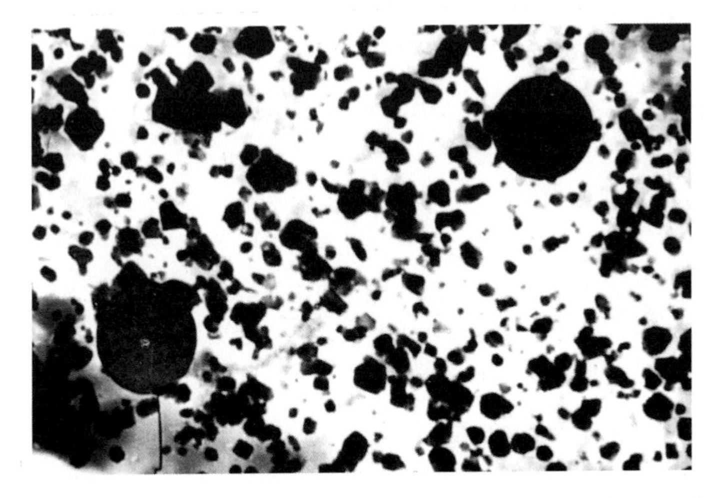

Figure 4.1. Paint is a composite material made up of an organic binder holding together a mixture of pigments and powders. In the relatively simple electron microscope picture of a section through a paint film, the large opaque circles are color pigments, while the smaller ones (which are about 1 micrometer in size) scatter light to make the paint opaque to light.

Since the fracture patterns for on and off in Figure 4.2 are obviously very different, we can establish whether a motorist had his headlights on or off when involved in an accident.[6]

Commenting on the problem of characterizing paint fragments, John A. Cooper, a forensic scientist at the Oregon Graduate Center at Beaverton, Oregon, states:

> *the initial examination of the paint fragment by microscope comparison, attempting to match physical characteristics such as color, layer structure, gloss, textural features, is only rarely successful.*[7]

He recommends an alternate technique called **pyrolysis gas chromatography**. The word pyrolysis means freeing by fire (the Greek word *pyra* has given us the term **pyrotechnics** for the art of making fireworks). In a pyrolysis investigation, a paint film is be heated in an oxygen-free piece of equipment, which results in the breakdown of the organic matrix of the paint into its constituents. These can then be analyzed by means of gas chromatogra-

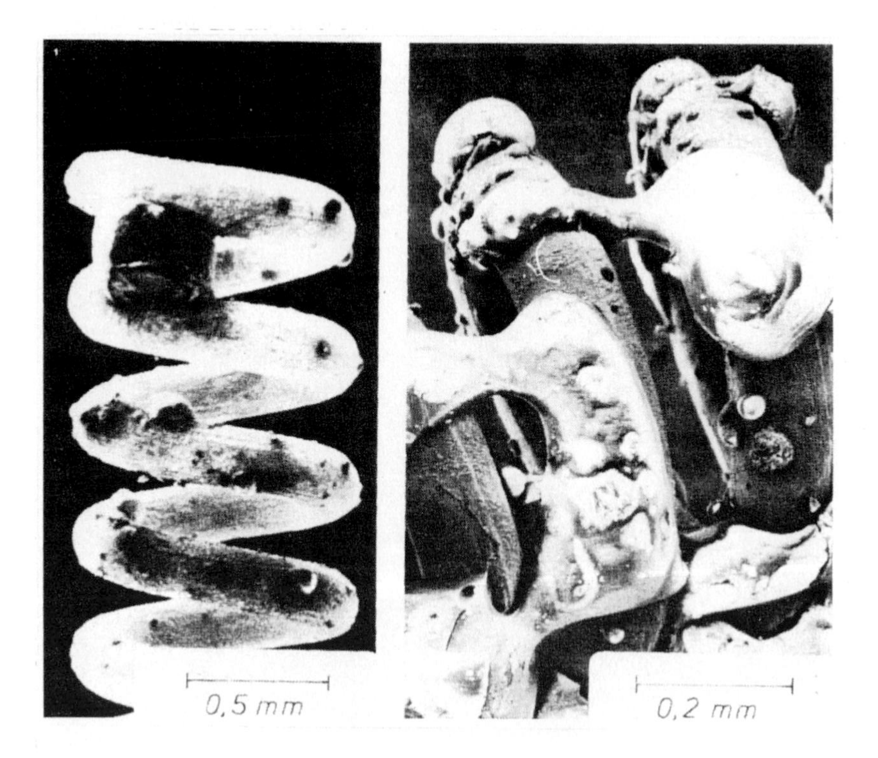

Figure 4.2. The condition of the filament in a broken automotive head light can indicate whether the lights were on at the time the light was broken.[6] a) Filament of a lamp that was off when it was broken. b) The filament of a lamp that was on when the lamp was broken shows splashes of molten metal.

phy. A full discussion of gas chromatography is delayed until Chapter 7 when we look into the problem of building synthetic bloodhounds to search for explosives and drugs.

Cooper seems to favor a technique known as **energy dispersive x-ray fluorescence**, referred to as **EDXRF** for short, for studying paint chips. X-rays are electromagnetic waves with wavelengths much smaller than ultraviolet light. In Figure 4.3(b) the whole range of electromagnetic waves found in the universe is illustrated. As can be seen from this diagram, x-rays can have wavelengths as small as a thousandth of a nanometer, while longer x-rays overlap with shorter ultraviolet wavelengths. Sometimes very short wavelength x-rays are called **hard x-rays**, whereas the longer wavelength x-rays are called **soft x-rays**. When these rays were first discovered, the scientists named them rays x-rays, because they did not know what type of radiation they had found ("x" signifies the unknown).

In some parts of the world, x-rays are called Röntgen rays, after the German scientist who first discovered these energetic electromagnetic waves.[8] For his discovery, Röntgen was awarded the Nobel Prize in Physics in 1901;

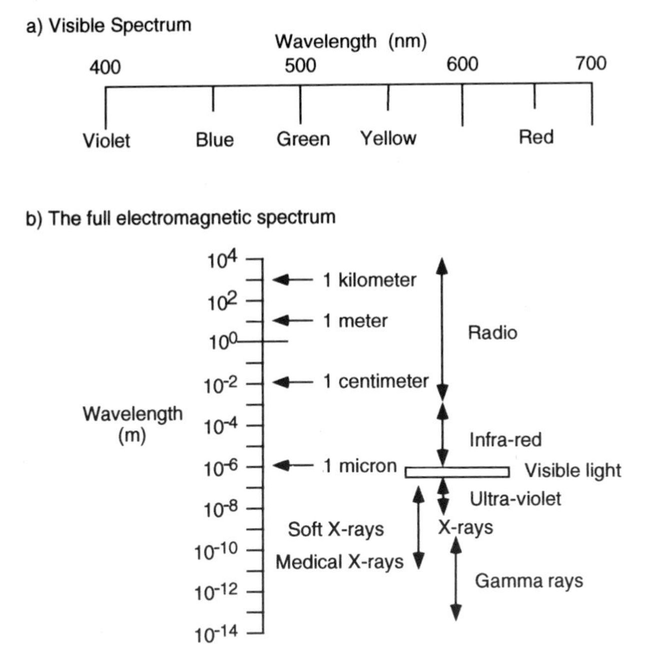

Figure 4.3. The wavelengths of electromagnetic radiation in the Universe range from less than one ten-thousandth of a nanometer to many kilometers in wavelength. a) The range of wavelengths of visible light. b) The complete spectrum of electromagnetic radiation with common names for particular ranges.

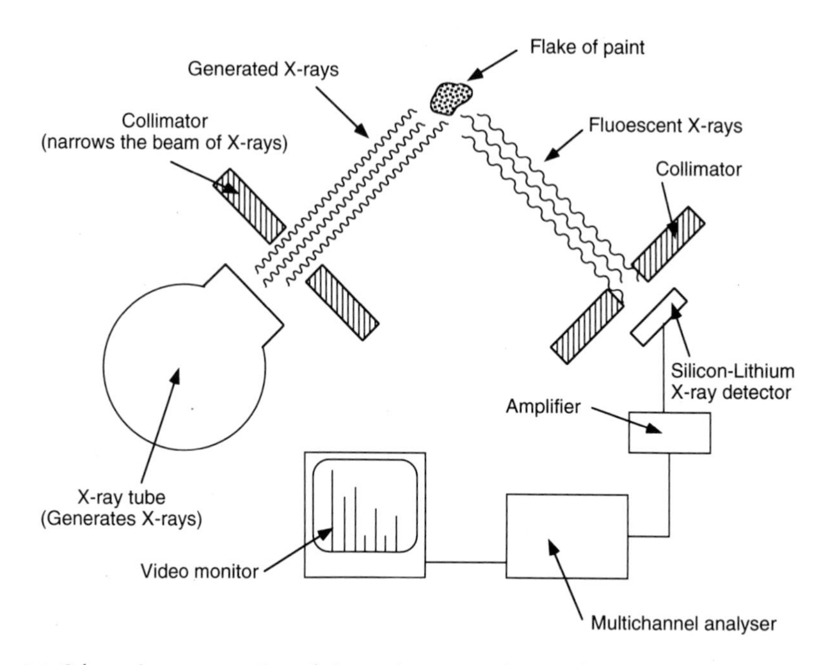

Figure 4.4. Schematic representation of the equipment used to study energy dispersive x-ray fluorescence.

he was the first recipient of this prize. He also found that these rays can pass through human flesh to generate photographs of the body's bone structure.

Just as ultraviolet light will stimulate an atom or a molecule to emit visible fluorescent light, hard x-rays beamed at a body or an object will stimulate fluorescence of longer wavelength x-rays. One can use a piece of equipment similar to the one shown outlined in Figure 4.4 to look at the energy and wavelength distribution of fluorescent x-rays. We now see the meaning of the term energy dispersive x-ray florescence. The typical output from an investigation of the structure of a ten microgram paint chip (a **microgram** is a millionth of a gram, a quantity written symbolically as 10^{-6} g) is shown in Figure 4.5. In this diagram the energy of the x-rays is described in terms of the number of bursts of energy measured by the counter. Therefore, the label "channel number" shown in Figure 4.5 corresponds to a series of channels each of which measures a particular wavelength x-ray.[7] Any particular element in the paint film generates fluorescent x-rays of more than one wavelength. Thus, in Figure 4.5 there are three peaks showing fluorescent x-rays from the element lead, which has the chemical symbol Pb. This symbol is used because in Latin lead is called plumbum. This same word gave us the word plumber to describe a person who worked with lead pipes. Although plumbers no longer use lead pipes, the name stuck.

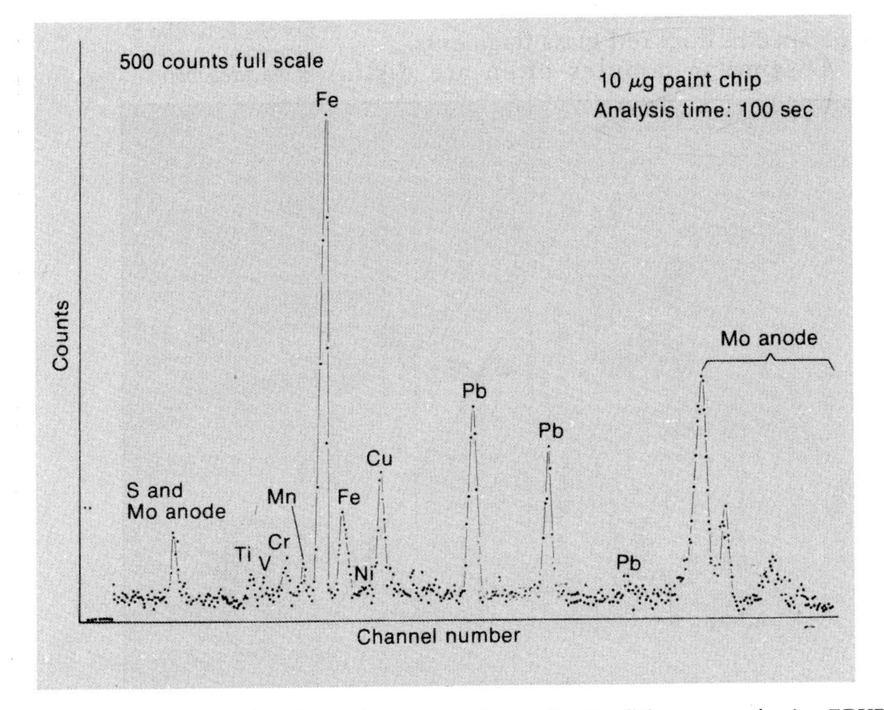

Figure 4.5. Information on the chemical structure of a small paint flake generated using EDXRF. (Reproduced with permission of *Industrial Research*.[7])

Lead is often found in paints. **Lead oxide** is added to paint to protect surfaces against the weather. In older paint, the yellow and white pigments are often compounds of lead. Iron is present in the spectrum because **iron oxide** is a widely used red pigment. Titanium dioxide is a white pigment added to paint films. The chromium in the paint chip probably came from a yellow pigment in the paint. All of the information shown in Figure 4.5 was generated in less than two minutes. The differentiation between two paint flakes that look alike visually, using this technique, is demonstrated by the data of Figure 4.6. It is obvious from the x-ray fluorescence spectra of the two paint films that paint sample B did not come from the same source as paint sample A: the elements antimony, vanadium, and cadmium are absent from the second paint fragment. The power of this technique to differentiate between sets of paint fragments is also illustrated by the work of Smale.[9] Smale studied 400 unrelated paint samples, and he was able to find unique spectra for all of them except for four, three of which were white paint fragments.

Another technique for studying paint film involves studying the way in which it absorbs various infrared waves. Thus, in Figure 4.7 data reporting

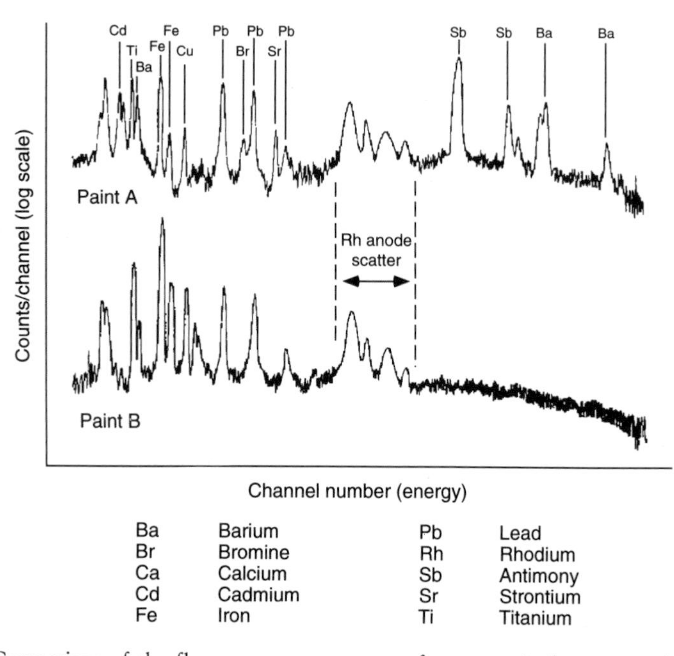

Figure 4.6. Comparison of the fluorescent x-ray spectra from two similar green paint flakes. The two spectra show quite clearly that the two flakes were not from a common source. (Reproduced with permission, *Industrial Research*[7]).

on the transmission of infrared waves through paint films by Compton and Powell are summarized.[10,11]

Other fragments frequently left at the scene of a hit-and-run accident are tiny pieces of glass and/or plastic from the clear and red lights on cars or bicycles. An important property of any glass or plastic is its **refractive index.**[9] If one looks at a stick placed in a bucket of water at an angle, the stick appears to be bent at the surface of the boundary between the air and the water. This is because light entering substances such as water or glass is bent at the surface, as illustrated in Figure 4.8. Scientists call this refraction. The refractive index of a substance measures how much the light will bend in a situation such as that of Figure 4.8 (a). The refractive index – a measure of bendability – is defined using a mathematical quantity known as the sine of the angle. Technically, it is the ratio of the sine of the angle at which light strikes the surface of a substance to the sine of the angle that light makes with the surface after it has entered the substance.

The reason for using the sine of the angle in this ratio does not need to concern the reader further. It is sufficient to recognize that the experts can measure and use this quantity in their search for truth. For our purposes we just have to know that this quantity measures the way in which light is **refracted** or bent at a surface such as the one shown in the diagram. The

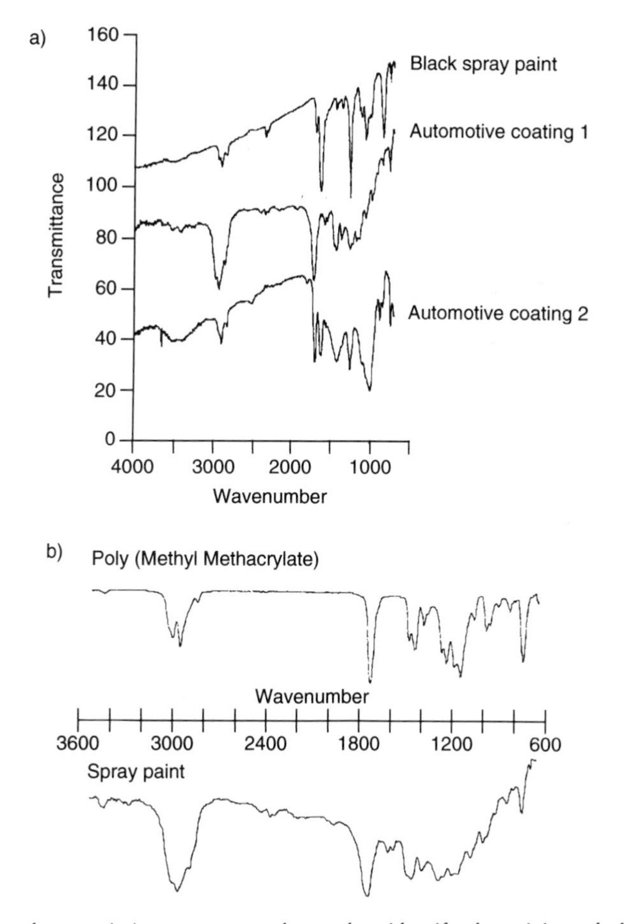

Figure 4.7. Infrared transmission spectra can be used to identify the origin and chemical structure of paint films. (Used by permission of *American Laboratory*[11]) a) Infrared transmission spectra of two black automotive paints and a black spray paint. b) Comparison of the spectrum of the spray paint in (a) with the transmission spectrum of poly(methyl methacrylate) shows the paint to be an acrylic based paint.

actual value of the refractive index varies from substance to substance and is dependent upon the wavelength of the light entering or leaving the substance. Typical values for many of the materials around us are listed in Figure 4.8(b).[8,9] The higher the refractive index of glass, the more it appears to be brilliant when it reflects and refracts light. Thus, what is known as good quality **crystal glass** is actually a glass with a high lead content, which has a large refractive index. **Quartz** is crystalline **silicon dioxide,** a compound also known by the name **silica**. Although it may seem strange to the reader, glass is actually not a crystalline body: it is a very sticky viscous liquid. Although no one notices it around the home, a glass object left in one position will

a)

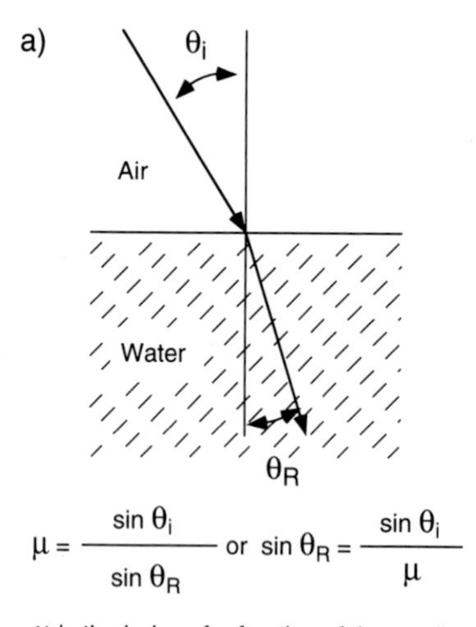

μ is the index of refraction of the medium (in this case water)

b)

Material	Index of Refraction
Crown glass	1.52
Light flint glass	1.58
Medium flint glass	1.62
Dense flint glass	1.66
Lanthanum flint glass	1.80
Ice	1.309
Fluorite	1.434
Rock salt	1.544
Quartz	1.544
Zircon	1.923
Diamond	2.417
Methanol	1.329
Water	1.333
Ethanol	1.360
Carbontetrachloride	1.460
Turpentine	1.472
Glycerine	1.473
Benzene	1.501
Carbon disulphide	1.628

Figure 4.8. Measuring the refractive index of a piece of glass or plastic is an important method of characterizing the fragment. a) The refractive index of a medium can be determined from the degree by which light is bent upon entering the medium. b) Some refractive indices of common media. c) Cargille Laboratories produces kits containing several liquids of known refractive index for use in the identification of various materials. (Used with the permission of Cargille Laboratories.)

deform slowly. This deformation can actually cause problems in terms of accuracy when measuring temperature with glass thermometers. If we were to take a piece of glass of a given refractive index and place it in a liquid of the same refractive index, the glass would apparently disappear. Thus, if one were to take crown glass beads (see the Table in Figure 4.8 (b)) and put them into water, they would still be faintly visible. But, if one were to put them into liquid benzene, they would virtually disappear from sight. I once used this fact to study the behavior of some red glass spheres in a suspension of clear glass spheres in mineral oil. The mineral oil made the clear glass spheres disappear from sight, thus, we could watch the red spheres without interference, as they fell down the middle of the column through the barely visible suspension of clear glass beads.[12]

Paper is a mass of cellulose fibres. Newspaper appears white because it contains many pockets of air trapped in the fibrous mass of the paper. This causes the light falling onto the surface to bend backwards. The turning back of the light is caused by the fact that the air and cellulose fibers have different refractive indexes. (In many cases, paper actually contains pigments to make it whiter. We will be discussing this fact later in Chapter 9, where we look at fraud involving paper documents.) Cooking oil has a refractive index

c)

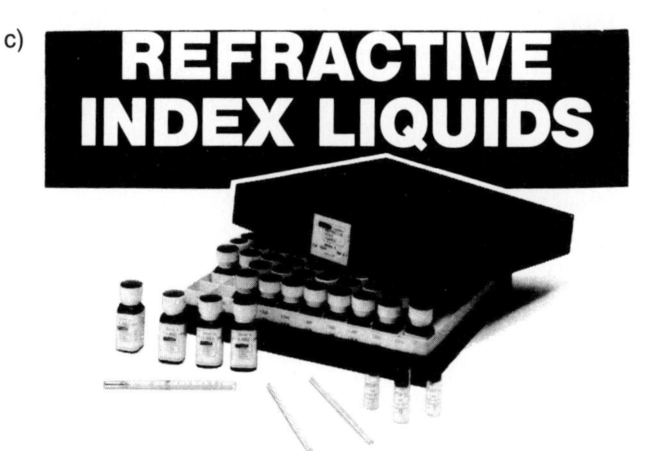

close to that of cellulose. When a drop of oil is placed on a mat of cellulose fibers, the air gaps fill with oil, changing the refractive index at the boundary of the cellulose fibers. As a consequence the light is no longer strongly reflected away from the mat of fibers and one can see through the paper at the site of the oil drop.

If one wishes to read a letter through an envelope without breaking the seal, one can place a liquid of a similar refractive index on the envelope to make it transparent. However, using oil for the purpose would permanently damage the letter and make it obvious that the letter had been tampered with. Alcohol (pure ethanol, C_2H_5OH) can be used to achieve the same effect, and the subsequent evaporation of the alcohol will leave the paper seemingly untouched. In some cases, however, the alcohol may dissolve some components of the ink and spread it around, damaging the handwriting on the letter inside the envelope. One can, however, make most paper transparent by putting the envelope in a tray of **liquid air**. The liquid air fills the gaps in the fibers, thus making it possible to read the hidden writing. The subsequent evaporation of the liquid air does not leave any trace of the fact that another person has tried to read the material inside. One should, however, be very careful if attempting this technique since if one accidently puts a finger in the liquid air, it could snap off with the slightest touch. At the very least one would certainly suffer from severe frostbite! (The typical temperature for liquid air is about –180 °C.)

Fragments of glass or plastic left at the scene of a crime have characteristic refractive indexes. If one wanted to match fragments from headlights on a car with glass fragments found on a victim's body, one could measure the refractive index of the fragments from the headlight and those found at the scene of the hit-and-run accident. Again, if a broken window generated fragments that were then found on a burglary suspect, it would be important to be able to match such evidence. A set of liquids, called Cargille liquids, contains varying refractive indexes and can be used to ascertain the refractive index of a substance. To use these liquids, one drops the fragment into a sequence of jars and observes whether it disappears at a particular refractive index (see Figure 4.8 (c)).

Many glasses contain small amounts of different chemicals and, therefore, we can look at glass fragments using energy dispersive x-ray fluorescence. In Figure 4.9 some data reported by Cooper, who used this method to look at two apparently similar samples of red glass, demonstrates this fact. The fluorescent x-ray spectra of the two glass samples indicate that they are unrelated. The absence of x-rays generated by fluorescent manganese was a particularly strong reason for rejecting any common source for the two samples. Cooper tells us that in one study 80 glass fragments with similar physical characteristics were uniquely identified using EDXRF.

4.3 Tell-Tale Dust

One of the world-famous laboratories involved in studying dust and fibers is McCrone Associates of Chicago. To help determine the origin and nature of dust and fibers found at the scene of a crime as well as in other situations, this organization has published a book known as "The Particle Atlas."[13] Several illustrations from that Atlas are shown in Figure 4.10. One of the workers at the McCrone laboratories, Skip Palenik, has written an interesting review of the use of microscopy to support the law. In one case, Palenik recalls how he was approached by the Pennsylvania state police to determine where a murder was committed so that jurisdiction could be

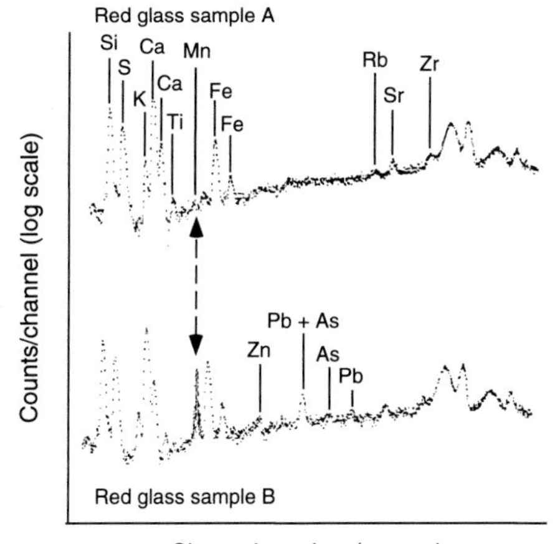

As	Arsenic	Rb	Rubidium
Br	Bromine	S	Sulphur
Ca	Calcium	Sb	Antimony
Cd	Cadmium	Si	Silicon
Fe	Iron	Sr	Strontium
K	Potasium	Ti	Titanium
Mn	Manganese	Zr	Zirconium
Pb	Lead	Zn	Zinc

Figure 4.9. EDXRF can be used to study tiny fragments of a substance. In this case a comparison of two samples of red glass shows that they are unrelated. (Reproduced with permission of *Industrial Research*.[7])

assigned. In this case a young girl was found dead in the trunk of a car parked on the side of a road in Ohio. Several miles away in Pennsylvania, just across the state line, her handbag and a pistol were found. Palenik's study showed that the soil on the victim's body could not be called soil in the normal sense of the word; it was crushed steel-mill slag mixed with a small quantity of soil containing characteristic heavy minerals. This material matched, practically particle for particle, the material on the ground where the gun and handbag were recovered. Apparently, crushed slag from a local steel mill had been used as landfill for a small portion of the road at that location. Therefore, it seemed almost certain that the murder had been committed in Pennsylvania. Later, when the police were able to convict a suspect of the murder, he confessed to having moved the body after committing the crime.[14]

Bisbing of McCrone Associates describes an interesting application of dust characterization involving latent fingerprints.[15] In the course of investigating a crime, a fingerprint was developed on a windowsill and subsequently identified as that of the defendant. In cross examination, the following question

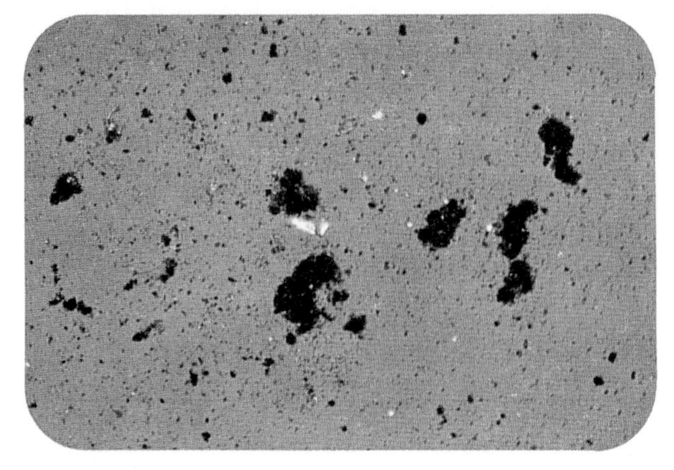

Gasoline Engine Test Filter

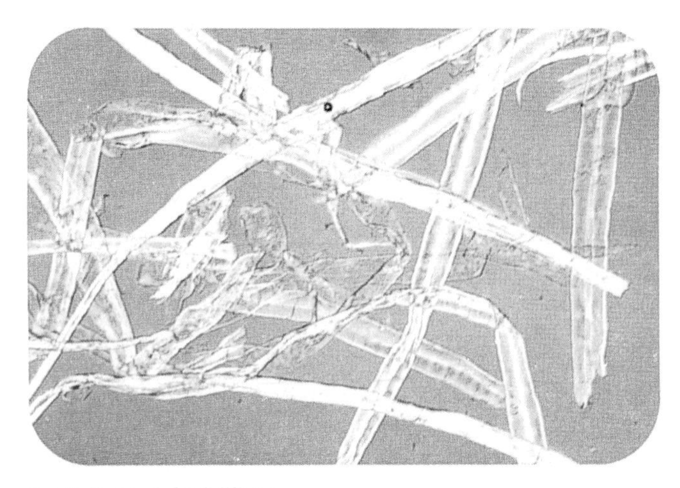

Kraft Paper (pine) Fibers

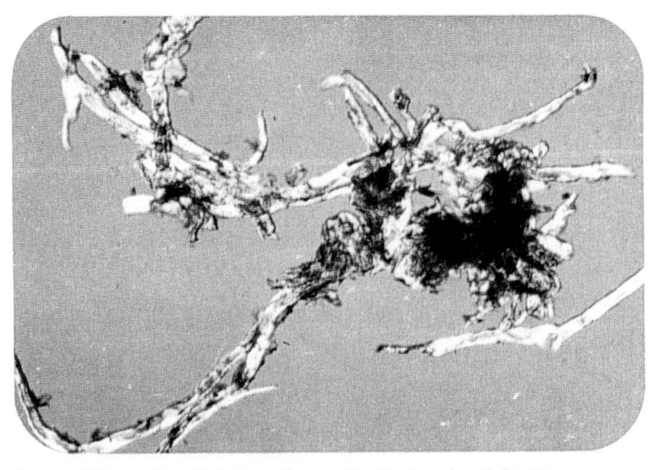

Paper Fibers Bonded together with Vulcanized Rubber

was raised: had the fingerprint been lifted from somewhere else? The windowsill was made of marble. The other surfaces suggested as possible origins for the fingerprint were painted. The developed fingerprint was lifted from the windowsill using a transparent tape. In further studies carried out by the McCrone laboratories, portions of the transparent tape were separated from the paper backing and placed sticky side up on a microscope slide. It was then covered with a thin glass top. No paint particles were found anywhere on the tape. However, on a portion of the tape, where powder had accumulated in a crack at the surface of the windowsill, calcium carbonate crystals were found. These crystals were shown to be similar in all respects to the marble dust on the window sill.

Broad reviews an unusual example of fragmentary evidence, which tells the forensic scientist an interesting tale. A tiny micron-sized sphere of metal was found trapped in the cellulose fiber net of a bank note. The bank note was recovered after a robbery in which a safe had been opened with an oxyacetylene torch. The metal sphere was examined using x-ray fluorescence and the experts were able to show that it came from the safe which had contained the stolen bank notes.[4]

Palenik describes a civil liability suit in which fragmentary evidence played an interesting role. A woman had apparently slipped and fallen in the supermarket. Subsequently, it turned out that she was more seriously injured than anyone realized at the time. The attorney representing the woman in her law suit against the supermarket approached McCrone Associates to help prepare the legal case. They were asked to look at stains on the upper leg and the back of her pant suit, and to see whether the stains were related to her fall. The stains had not been present when she dressed in the morning. Palenik reports the following:

> *portions of the stains were scraped onto microscope slides and soaked in distilled water. Characteristic spongy Prenchma cells of a fruit were noted at once. Microchemical tests for reducing sugar also were positive. Finally a few starch grains were detected but these were not immediately identifiable. Other microscope slides showed these distinctive grains in great quantity. It took a little time to identify the starch by reference to our collection of known starches. When we did the answer was obvious. The broad grains with distinct concentric rings surrounding the hilum were banana starch.*

The woman had slipped on a banana.[14]

◀ **Figure 4.10.** The discovery of the origin of dust found at the scene of a crime, or constituting a nuisance or health hazard in a civil law case, has been greatly facilitated by the publication of *The Particle Atlas*, by McCrone and Associates, a "Who's Who of Dust." A typical set of illustrations from this book demonstrates its usefulness. (Photomicrographs courtesy of McCrone Research Institute.)

4.4 Hairs and Fibers

Hairs and fibers found at the scene of a crime have played a significant role in solving crimes. The appearance of hairs and fibers under a microscope differs very widely, as can be appreciated from the pictures shown in Figures 4.11 and 4.12. Wool fibers from sheep have a very distinctive scaliness, as shown in Figure 4.12(a). These scales help move fibers to interlock with each other when a wool fabric is manipulated. This explains the well-known fact that older wool garments mat and contain little bobbles on the surface of the wool.

Bisbing reviewed a case in which fibers were transferred to an assault victim from the attacker, and vice-versa. This provided significant evidence for conviction, as summarized in Figure 4.13. Bisbing points out that the rayon fibers contained small yellow pigment fine particles. He tells us that these particles were characterized by microscopy, after being separated from the fibers. The pigments were collected by first dissolving the fibers, and then carefully picking up the pigment particles from the solution and placing them on a microscope slide.[15]

McCrone, in a review of the use of microscopes to solve various types of problems, points out that viewing fibres under an optical microscope played an important role in convicting Wayne Williams for a series of murders in Atlanta.[16] He praises the work of Larry Peterson, of the Georgia Bureau of Investigation. Peterson was able to show the presence of unique hairs and fibers on many of the 28 Atlanta murder victims and in Wayne William's apartment and various automobiles. One of the incriminating fibers is shown in Figure 4.14.

Fibers have proved to be the source of important information not only in criminal cases but in several liability investigations. Thus, McCrone and Palenik cite an investigation in which they were asked to look into what was causing the health problems of a woman after she moved into a newly built home.[17] Upon taking up residence in the house, she complained of an itch. The itch grew steadily worse until it affected her whole body and she had to be hospitalized. In hospital, her health improved but deteriorated quickly again after she returned home. Her doctor suspected that her rash was similar to that caused by glass fibers found in the insulation installed in many houses. McCrone and Palenik collected dust in the house from the top of the refrigerator and from behind the stove. They then sampled all known sources of fibers inside the house, including draperies, carpets, the attic, and other heating insulation material. Examination by microscopy of the dust that they collected showed the presence of a large number of glass fibers in every sample, yet none of the fibers matched those from the known sources in the home. Investigators went back to the house to look for pink glass fiber insulation. Two weeks later when investigating more samples of insulation from

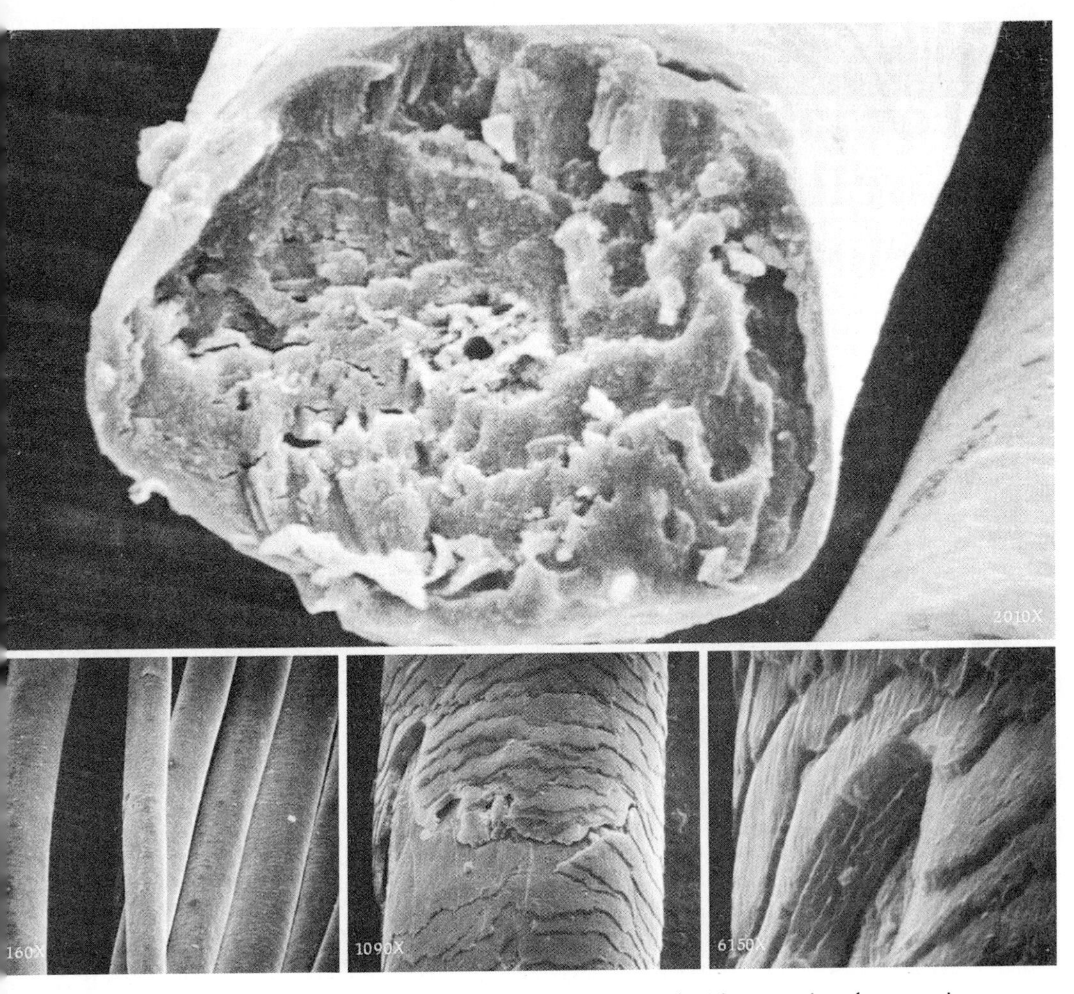

Figure 4.11. Human hair has a distinctive structure when viewed with a scanning electron microscope. (Reproduced with permission from *Science Spectrum*.)

the home, they found one sample that consisted of glass fibers identical in all respects to the fibers isolated from the dust samples around the house. After McCrone and Palenik submitted their report, the contractor pulled out all the pink fiberglass insulation and discovered a flaw in a heating duct. This flaw allowed fragments of insulation fiber to be circulated throughout the house, causing irritation to the inhabitants.[17]

The way in which the fibers in a piece of fabric tear can also give evidence to the truth or falsehood of statements involving the way in which the fabric came to be torn. In a case described by Palenik, a policeman interrupted a burglar in the middle of breaking into a house in a suburb of Chicago. The

a - Resin wool ;
——— 100 μm

b - Electret I ;
——— 100 μm

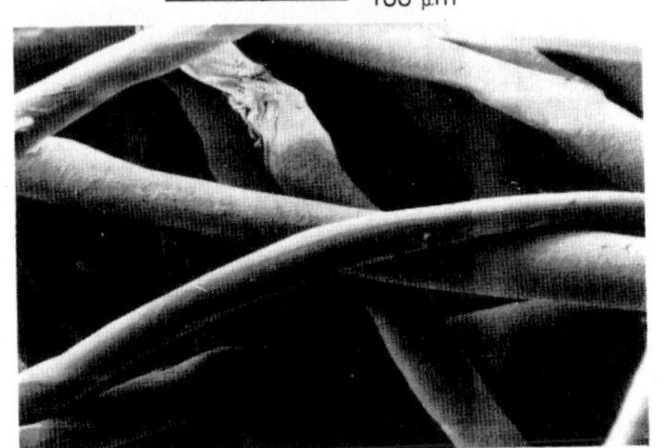

c - Electret II ;
——— 100 μm

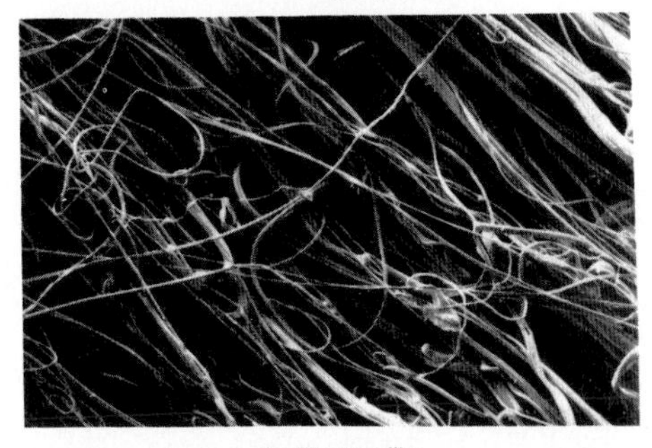

d - Electret III ;
——— 100 μm

e - Mixed fibre material ;
——— 40 μm

f - Polycarbonate extruded from a solution ;
——— 20 μm

g - Polypropylene extruded from a melt ;
——— 10 μm

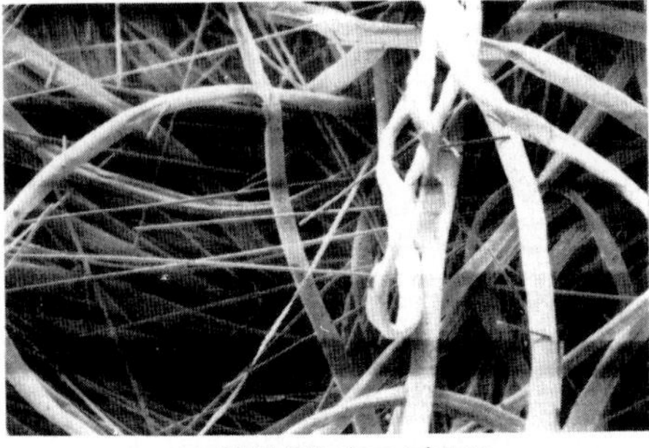

h - TEFLON glass mixture.
——— 200 μm

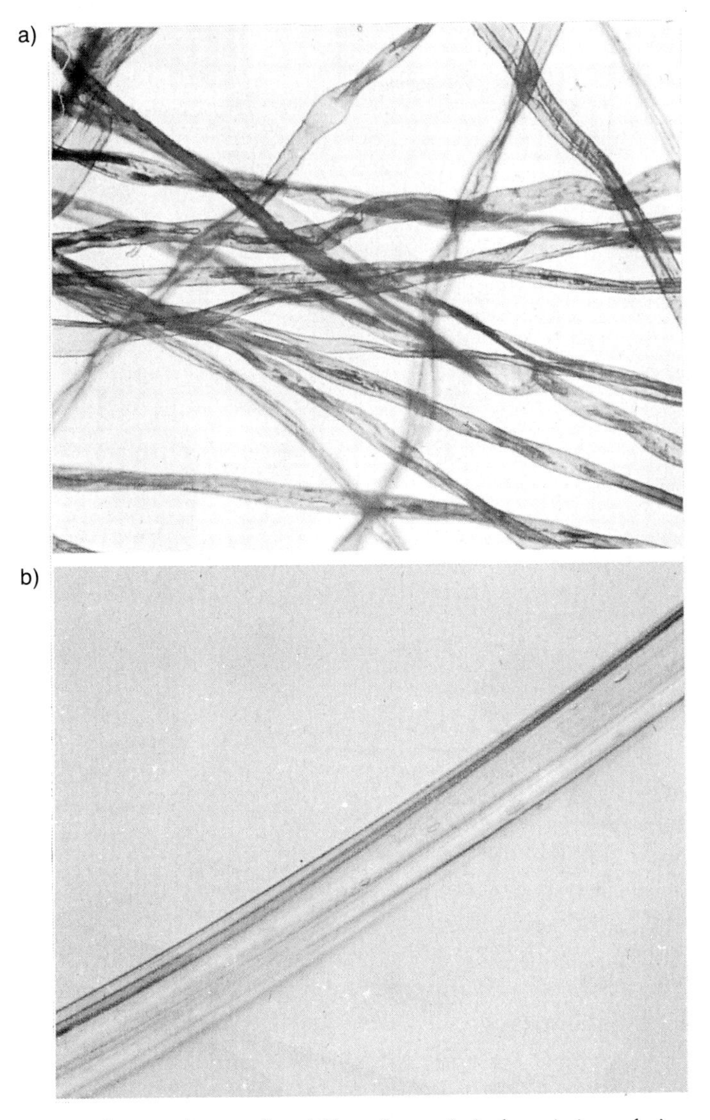

Figure 4.13. In assault cases the transfer of fibers from criminal to victim and vice versa can result in significant evidence leading to conviction. (Photomicrographs courtesy of McCrone Research Institute.) a) Purple cotton fibers from a victim's pants found on a suspect. b) Yellow rayon fiber from an assailant's clothing found on a victim's hands.

◀ **Figure 4.12.** Various types of fibres can be recognized by the expert microscopist. (Crown copyright. Reproduced with the permission of the Controller of Her Majesty's Stationery Office.)

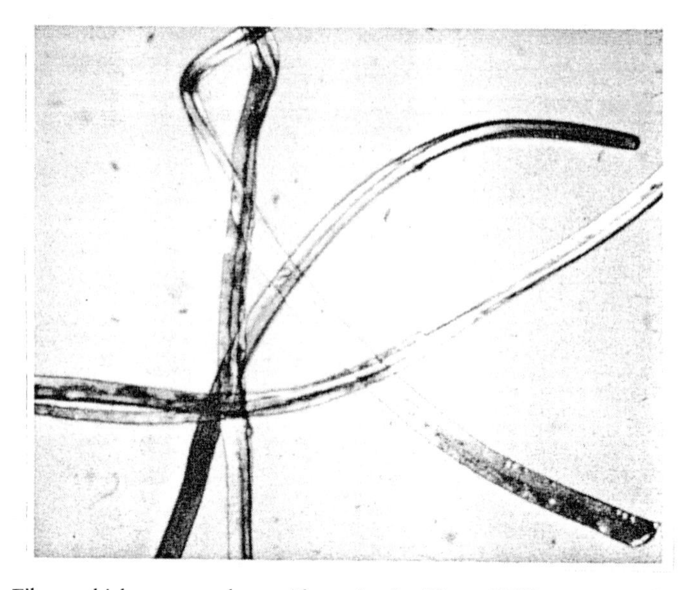

Figure 4.14. Fibers which were used as evidence in the Wayne Williams case. (Photomicrographs courtesy of McCrone Research Institute.)

policeman claimed that when he finally stopped the burglar, the suspect tried to slash at him with the knife he was carrying. The suspect denied doing this, and claimed that the tears in the policeman's shirt must have been caused when he caught his shirt on some thorn bushes which they had run through.

To ascertain who was telling the truth, McCrone laboratories was presented with the policeman's shirt, which had two short tears in the breast pocket. They were also given two kitchen knives recovered from the suspect. In addition, they studied an identical uniform shirt and a sample of the thorn bushes, which the men had run through. An examination of the tears in the first shirt showed that the edges were cleanly cut. Fibers from the cloth were removed and identified as blue dyed cotton and polyester. An examination of the blade of the first knife showed nothing but some protein-based fragments and remnants of fingerprint powder. The study of the second knife showed two minute fibers at the tip of the blade. These were carefully removed and viewed under a microscope. Both fibers measured one millimeter in length and were identified as blue dyed cotton. Viewed side-by-side with fibers from the pocket of the shirt, their colors were found to be indistinguishable. In a series of microchemical tests, the fibers from the shirt and knife were also found to be similar in all respects. Finally, comparative tests were made on an identical police shirt. Cuts were made in the fabric with the suspect's knife. These cuts had exactly the same appearance as those in the shirt worn by the policeman at the time of the crime. McCrone and Palenik pointed out that after a cut was made, the knife invariably carried numerous cotton fiber fragments. Apparently, no polyester fibres were transferred to the knife because of the physical properties of that type of fiber. Palenik tells us:

As an additional test, we tried to cut the shirt with the thorn bushes. This proved quite difficult. When we were successful in penetrating the fabric, all that we could do was pull out a tuft of material. In most cases, the thorns came lose and frequently embedded themselves in the fabric. We felt certain at this point in concluding that the second knife had been used to cut the policeman's shirt and that it was highly unlikely that the thorn bushes could cause such a clean cut. Presented with this evidence the suspect finally confessed.[14]

The distribution of different chemicals in hair can give useful information to the forensic scientist. The term **autopsy** is familiar to most people. It describes the act of examining a body after death to determine the cause of death. It is a rather curious word since it literally means self-inspection. The root of the word is *auto*, a Greek word meaning, "self" (an automobile is self-mobile), and *opsis* meaning "a sight or an inspection." It is hard to imagine how a corpse could be involved in self-inspection, but the term is widely used. One dictionary implies that the term means an inspection that you make yourself to ensure that no-one else is deceiving you. It is from the point of view of the examiner, not the victim – really the act of seeing with one's own eyes.

Another word used in the medical profession is **biopsy**, meaning an inspection of the tissue taken from a living system. A medical specialist wishing to determine whether a cancer is malignant or benign will take a small sample from the tumor for biopsy. The reason why hair contains useful information for a biopsy or autopsy is because of the way in which it is generated. To enable the reader to appreciate the technical terms used when studying a hair, it is worthwhile giving the definition of hair presented in a medical dictionary:

*Hair; a thread-like keratinized outgrowth of the epidermis of the skin. It develops inside a tubular structure known as follicle. The part above the skin consists of three layers: an outer cuticle; a cortex forming the bulk of the hair and containing the pigment that gives the hair its color; and a central core (**medulla**) which may be hollow. The root of the hair beneath the surface of the skin is expanded at its base to form a bulb, which contains a matrix of dividing cells. As new cells are formed the older ones are pushed upward and become keratinized to form the root and the shaft.*[18]

The term keratinized also needs some explanation. The same dictionary gives the following definition: "**Keratin**; a fibrous protein that forms horny tissue such as fingernails. It is also found in the skin and hair." What the dictionary definition tells us in simple terms is that hairs growing out of the skin are essentially dead strands of protein, which have turned into horny type tissue. This means that as each portion of a hair grows out from the scalp it incorporates chemicals that happen to be in the body when that particular portion of the hair emerges from the scalp. Therefore, as the chemicals in the hair change, one has a record of the changing chemical composition of the body over a period of time. This record is contained in the length of the hair.[19] In one study Berg and colleagues were able to show that the distribution of mercury along the length of a hair could be related to the beginning and ending of the period when a group of people were eating bread made from grain, which was contaminated with a fungicide containing mercury compounds. (**Fungicide** means something that kills **fungi** such as mushrooms and molds.)

Energy dispersive x-ray fluorescence can be used to measure the presence of metals in hair. Two other techniques that have been used to study the chemical composition of hair are known as **neutron activation analysis (NAA)** and **photon activation analysis (PAA)**. The physical basis of neutron activation analysis is the following: the material to be studied is placed in a strong flux of neutrons in an atomic reactor. As the neutrons bombard the atoms in the hair, some of the atoms are changed into radioisotopes, which have a short life. Radioactive materials that decay quickly are known as **short-life radioactive material** or short-lived isotopes. Other materials can take centuries to completely decay back to their non-radioactive forms and these are called **long-lived isotopes**. Very small amounts of an original element, **nanograms** or less (a nanogram is a billionth of a gram written symbolically as 10^{-9} g or 1 ng) can give rise to detectable levels of radioactivity, which can then be measured using very similar equipment to the x-ray fluorescence equipment described earlier.

Steinberg has reviewed the use of neutron activation analysis in forensic science.[20] He points out that the technique was first used in a murder investigation in May 1958. The body of a sixteen-year-old girl, Gaetane Bouchard, was found in the Canadian city of Edmunston near the Canadian–American border. An American, John Follman, was the prime suspect in the case, but he denied having anything to do with the crime. A careful examination of the girl's body resulted in the recovery of a single hair in the girls tightly clenched fist, which could belong only to her or to the murderer. The police asked an NAA specialist, Robert Gervie, to take part in examining the hair. In the course of his investigation, Gervie inspected the hair of hundreds of people. He looked for concentrations of materials such as arsenic, sodium, copper, zinc, and bromine. As a result of his work, it was proved that the hair in the victim's hand belonged to Follman and this became the decisive piece of evidence establishing his guilt.

G. Seaborg, in his review of the use of neutron activation analysis in forensic science, states that the technique was first used by the Internal Revenue Service in 1964. Within six years the technique had been used in 1,500 criminal cases to examine a large variety of substances. These substances included paint, soil, bullets, metal, putty, adhesive tape, grease, hair, wire, and gunshot residues. In particular, he reviews a case from the late 1960s in which hair samples were taken from the head of man injured in a hit-and-run accident, and matched with hair caught in the windshield wiper of a car owned by the primary suspect. Seaborg comments that:

> *It was a difficult case because hair taken from different parts of the same head may differ slightly in composition. Highly efficient analyses were required to remove the question of reasonable doubt concerning the identity of the hair from the jury's mind in order to establish the guilt of the suspect.*[21]

In another case reviewed by Seaborg a man was convicted of operating an illegal distillery in Georgia. The crucial evidence was mud samples from the defendant's truck picked up in Brooklyn, New York, that exactly matched other mud samples taken from a Georgia road leading to the distillery that produced the alcohol.

Earlier in this chapter we showed that gamma rays are electromagnetic radiation similar to light, but having much shorter wavelengths than either x-rays or visible light. The smallest packet of energy present in any electromagnetic wave is known as a **photon**. If the hair being studied is placed in a strong flux of **gamma rays**, it is said to be exposed to **high-energy photons**. Again, the elements present in hair can be changed by high-energy photon flux into short-lived radioisotopes. Arsenic atoms exposed to a neutron flux can be changed into short-lived isotopes which are easy to measure. Using this technique, a few parts per million of arsenic in human hair can be measured. Katz has given a review of the way in which different techniques can be used to study different metals in human hair.[19] In one court case against a refinery, which was recycling automotive electric batteries, investigators were able to show that children in the streets around the refinery were exposed to fumes and fine aerosols of lead (any cloud of dust is an aerosol). Scientists were able to estimate the level and length of exposure of the children to the lead fumes by measuring the amount of lead in samples of hair. Anyone that suspects he has been subjected to slow poisoning could have his hair examined by this technique.[19,22]

4.5 Who Killed Napoleon?

Napoleon I, Emperor of France, was sent into exile when he was captured following his defeat at the battle of Waterloo. In October of 1815, he arrived at the island of St. Helena in the south Atlantic, 1,200 miles from the West Coast of Africa. Napoleon died in May 1821. Three weeks before he died, on May 5th, Napoleon Bonaparte wrote the following words in his last will and testament: "I die prematurely, murdered by the English." Before his death, Napoleon showed all the symptoms of arsenic poisoning. In a medical dictionary we have the following entry defining arsenic and its toxic effects:

> ***Arsenic;*** *a poisonous grayish metallic element which when ingested produces the symptoms of nausea, vomiting, diarrhea, cramps, convulsions and the individual eventually falls into a coma after ingesting large doses. Arsenic was formerly used in medicine, the most important arsenical drug was used in the treatment of syphilis and dangerous parasite diseases.*[23]

The fact that arsenic was used to treat various diseases must be born in mind when considering evidence of possible arsenic ingestion by historic figures who manifest high arsenic levels in their hair.[18,23]

Napoleon asked that, when he died, locks of his hair be given to various individuals. The British Governor of St. Helena, Governor Lowe, sent snippets of the ex-emperor's hair to several families including that of Louie Marchand who had been Napoleon's devoted valet for many years. In 1955 the autobiography of Marchand was published by his grandson. This book was read by Sven Forshufvun, a Swedish toxicologist and a student of Napoleon's career. Forshufvun obtained some of Napoleon's hair from Marchand's descendents and had it studied by means of neutron activation analysis. It showed relatively large amounts of arsenic. As various people have studied the possibility that Napoleon was deliberately poisoned, forensic experts focussed on the three possible suspects identified by historical researchers. Napoleon himself pointed the finger at Governor Lowe and his assistants. He claimed in his diaries that the British were systematically poisoning him with arsenic administered in his food by the servants of the British.

After an exhaustive study of diaries and evidence, Forshufvun concluded that the most likely candidate for the deliberate poisoning of Napoleon was Count de Montolon, who, in spite of his ties with the Bourbon King's Monarchy, which was restored to the French throne after Napoleon's defeat, accompanied him to St. Helena. It appears that Napoleon liked a South African wine imported in casks and bottled at St. Helena. Only he drank this particular wine, and the wine steward, Montolon, kept it under lock and key. On two occasions the wine was given to other people by mistake and on both of these occasions the people became ill after drinking it. A Swedish toxicologist believes that the Bourbon Monarchy would be very pleased to see the death of Napoleon, whose memory was still cherished by the French people. Therefore, while he still lived he constituted a threat to the stability of the Monarchy.

If the Count did poison Napoleon, it is not clear whether he was acting on his own to curry favour with the Bourbon King, or whether he actually acted as their agent. We also know that Count de Montolon's wife left the island before Napoleon died and it is rumored that Napoleon was the father of a child born to the Countess. It could be that, if Count Montolon was the main instigator of any poisoning, the murder of Napoleon was simply an act of revenge.[24,25]

In 1982, Dr. David Jones of the Physical Chemistry Department of the University of Newcastle in Great Britain came up with a very interesting third suspect: Napoleon's wallpaper![26] In his article, Jones points out that during the 19th century a pigment then known as, **Scheeles' green** was widely used. This pigment was a compound of arsenic. In the words of Dr. Jones, "this cheap, vivid, stable pigment was welcomed by the color industry of the day and, by 1800, copper arsenite and its relatives were widely used in paints, fabrics and wall papers." Dr. Jones states:

> *The arsenic in the new pigments seemed able to leave the wallpaper of the room and get into occupants. Hundreds of luckless householders developed the various symptoms of arsenite poisoning and quite a number died. After 1900 the copper arsenic was replaced by safer pigments. An Italian Biochemist, Gossio, in 1893 was able to show how the wallpaper attacked the people living in the room decorated with the wallpaper. Apparently, the inorganic pigment was safe as long as the wall remained dry, but if the walls became wet either from condensation or rising damp, and this was very probable event in the badly heated homes of the 19th century, then the wallpaper became moldy.*

Molds which are related to mushrooms and fungi are fascinating materials that can manufacture some very surprising chemicals. The mold that grows on melons or cheese can produce the drug penicillin. There are reports that during the Chaldea crusades in the 12th century doctors applied moldy bread poultices to wounds. If this story is true, it would constitute a very early use of the curative powers of penicillin. Apparently, a mold known by the name Scopulariopisis Bevicaulis loves to grow on wet wallpaper. It could get rid of the arsenic compounds in the wallpaper by making a vapor known by the name **trimethoxy arsenic**. As the mold broke down the wallpaper, this chemical was given off into the air of the room and the inhabitants breathing in these fumes would suffer accordingly. On a program broadcast by the British Broadcasting Corporation (BBC), Dr. Jones suggested that if Napoleon's room had been decorated with wallpaper containing this dye, then the arsenic found in Napoleon's hair could have come from his wallpaper. Apparently, Napoleon had complained continuously about the fact that the house he was living in at St. Helena was cold and damp. After the broadcast, Dr. Jones received an interesting letter from a Shirley Bradley of Norfolk. In her letter, Ms. Bradley advised Dr. Jones that:

> *I have in my possession an old book rather like a common place book (scrapbook of 1823–1829) in which there is a sample of wallpaper with the following words beside it: 'This small piece of paper was taken off the wall of the room in which the spirit of Napoleon returned to God who gave it.'*

Later, Ms. Bradley permitted Dr. Jones and Dr. Ken Ledingham, physicists at the Glasgow University, where the original analysis of Napoleon's hair was performed, to study this piece of wallpaper. The wallpaper was analyzed by means of x-ray florescence, neutron-activation analysis, and photon-activation analysis. All of the studies revealed that the wallpaper did indeed contain arsenic. However, the arsenic content was relatively low compared to some levels found in other wallpapers from that period. Nevertheless, Dr. Jones still concludes that one very real source of arsenic for Napoleon could have been his wallpaper. He tells us that one of the problems with trying to assess the effects of the wallpaper is that one must expect wild fluctuations both in the output of arsenic, due to the level of dampness in a room, as well as in the uptake of the poison by the inhabitants. Dr. Jones also pointed out that Napoleon's imperial colors were gold and green. In other words, he may well have been exposed to flags, drapes, curtains, and even clothing, all dyed green with copper arsenite.[26]

At the moment it is impossible to decide who was the real culprit in the poisoning of Napoleon, but mold growing on green wallpaper could have been the "sinister poisoner" silently attacking Napoleon. No matter who or what was responsible for the poisoning, one thing remains certain: Napoleon was the victim of arsenic poisoning. When his body was exhumed for return to burial in France, by his nephew who became Napoleon III, workers found that Napoleon's body was fully preserved in death. This is an important piece of evidence indicating arsenic poisoning as a cause of death. However, it should not be forgotten that, at the time, arsenic was used by morticians for embalming bodies.

4.6 Fragmentary Evidence and the Saga of the Great Fiddle

This section does not deal with a great robbery or confidence trick. Instead, it concerns the application of forensic technology to an investigation into possible reasons why the famous violins, and other musical instruments, made by Antonio **Stradivari** (1644–1737), produce such wonderful sounds when they are played. Legend has it that Stradivari used to go through the forests of Lombardy, Italy, rapping on trees to find the finest spruce available for the top surface of his violins and the finest maple for the back.[27]

Scientists have always suspected that the varnish applied to the Stradivarius instrument played an important role in its ability to generate superb musical sounds. In 1988, chemists at the University of Cambridge, England, studied the varnish using modern techniques available to forensic scientists. They carried out these studies on fragments of varnish taken from a 1677 Stradivarius cello. Using a transmission electron microscope, they discovered that beneath the surface layer of orange-red varnish was another coating 50 micrometer thick. This previously unknown hard layer of varnish was studied using **energy dispersive x-ray spectroscopy (EDAX)**. In this technique, the sample is bombarded with a beam of high-energy electrons which then give out x-rays. This is not exactly the same as EDXRF, since in x-ray fluorescence the same type of radiation of a shorter wavelength is used to stimulate emissions at a longer wavelength. In EDAX, the stimulating beam of electrons is not electromagnetic radiation. The hard layer underlying the orange varnish was found to have the x-ray spectrum shown in Figure 4.15 (b). The scientists were able to recognize that the EDAX pattern shown in Figure 4.15 (b) was similar to that generated by a volcanic ash found in the area where Stradivari lived, shown in Figure 4.15(c). This volcanic ash is called **pozzolana**: it is named for the town where it was first discovered. This ash is used in modern Italy to manufacture high-quality cement. When mixed with lime, it produces a very hard and extremely durable material that can be used as building stone. The EDAX spectrum of the volcanic ash closely resembles the hard paint layer under the orange varnish. The Cambridge scientists suggest that Stradivari made a paste of pozzolana powder and water by adding a binding agent such as egg white. Apparently, he then applied this material to seal the wood of the instrument before adding the orange varnish. In their studies, the Cambridge scientists ruled out the possibility that this hard layer was a residue of pumice used by Stradivari to smooth the surface of the wood. When they analyzed a sample of pumice, which is also of volcanic origin, they were able to show that pumice had much higher levels of silicon than pozzolana. In Figure 4.15 (a) and (d) the EDAX spectra for the orange varnish, and for the wood of the musical instrument are shown. The Cambridge scientists believe that the secret layer, based on the use of pozzolana powder, was important in producing the powerful and penetrating tones that have made the Stradivarius a much prized instrument. They also speculated that an added bonus was the strength it gave to the instrument, and that this has helped them to function efficiently three hundred years after they were made.[28]

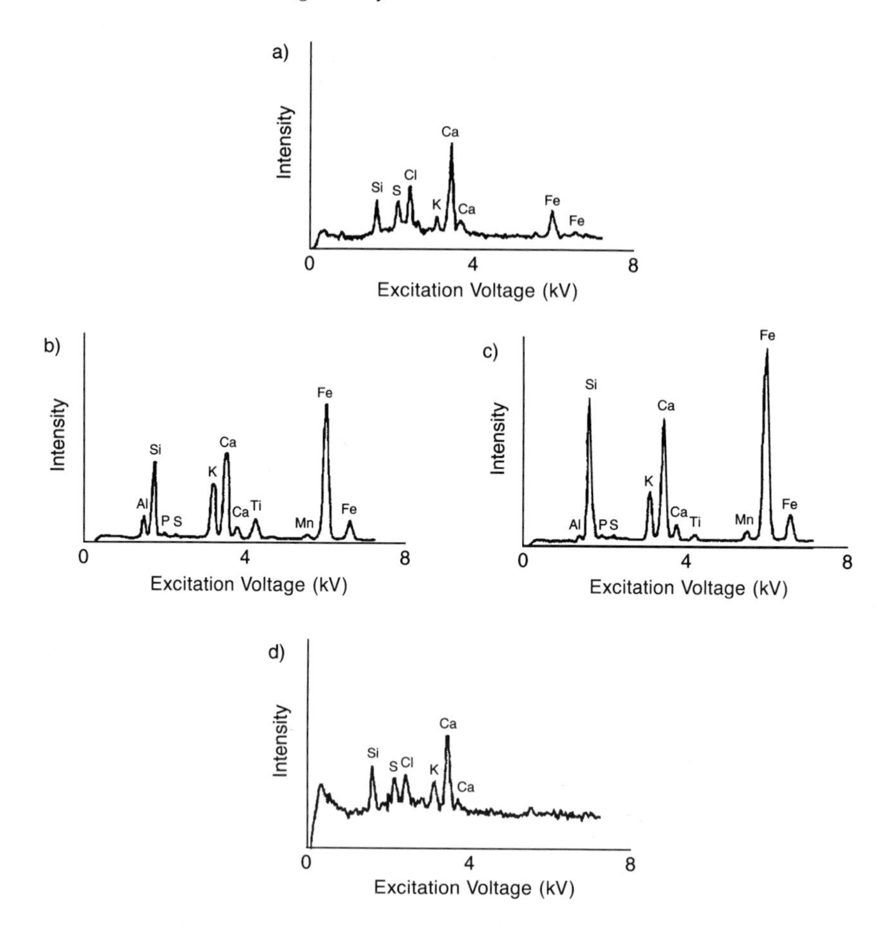

Figure 4.15. It may be that the origin of the beautiful tones of the family of Stradivarius musical instruments, such as the violin and cello, lies in the structure of a primary coating containing volcanic ash applied to the instrument before finishing with orange varnish.[28] a) EDAX analysis of the varnish layer. b) EDAX analysis of the "Stradivarius Layer." c) EDAX analysis of volcanic ash found in the area of Stradivari's home. Note the similarity to the "Stradivarius Layer." d) EDAX analysis of the wood of the instrument.

4.7 Fragmentary Evidence Points to Modern Fabrication of Supposedly Antique Documents

McCrone's discovery of the fraudulent nature of a document known as the **Vinland map** is a modern case where fraud was exposed by studying tiny fragments of a material. Part of the Vinland map is shown in Figure 4.16 (a). The map was purchased in 1957 from a European dealer by a book seller

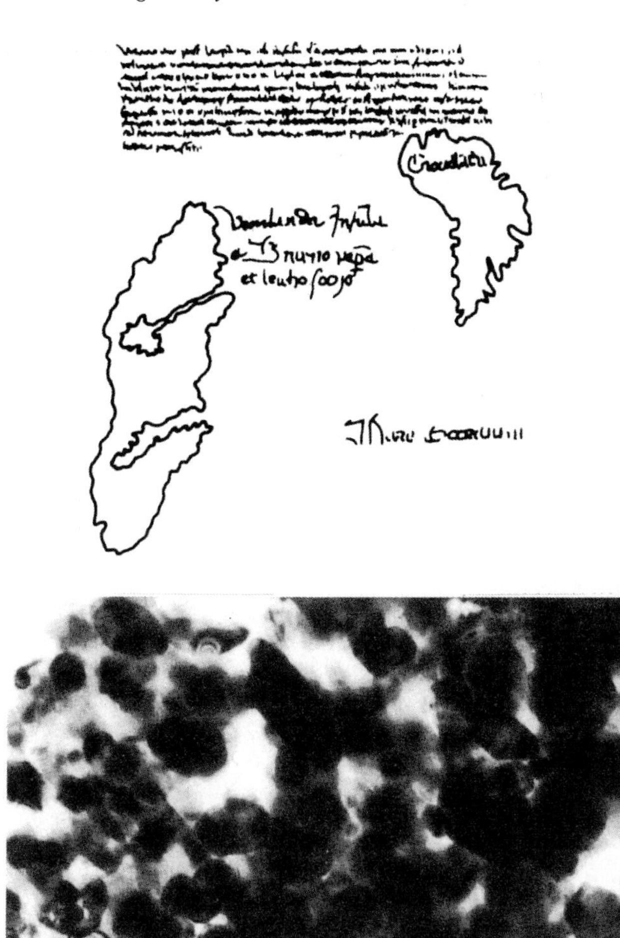

Figure 4.16. Detailed examination of the pigments in the so-called Vinland map indicated that it was a fraudulent document because the crystalline white pigment found in the ink was not manufactured until 1920.[29,30] (Photos courtesy of McCrone Research Institute.) a) Appearance of the "Vinland Map" b) Transmission Electron Micrograph (TEM) of an ink fine particle from the Vinland map. The image is magnified approximately 50 thousand times.

named Laurence Witten. The map was bound together with another document, known as the Tartar relation, dating from 1440, and a fragment of a medieval encyclopedia known as the "Speculum Historical." Some scholars were suspicious of the authenticity of the Vinland map from the moment of its "discovery," because it showed Greenland as a separate island. At the time of Lief Erikson's voyage to North America, however, Greenland was not known to be an island, since most of its shoreline was buried under arctic ice.

In 1959, Witten sold all three documents to an anonymous buyer who donated them to Yale University for a price reported to be close to one million dollars. Scholars at Yale determined that the map had been drawn about 1440. To check up on the authenticity of the map, McCrone and his associates took samples of ink from all three documents. They took 29 samples from the map, 18 from the Speculum, and 7 from the Tartar relations. In McCrones words, "All 54 samples were so small that if collected into a single pile, it would hardly be visible to the unaided eye." Microanalysis established that the inks used for the Tartar relation and the Speculum were composed of similar constituents, basically of iron gallotannate. The conclusion that these manuscripts were almost certainly genuine was supported by an earlier study of the paper on which they were written. The ink on the Vinland map, however, was discovered to be different from the ink on the other two documents. The same ink was used on both map boundaries and legends. However, the map itself was drawn with a firmly adhering brownish-yellow ink on parchment. This ink was found to be overlaid one time with a thin shiny black ink that had almost disappeared. Polarized light microscopy and other techniques showed a high concentration of **anatase** in the brownish-yellow ink but not in the black fragments. Anatase is the popular name for a chemical known as titanium dioxide, a white pigment used widely in cosmetics and paint. As McCrone states:

> *Further electron microscopy revealed that the anatase crystals are inconsistent in shape with a pulverized material, but are characteristic of a precipitated product and indistinguishable from commercial pigment anatase first made during the 1920s. The conclusion that the Vinland map was skillfully constructed during the 20th century was inescapable.*[29]

One wonders how the 10,000 people, who paid $15 dollars each for copies of the map, felt about the fraud![29,30]

References

1. S. Palenik, "Microscopy and the Law," *Industrial Research Development*, March 1979.
2. E. Locard, *Revue Internationale de Criminalistiques 1* (1929), 176.
3. E. Locard, "The Analysis of Dust Traces," *The American Journal of Police Science 1* (1930), 276.
4. J. Broad, *Science and Criminal Detection*, Macmillan, London, 1988.
5. H. J. Walls, *Forensic Science*, Frederick A. Praeger, New York, 1968.
6. O. Johari, I. Corvin, "Scanning Electron Microscopy and the Law: A Review," *Canadian Research 24* (1976), 24.
7. J. A. Cooper, "The Criminal Leaves His Card," *Industrial Research*, 15 November 1977, 22.
8. P. A. Tipler, *Physics for Scientists and Engineers*, Third Edition, Worth, New York, 1991, p. 985.
9. D. Smale, *J. Forensic Sci. Soc. 13* (1973), 8.
10. J. R. Ferraro, L. J. Basile, "The Diamond Anvil Cell as a Sampling Device in the Infra-Red," *American Laboratory*, March 1979, 31.
11. S. Compton, J. Powell, "Forensic Applications of Infrared Microscopy," *American Laboratory*, November 1991, 42.
12. Readers interested in this particular experiment can find a review of the work in B. H. Kaye, *Direct Characterization of Fineparticles*, Wiley, New York, 1981.
13. W. C. McCrone, *The Particle Atlas*, Ann Arbor Scientific Publishers, Ann Arbor, MI, 1980.
14. S. Palenik, "Microscopy and the Law," *Industrial Research Development*, March 1979.
15. R. E. Bisbing, "Clues in the Dust," *American Laboratory*, November 1989, 19.
16. W. C. McCrone, "The Renaissance of Light Microscopy," *American Laboratory*, October 1988, 42.
17. W. C. McCrone, S. Palenik, "The Solids We Breathe," *Industrial Research*, April 1977.
18. R. W. Pease Jr. (Ed.), *Webster's Medical Dictionary*, Merriam Webster Inc., Springfield, MA, 1986.
19. S. A. Katz, "The Human Hair as a Biopsy Material for trace Elements in the Body," *American Laboratory*, February 1979, 44.
20. A. S. Steinberg, "Neutrons Seek the Murderer," *Quantum*, May/June 1992, 20.
21. G. T. Seaborg, "Expanding the Role of the Atom in the Humanities," *The Physics Teacher*, November 1970, 422.
22. T. Giovandi-Jukubczak, "Measurement of Mercury In Human Hair," *Arch. Environmental Health 28* (1974), 139.
23. E. A. Martin (Ed.), *The Bantam Medical Dictionary*, Bantam Books, New York, 1982.
24. J. H. Gardner (Ed.), *Great Mysteries of the Past*, Reader's Digest Association, Montreal, 1991, p. 124.
25. S. Forshufvud, *Who Killed Napoleon?*, Hutchinson, London, 1962.
26. D. Jones, "The Singular Case of Napoleon's Wallpaper," *New Scientist*, 14 October 1982, 101.
27. WGBH Educational Foundation, *Nova, Adventures in Science*, Addison-Wesley, Boston, 1983.
28. J. Emsley, "Stradivari's Musical Varnishing Act," *New Scientist*, 24 March 1988, 36.
29. W. C. McCrone, "The Vinland Map," *Analytical Chemistry 60,* (1988), 1009.
30. H. Wallis, F. R. Maddison, J. D. Painter, D. B. Quinn, R. M. Perkins, J. C. Crone, A. D. Baynis-Cope, W. C. and L. B. McCrone, "The Strange Case of the Vinland Map," *The Geographical Journal Part II 140* (1974), 183.

Chapter 5

Bullets, Bombs, and Body Armor

Chapter 5

Bullets, Bombs, and Body Armor

5.1 Gunshot Residues

On the title page of this chapter there is a remarkable photograph of a bullet speeding towards its target.[1] This picture dramatically illustrates that whenever a gun is fired a cloud of smoke is generated by the burning propellant that drives the bullet out of the gun. Some of this cloud of smoke inevitably ends up on the hands of the person firing the gun. (Note that this cloud of gun smoke is usually rich in lead fine particles which can cause lead poisoning in police who do target practice in poorly ventilated target ranges without respiratory protection.) Sometimes it is possible to check the hands of the person suspected of firing the gun to see if the fine particles trapped on the skin of the suspect can be matched to the known characteristics of the propellent used in the round of ammunition fired from the gun. In Figure 5.1 components of a typical round of ammunition are shown. In terms of explosives it is useful to talk about high and low explosives. The gun powder used to drive a bullet from a gun is typically known as a **low explosive** because of the rate at which it burns, whereas a substance such as TNT (trinitrotoluene) is known as a **high explosive**. In Figure 5.2 the chemical formulae of some well-known explosives used in ammunition and the making of terrorist bombs are shown. Basically, they are compounds high in carbon, hydrogen, and nitrogen, which burn to generate large volumes of gases at high pressure. It is the buildup of high pressure gas as the propellant burns that drives a bullet out of the gun. This has to be initiated by a small detonator cap contained in the end of the case, as shown in Figure 5.1. Gunshot residue contains the by-products from the burning of the detonator, the basic propellent used in that particular shell, and the metal fume given off by the bullet as it passes down the barrel of the gun. The bullet itself is usually slightly larger than the barrel of the gun. The barrel contains spiral grooves, the raised areas between the grooves are known as **lands**. When the bullet is pushed into the barrel by the burning propellant it is deformed as the lands bite into the metal, causing the bullet to spiral as it leaves the gun. This spin stabilizes the bullet's path as it flies towards its target. A gun barrel containing lands is said to be a rifled. Long barrelled guns which generate spinning bullets have also come to be known as rifles. In a modern gun, the

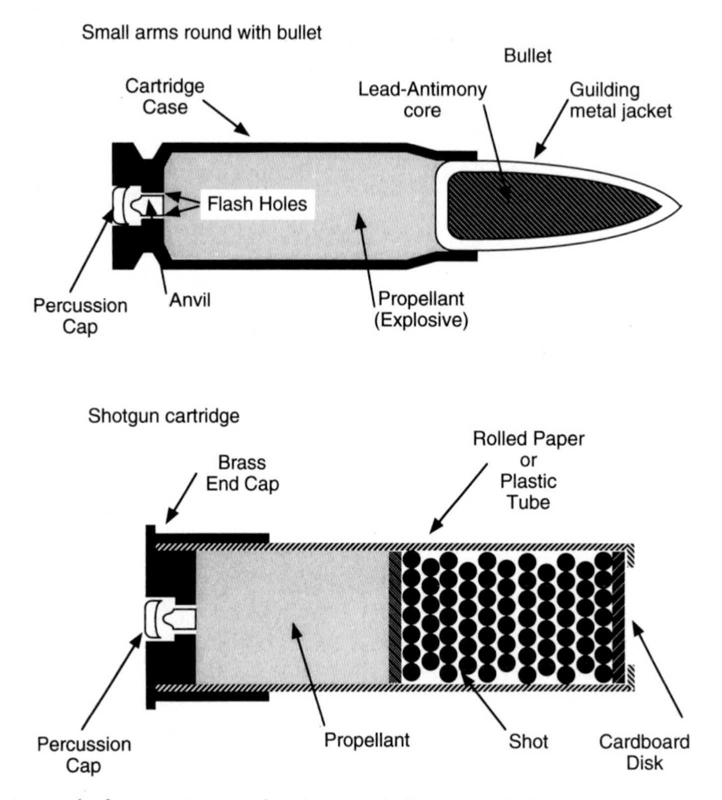

Figure 5.1. A round of ammunition and a shotgun shell consist of the missile (bullet or lead shot), the explosive propellant, and the detonator.

bullet leaving the muzzle can be travelling at more than twice the speed of sound and be spinning at 4,000 revolutions per minute.[2]

In Figure 5.3 two remarkable pictures taken by Edgerton illustrate how scientists gain knowledge about the dynamics of a bullet. Figure 5.3 (a) shows a bullet passing through a playing card. The grooves and scratches on the bullet made by the gun from which it was fired are clearly visible in this photograph. In Figure 5.3 (b) the lines moving from the top and the bottom of the bullet to form a triangular shape are the pathway of the shock waves created by the bullet. These shock waves travel out from the axis of the bullet's pathway at the speed of sound. They are visible because shock waves consist of compressed air, which has a different refractive index from that of ordinary air. (This is why the air above a hot road appears to shimmer. Its density is being changed by the heat of the road and hence its refractive index is different, a fact which sometimes creates mirages along a hot highway.) If one measures the triangle for the shock waves of the bullet of Figure 5.3 (b), one can calculate that the bullet is travelling at 2.5 times the speed

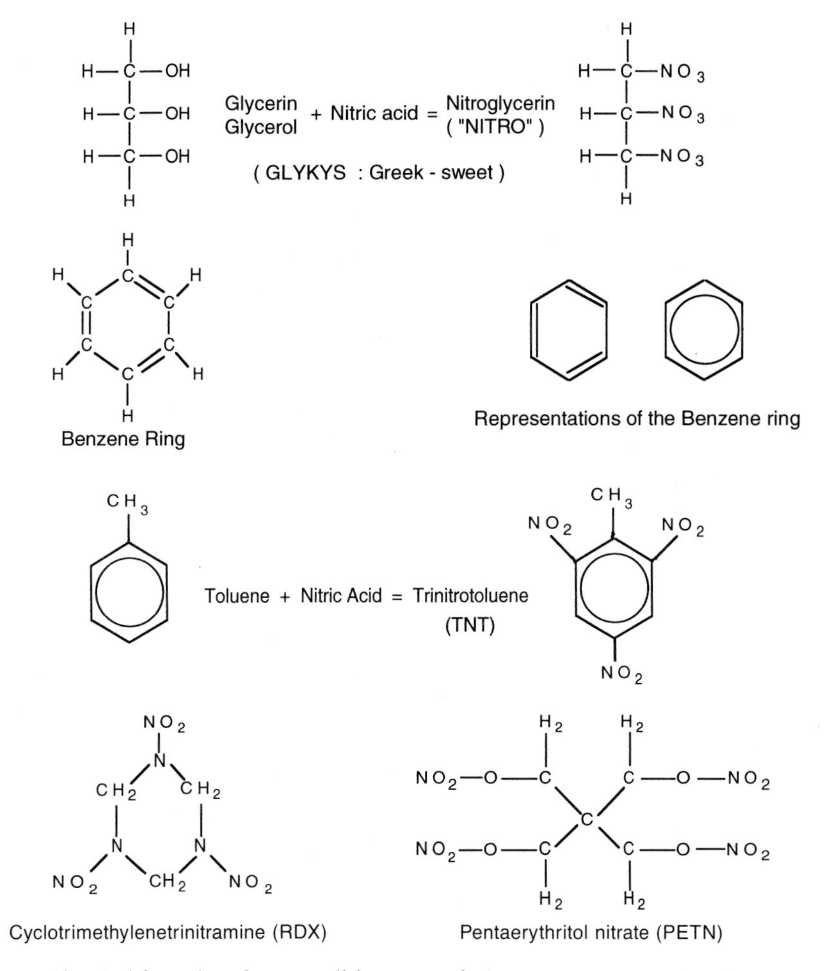

Figure 5.2. Chemical formulae of some well-known explosives.

of sound. The fact that a bullet is travelling faster than the speed of sound is enshrined in the famous statement made in many cowboy movies that "you never hear the one that gets you." This folk statement suggests that the bullet arrives at its target before the sound wave created by the firing of the gun. Sonic shock waves were first brought to the attention of the general public when aircraft started to travel faster than the speed of sound. When an aeroplane has broken the sound barrier (traveling faster than the speed of sound), the aircraft can be seen passing overhead before the sound waves from its engine reach the person on the ground. Because the average citizen is unaware of the problems of shock waves, he is not aware of the fact that in a confined space the shock waves travelling with the bullet can create very

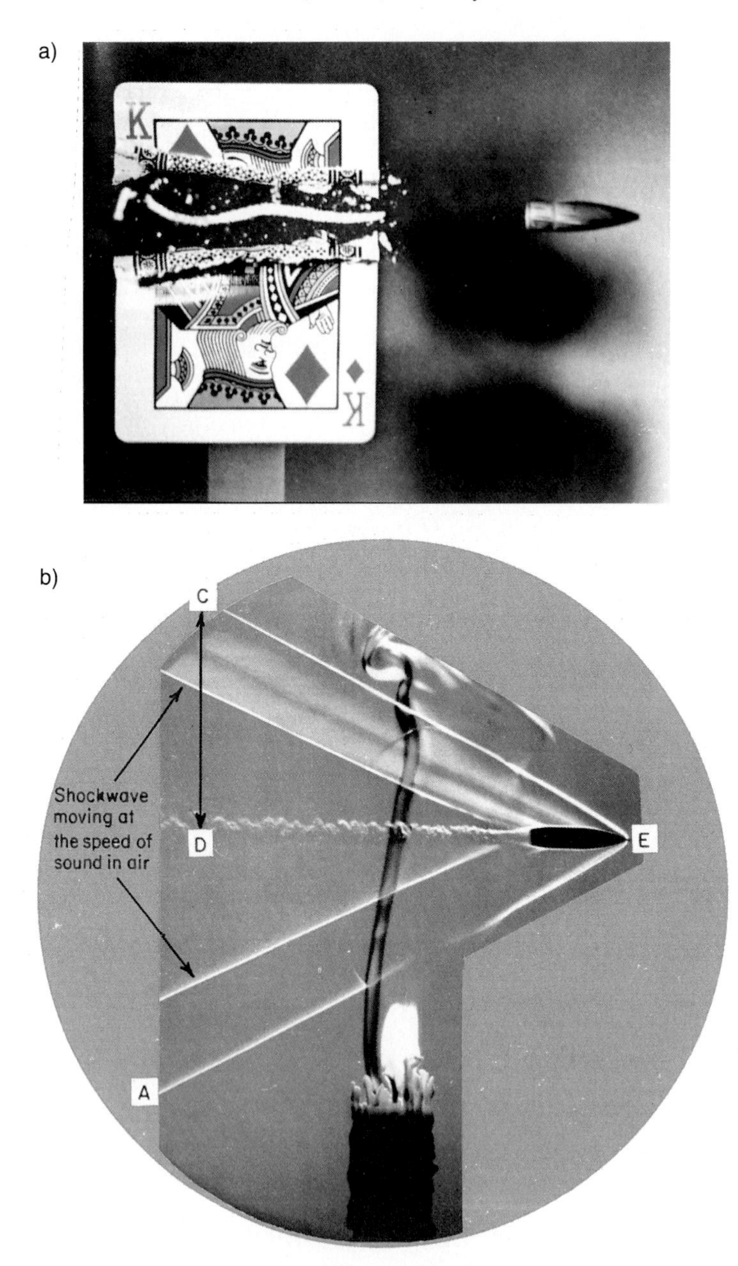

Figure 5.3. Modern photographic methods make it possible to study the dynamics of bullets in detail. a) High speed photography can capture the spinning of the bullet created by grooves (rifling) in the gun barrel. Note the curling of the paper. b) High speed photograph of a bullet moving through the heat of a candle flame displays the shock front created by the supersonic bullet. (Exposure time 0.33 microsecond.) The photographs were taken by Dr. H. E. Edgerton. Reproduced from E. Zwingle, "Doc Edgarton. The Man who made Time Stand Still", *National Geographic*, October 1987, 464.

strong echoes. A famous case in which the echoes generated by reflective shock waves may have confused witnesses is the assassination of President Kennedy in 1963. While discussing this case, F. Smythe comments that the sound waves caused by the discharge of a gun can cause echoes in a confined space, which sound like two or more shots. In the case of firing a bullet at more than the speed of sound, a second shot will be heard as the bullet breaks the sound barrier. A ricochetting bullet may also whine, similar to how they sound in westerns, further confusing the witness.[3]

Figure 5.4 illustrates the way in which gunshot residue can lodge in the wrinkles of the skin and around the wound on the body, if the gun is fired close to the victim.[4] It is very difficult to remove gunshot residue from inside the wrinkles of the skin, but one widely used technique, similar to the method known as a nitrate test, involves the use of paraffin wax. It is outlined by Nickolls in his text written in 1956:

> *The method of testing hands of the suspect for nitrates derived from the blow back of the ignition of a cartridge fired from a weapon was developed many years ago. The method has been the subject of considerable work in the United States. It consists of a casting in paraffin wax of the forefinger, thumb, and other parts of the hand subject to blow back and the treatment of the cast with a solution of diphenylamine in sulfuric acid. The spots of nitrate are shown as deep blue spots. In practice, I have found this test quite unreliable since it is common to obtain positives when no gun has been used and negatives when a gun has been used. In these circumstances I do not consider the test of forensic value.*[5]

In a more modern version of the paraffin test, the melted paraffin is poured over the suspects hand, allowed to set, and peeled off to form a kind of thin skinned glove. This wax replica of the hand is then placed into a nuclear reactor and bombarded with neutrons (this is an example of neutron activation analysis discussed in Chapter 4).

By studying the radioactive material produced by the neutron bombardment, one can look for elements such as mercury, lead, antimony or barium, which are typical metals found in gunshot residues. If these metals are found, in the words of F. Smyth:

> *The chances are that the suspect has recently fired a gun but it is only fair to say that when this process was tried on Lee Harvey Oswald, President Kennedy's assassin, all the results were negative.*

Smyth also reminds us that a popular feature of detective fiction shows the examiner lifting a pistol to his nose, sniffing it, and pronouncing that it has been recently fired. Again, in the words of Smyth: "Unfortunately, unless this is done within minutes of the firing, in real life it tells the investigator nothing".

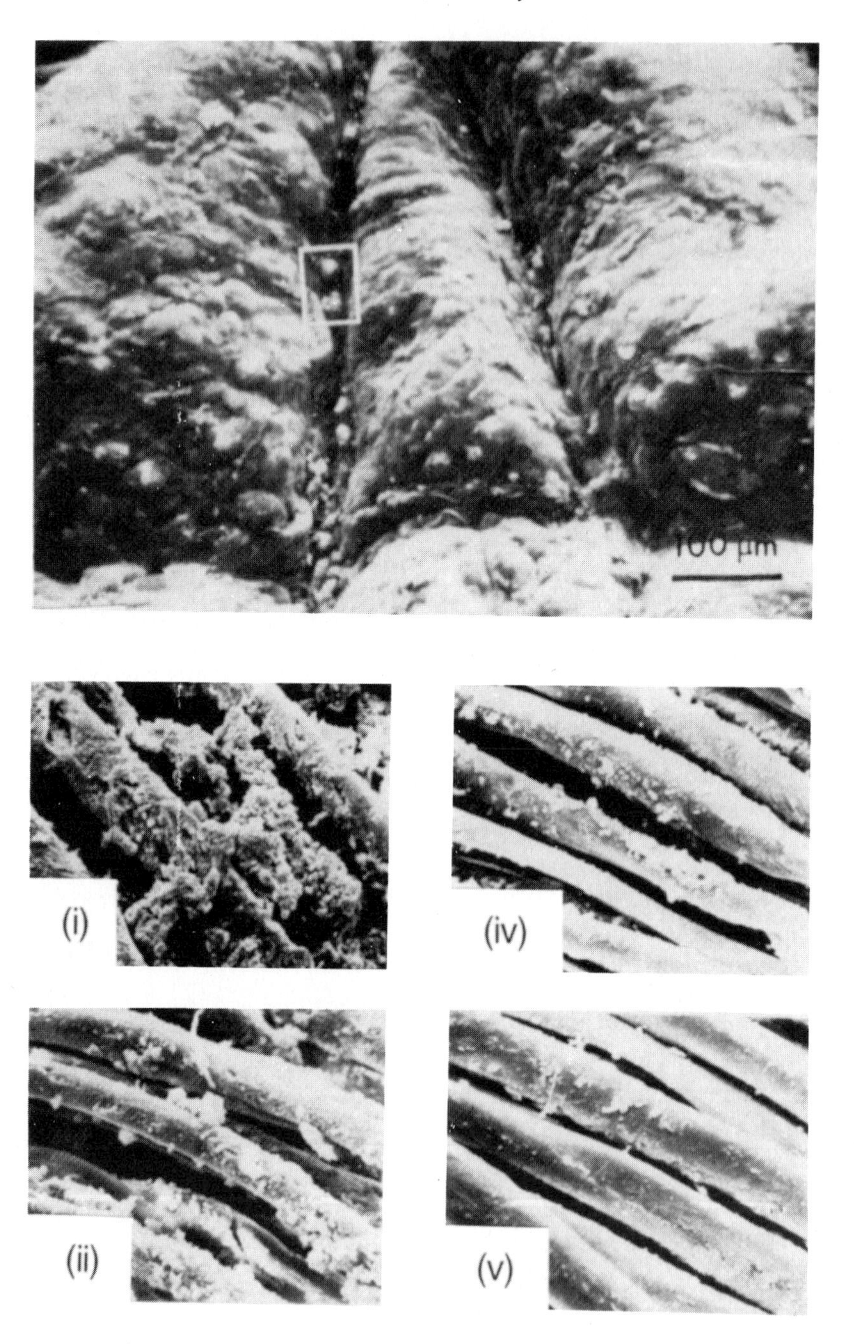
100 µm
(i)
(iv)
(ii)
(v)

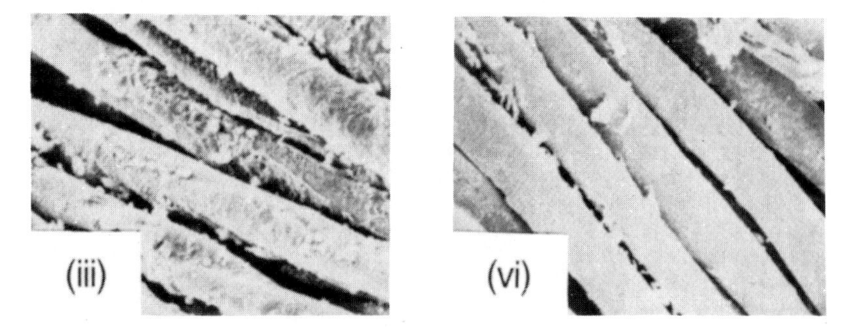

Figure 5.4. Scanning electron micrographs showing the deposition of gunshot residue on the hand of the assailant and the clothes of a victim at close range.[4] a) Gunshot residue in the wrinkles of the skin on the back of a hand. b) Micrographs of the first untorn strands of yarn near a bullet hole in cotton fabric caused by shots fired from (i) 5 cm (ii) 10 cm (iii) 15 cm (iv) 30 cm (v) 50 cm (vi) 80 cm.

In studies of gunshot residues it has been found that the fine particles from the smoke are often spherical and range in size from below one micron to tens of microns in diameter. To use more modern methods of studying residual fine particles, one can use an adhesive tape to lift the gunshot residue from a suspect's hands, clothing or vehicle. Fine particles removed in this way are studied by modern techniques called back scattered electron imaging (BEI) and secondary electron imaging (SEI), see Figure 5.5 (a). Readers interested in these techniques can consult the article by Germani of McCrone Associates, and the references that he quotes.[6]

Germani tells us that analyzing gunshot residue using the advanced techniques that he describes has been accepted as evidence in several murder cases.[6]

Energy dispersive x-ray fluorescence has also been used to study gunshot residue. Cooper tells us that the method is rapid, non-destructive, and capable of simultaneous multi-element analysis. Cooper also tells us that lead, antimony, and barium can be measured in gunshot residue samples in as little as five to ten minutes after wiping both hands of the suspect with a filter paper moistened with a diluted acid solution. (Cooper comments that neutron activation analysis requires several days for information to be fed back to the investigation officer.) In one study of gunshot residues using EDXRF, Cooper reports that one of two bullets was suspected of having made a hole in a white shirt. A comparison of the elemental content of the bullets and the debris around the bullet hole revealed the presence of tin around the bullet hole and in one of the bullets, but not in the other bullet.[7] An example of the sensitivity of this technique is shown in Figure 5.5 (b).

Another analytical procedure used to characterize gunshot residues is a method known as **atomic absorption**, often referred to as **AA**. This analytical procedure has many applications in forensic science and, therefore, we

a)

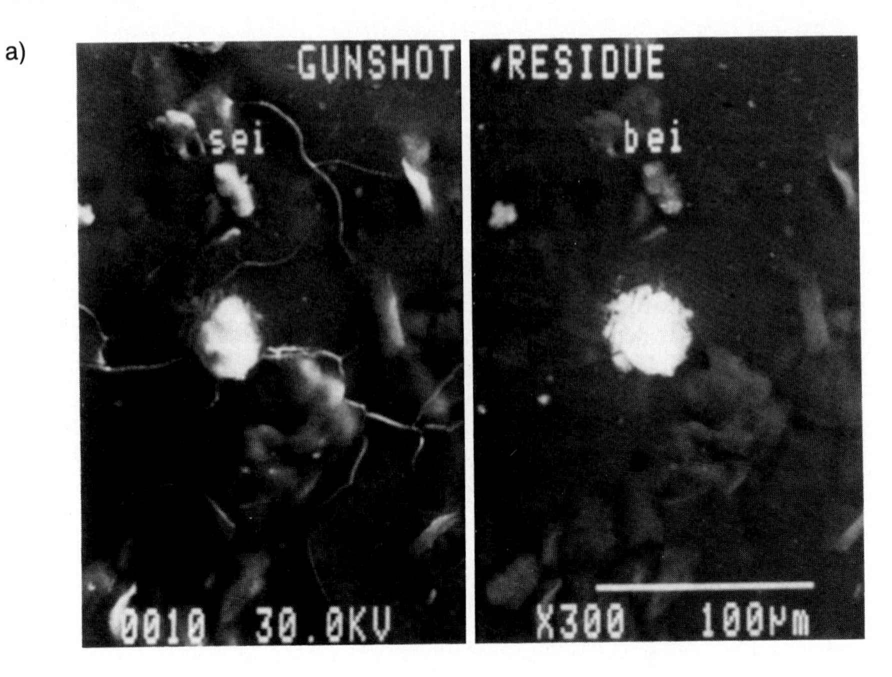
GUNSHOT RESIDUE
sei
bei
0010 30.0KV
X300 100μm

b)

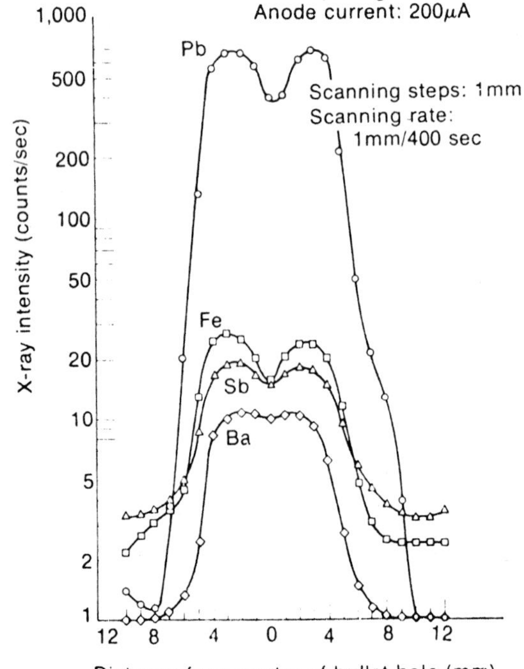
Excitation conditions
Tube anode: Mo; filter (none)
Anode voltage: 35 kV
Anode current: 200μA
Scanning steps: 1mm
Scanning rate:
1mm/400 sec
Pb
Fe
Sb
Ba
X-ray intensity (counts/sec)
1,000
500
200
100
50
20
10
5
2
1
12 8 4 0 4 8 12
Distance from center of bullet hole (mm)

will discuss it in some detail. If one takes a Bunsen burner and sticks a piece of table salt (sodium chloride) into the flame, the hot salt vaporized into the flame and generates an intense yellow flame (when I was a student of physics, in order to generate monochromatic sodium light, we used to mold table salt with a small amount of water to make suitable salt sticks). The yellow light emitted by the flame has a wavelength of 589 nanometers (0.589 micrometers). Some textbooks use a unit of length called the Ångström, Å to state the wavelength of light. One **(Å)ngström** equals 10 nanometers. Ångström was a Swedish physicist, who lived from 1814 to 1874. He was a pioneer in the study of the various components of the light spectrum given out by the sun and the aurora borealis (northern lights). In a technique known as flame photometry, the various elements present in a solution under test are detected by analyzing the light given out by a flame into which the solution is sprayed.[8] The basic concepts embodied in flame photometry are illustrated in Figure 5.6 (a). The inspection slit is movable, so that any particular part of the spectrum can be examined in detail; a photographic record of the various components of a spectrum can be created by moving the inspection slit over the entire spectrum.

If one applies high voltage to electrodes in a glass tube containing sodium ions, the pure yellow light of 589 nanometers is given out by sodium atoms when they collide with electrons speeding from the negative electrode (called the **cathode**) to the positive electrode (called the **anode**). Such an electrical system is used in yellow "sodium" street lights used in some cities.

In the analytical technique known as atomic absorption, an intense light source that emits the light characteristic of a substance, which may be present in the mixture to be analyzed, is shone through the flame into which the solution under investigation is sprayed. For example, if our lamp shown in the sketch of Figure 5.6 (b) were generating sodium light, any sodium atoms present in the solution being investigated would absorb the light from the electric discharge lamp. As a result, instead of the lines characteristic of sodium, dark lines would be present in the absorption spectrum. Similarly, if, in our examination we used a high-voltage lamp emitting light characteristic of lead atoms, and if there were lead present in the solution of material taken from a suspect's hand, we would have lead-absorption lines in the spectrum generated in the atomic absorption equipment. (The name atomic absorption obviously comes from the fact that the presence of materials in the flame is deduced from the adsorption behavior of the atoms being

◀ **Figure 5.5.** Gunshot residues can be studied using advanced imaging techniques and EDXF, to determine if they contain heavy metals vaporized from the bullet or present in the detonator and propellant materials. a) Modern techniques using beams of electrons can image gunshot residues. Left image is produced by SEI and the right image by BEI. (American Laboratory, April 1993, used by permission.[6]) b) Plot of the X-ray intensities obtained by scanning a bullet hole in a white shirt. (Used by permission of *Industrial Research*.[7])

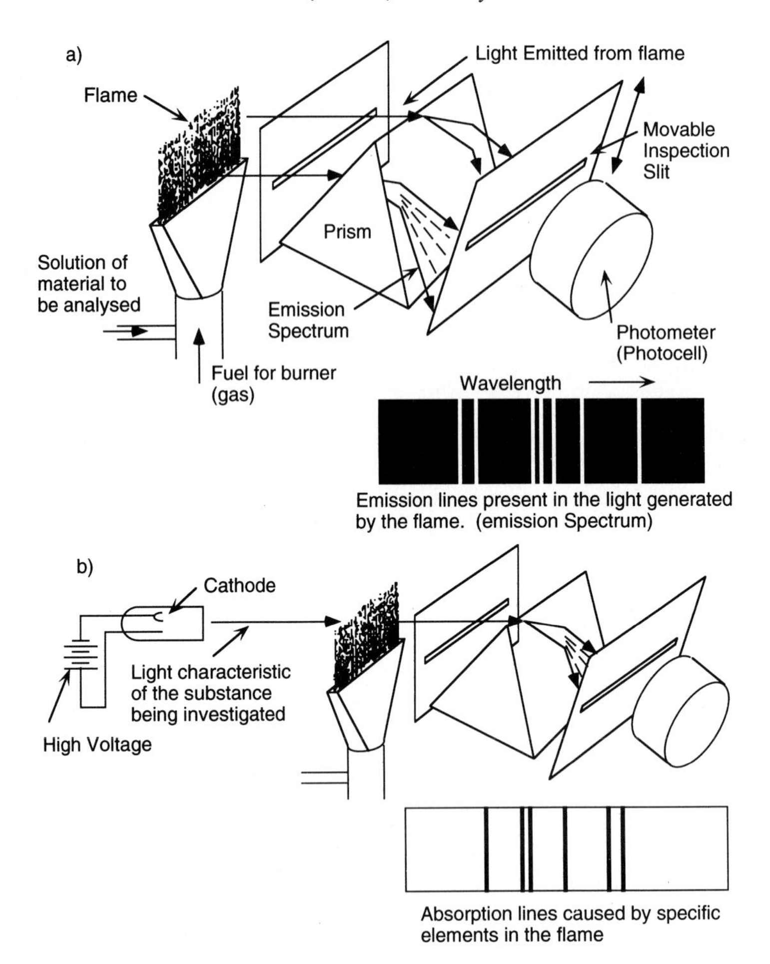

Figure 5.6. Flame photometry and atomic absorption can be used to study the constitution of gunshot residues. a) In flame photometry the material to be investigated is sprayed into a flame where the heated atoms emit light characteristic of their atomic structure, seen as intense lines in the spectrum of the flame. b) In atomic absorption, the presence of an element in the flame results in the absorption of particular wavelengths of light, characteristic of an element, from the light of a lamp viewed through the flame.

sprayed into the gas flame.) The use of atomic adsorption techniques enables the forensic scientist to look for specific elements such as lead, barium, and antimony, using reference light sources giving out the characteristic wavelengths generated by the excited atoms of these elements.

In the discussion so far we have concentrated on investigating the presence of metal contamination on the skin from the gunshot residue. One can look for unburnt particles of propellant and detonators on the hand using a technique known as **liquid chromatography**. Atomic absorption and EDRXF are not suitable for looking for low atomic weight elements such as carbon, hydrogen, nitrogen, and oxygen present in the propellants and explosives used in the manufacture of bullets and bombs. The term **chromatography** means "colored writing." However, if a visitor to a forensic laboratory were to look at a modern equipment for carrying out liquid chromatography, he would wonder where the name came from, since there is very little evidence of colored writing in the output from the analytical instrument. An instrument used to carry out chromatography is called a chromatograph. This name was given to the instrument by its inventor, because of the appearance of the device when he used it to separate the various pigments present in plant material. The inventor was a Russian botanist Michail Tswett, who lived from 1872 to 1919. In 1906, Tswett extracted the colored pigments from leaves using gasoline. He poured this material into a column of crushed calcium carbonate and he continued to pour the liquid through the material in the column, colored bands appeared in the calcium carbonate. It is now known that the pigment molecules have different sizes and different affinities for the surface of the calcium carbonate. As the liquid was poured down through the column, the pigment molecules moved at different rates, alternately clinging to and leaving the surfaces of the calcium carbonate. Therefore, they separated into groups according to their molecular size and properties. Tswett gave his colorful name to the technique because it seemed to him that dyes were writing their presence on the column of calcium carbonate.

One can carry out a simple experiment, which parallels the original work of Tswett, using a piece of filter paper and a drop of drawing ink. Many drawing inks have various components in them to give different properties to the ink. If one takes a circular filter paper and places a drop of ink in the middle of the filter and then continuously drops water onto the center of the spot of ink, one creates a front of liquid moving away from the center. As the movement of the liquid continues, the components of the ink move at different speeds to create a pattern as illustrated in Figure 5.7. The liquid carrying the various parts of the ink outward is called the **eluant**, and one states that the eluant liquid has **eluted** the various components of the dye outwards. The word "elute" comes from a Greek root word which means "to wash out;" it is related to the words dilute and pollute. When one uses a steady stream of drops to move the various components of the ink through the fibrous mat of the filter paper, one is using a technique which is described as **thin layer paper chromatography**. This technique can be used to study the structure of ink used in a particular pen and can be used to check on the authenticity of signatures claimed to have been written with a certain pen. (This aspect of chromatography will be discussed in greater detail in chapter 9) In modern chromatography equipment, instead of observing the separation of a sample into its components as it travels down a column, one monitors the rate of arrival of the components at the end of a chromatography column. If the eluant used in a piece of equipment is a gas

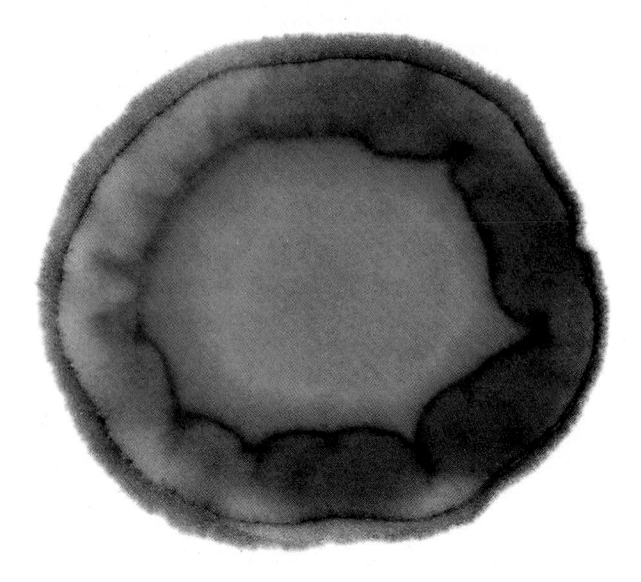

Figure 5.7. A drop of ink, eluted outwards from the center of a filter paper, separates into its components. This process is known as thin layer liquid paper chromatography (TLC).

(referred to as a **carrier gas**) the process is known as **gas chromatography**. In Figure 5.8, the basic structure of a gas chromatographic equipment used to generate what is known as a **chromatogram** is illustrated. In Figure 5.8 (b), the gas chromatogram of two different whiskeys are shown. If the eluant is a liquid, the process is known as liquid chromatography. In Figure 5.9, two applications of liquid chromatography to investigations of explosives are shown. So-called plastic explosive (shown in the figure) is a mixture of two explosive materials with a plasticizer. The data of Figure 5.9 are taken from an investigation reported by the Waters Corporation.[9]

Actual propellant mixtures and explosive material often contain small amounts of various additives. A liquid chromatography investigation of confiscated explosives or ammunition can often be used to track the suppliers of the explosives. In Figure 5.9 (b), various components of a particular explosive, and the amounts of material in a standard sample are shown. Note that μg and ng shown in these two diagrams denote **micrograms** and **nanograms**, respectively.

Figure 5.10 illustrates the use of liquid chromatography to look for nitroglycerin in gunshot residue collected from the back of the hand of a suspect.[10] The debris was collected with a cotton swab dipped in acetone. Using liquid chromatography, nitroglycerine was indeed found, as illustrated in Figure 5.10 (a). The nitroglycerine was detected even in unburned gun powder that was not visible to the naked eye. Investigators generating the infor-

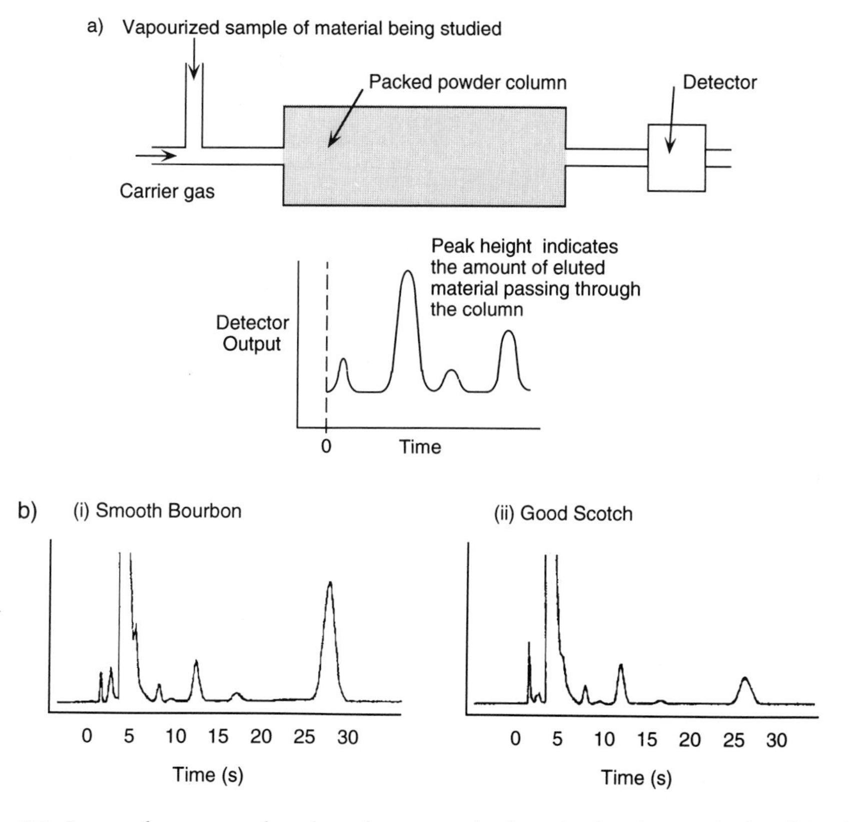

Figure 5.8. In gas chromatography, the substance under investigation is vaporized and carried through the chromatography column by a carrier gas. The chromatogram is a record of the arrival time of the various components present in the sample. a) Basic configuration of gas chromatography equipment and a typical chromatogram. b) Chromatograms showing the components present in two types of whiskey.

mation of Figure 5.10 noted that the nitroglycerine was not detected on the palm of the hand firing the gun.

To investigate the effects of a delay in swabbing the suspect's hand, several different persons fired the weapon and hands were swabbed at different times. Some hands were swabbed after 5 minutes, others after 3 hours. The result to the investigation is summarized in the Figure 5.10 (b). The investigators note the following:

> *Immediately after firing, nitroglycerine levels were fairly high probably due to visible fine particles on the skin. Later, as movement dislodged the fine particles the concentration on the hand subsided to lower but still detectable amounts.*[10]

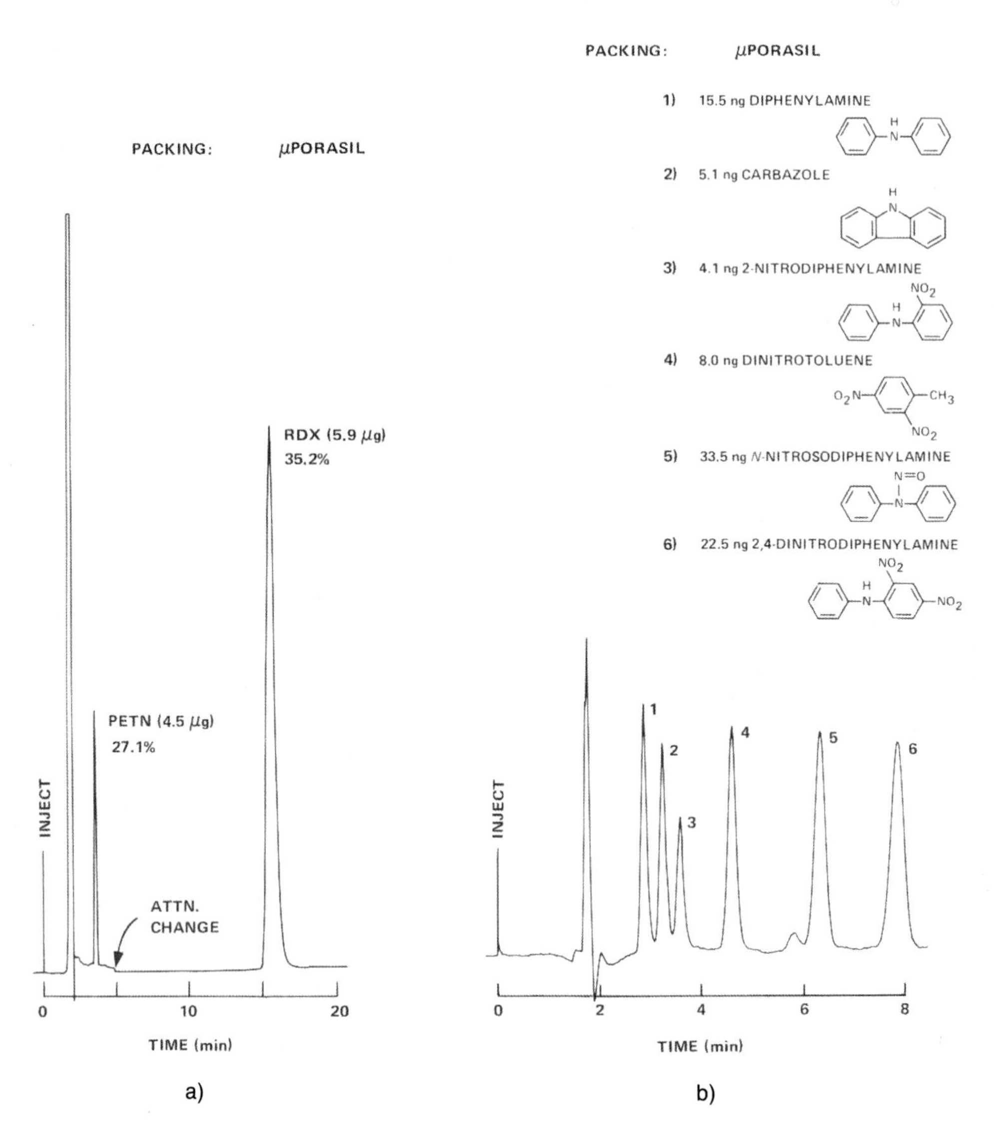

Figure 5.9. Liquid chromatography can be used to identify the components of an explosive.[10] a) Liquid chromatographic analysis of a confiscated plastic explosive. b) Liquid chromatography can identify various additives in an explosive or a propellant and may be used to identify the manufacturer.

So far in our discussion of techniques for studying gunshot residue, we have focussed on techniques where one swabs the hands of a suspect with acetone; however, other solvents can also be used. Dr. Robert Shaler of the chief medical examiners office in New York swabs the hands of suspects

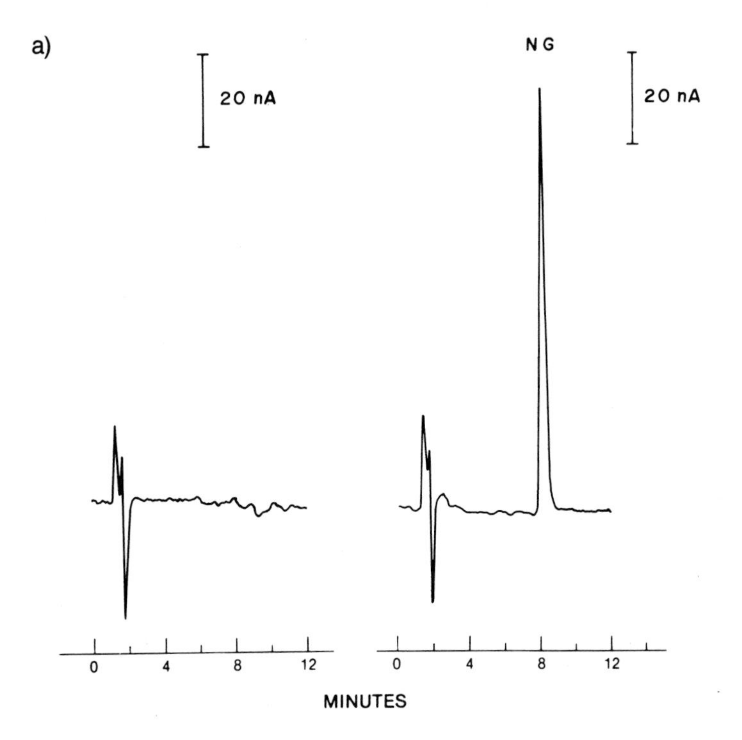

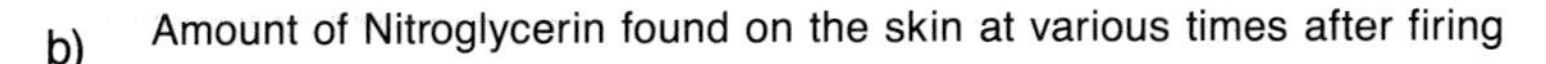
b) Amount of Nitroglycerin found on the skin at various times after firing

Sample	Swabbing Time after Firing	Visible Flakes	Amount of Nitroglycerine Detected
1	0 min	Yes	16.9 µg
2	0 min	Yes	9.0 µg
3	0 min	Yes	3.7 µg
4	34 min	Yes	11.7 µg
5	34 min	No	1.0 µg
6	55 min	No	0.8 µg
7	59 min	No	0.77 µg

Figure 5.10. Liquid chromatography can be used to look for unburned propellant on the hand of a person suspected of firing a gun. (© 1980, Bioanalytical Systems Inc., West Lafayette, Indiana.) a) Chromatograms obtained prior to and after firing a 0.22 calibre revolver. The nitroglycerin peak (NG) is quite apparent in the second chromatogram. b) Table showing variations in the amount of nitroglycerin detected on hands at various times after firing a weapon.

with a chemical that reacts with lead, **sodium rhodizonate**, to determine whether or not a person has recently fired a gun. If lead is present on the hand, the fine particles of lead will turn reddish purple. Commenting on this technique Shaler states:

Of course, it is merely a presumptive test and a fairly crude one at that. We cannot prove that the guy fired a gun at a particular moment only that it was there on his hand. Nevertheless, it can be an important piece of evidence; one more link in a chain establishing guilt or innocence.[11]

Not only does the hand of the person who fired the gun carry informative debris, but material to be found on a bullet removed from a body can yield useful information to the investigating biologist. One branch of modern biology is called **cytology**. The specialist in this area looks at the various cells that make up the body. An expert in bullet cytology (looking at the biological debris to be found on a bullet) is Maryanne Sens of the Medical University of South Carolina.[12]

When Sens receives a bullet from the scene of a crime, she washes it in a salt solution, filters the solution and examines the biological debris left behind. In discussing her technology, Sens states:

You can tell where the bullet has been. If it's been through the heart, there are little bits of heart tissue. If it went through the brain, there are traces of brain tissue.

Sens can match each bullet with a hole in the body and identify which bullets missed completely. She can also determine the victim's blood and tissue type and sometimes obtain a DNA fingerprint. Sens points out that if this technique had been available to the investigators of the John F. Kennedy assassination, the claim that the same bullet had struck both Kennedy and Texas Governor John Conolly could have been investigated.

5.2 Has the Suspect Handled Explosives Recently?

Closely allied to the problem of investigating whether or not a person has recently fired a gun is the problem of investigating whether suspects have handled explosives. This question assumed new prominence because of the overturning of 17 different convictions of alleged IRA terrorists in Great Britain in the early 1990s. (The Irish Republican Army, IRA, is an organization which is fighting the British with the ultimate aim of joining Northern Ireland, which is presently part of Great Britain, to Southern Ireland, known as Eire, forming an independent republic. Depending on your political point of view, the IRA can be seen as a terrorist organization or a political one.) The three cases in which there has been considerable controversy over the investigation into whether or not the suspects had handled explosives are known as the Birmingham Six, The Guildford Four, and the Maguire Seven.[13]

The Guildford Four were convicted in 1976 for planting a bomb. They confessed to the crime but these confessions have now been shown to have been baseless. At their trial the Guildford Four named Annie Maguire, an aunt of one of the four, as the bomb maker for the IRA. After the implication of Annie Maguire in the making of the bombs, four members of her family and three other people who were staying at the house were arrested. The Maguire Seven, as they came to be known, were convicted in 1976 of running a bomb factory. The only evidence against them was some forensic tests carried out by the Royal Armament Research and Development Establishment (RARDE) using thin-layer chromatography in an effort to establish that the Maguire Seven had handled nitroglycerin. At a later inquiry into the conviction of the Maguire Seven, conducted by a judicial official named May, a scientist from RARDE stated that never before had he seen so many positives on a plate at a reasonably high level of intensity. The evidence given at the inquiry produced four separate grounds for thinking that the Maguires had never been near nitroglycerin. The interested reader will find a full discussion of the May inquiry in Hamer's article.[13] Basically, however, it appears that a major problem was that the test kits (produced by the Royal Armaments Research and Development Establishment) used in the investigation probably contained a small bottle of ether which was already contaminated with explosives. The scientists who prepared these test kits had not only made the test kits but also made explosives at the same establishment. Obviously, it was difficult to avoid chance contamination of the test kits used by the police. Hamer concludes that Annie Maguire spent 11 years in prison with the most probable explanation being a contaminated test kit.[13] Hamer points out that we will never know for sure whether or not the police were using contaminated test kits. But in addition to this technical problem, Hamer criticizes the general use of doubtful forensic evidence by the judiciary officials at the original trial.

The Birmingham Six involved six Irishmen who were accused of planting a bomb that killed 21 people in November 1974. In the words of Hamer:

> *The case against the Birmingham Six had two main planks: the confessions of the six Irishmen and forensic evidence. The six men maintained that the confessions were beaten out of them and some of the written confessions showed the possibility that the original confessions had been altered as shown by the electrostatic deposition study of pages underneath the top page of the confession. Apart from these confessions of doubtful value, the forensic evidence that convicted the men was largely the work of Frank Scoose who worked at the home office forensics science laboratories.*

It is worth quoting Hamer's description of what happened at some length:

> *Scoose employed the* **Griess test** *to detect explosives in samples taken from the men's hands. In the Griess test caustic soda breaks down the molecule of nitroglycerine and produces nitrite ions. When sulfuric acid is added, the nitrites form a diazonium salt and a pink color develops when α-naphthyla-*

> *mine is added. The concentration of this caustic soda is crucial to the test. The Chorley laboratory said Scoose had used one percent concentration of caustic soda.*

In an investigative television report, Granada Television (one of the local television companies in Great Britain) asked Brian Caddy, head of the forensic science unit at the University of Strathclyde, to examine the evidence presented at the original trial of the Birmingham Six. In his investigations, as reviewed by Hamer:

> *Caddy showed that a 1 % concentration of caustic soda would also produce positive results for other substances, notably nitrocellulose, which is used to coat products such as record sleeves (album covers) and playing cards.*

Hamer tells us that the men had been playing cards on the night that they were arrested and this could have been how their hands were contaminated. Instead, the presence of caustic soda was interpreted as a positive Griess test. In a later investigation, Scoose backtracked and said that he had only used 0.1 % caustic soda. At this concentration the caustic soda would not have produced a false positive for nitrocellulose. It would have been specific for nitroglycerin. However, Caddy pointed out that at this dilution the Griess test is far less sensitive, and he said that, at this concentration of caustic soda, the police would have been able to visually detect nitroglycerine on the men's hands. It is reported that Scoose had not kept any notes, and he told the court that he did not think it necessary to keep a record of his experimental techniques. Due to the confusion over the validity of the test and other evidence, the Birmingham Six were eventually released from prison and their conviction overturned. It appears that these cases were instances where competent challenging of the forensic evidence would have greatly weakened the cases of the prosecution. But, the evidence went unchallenged because of a lack of knowledge by the defense lawyers. (For a discussion of the implications of the reversal of these decisions see the article by J. Lloyd.[14])

5.3 What Can Ballistics Tell Us?

The use of information contained in scratch marks made by a gun on a bullet has recently been reviewed by Knight.[15] He tells us that ballistic evidence in which scratch marks were used to link bullets found in a victim to a particular gun was first used in Great Britain in a case that involved the shooting of a police constable in Eping Forest in 1927. Four Webley 0.445 revolvers were taken from the suspects, F. Browne and W. Kennedy. Knight notes that the constable had been shot through both eyes; a macabre illustration of the myth that a dead man's retinae, the networks of nerves in the back of the eye, retain the last images of his life. Robert Churchill, a gunsmith called to give evidence at the trial, stated that the policeman's wounds contained black powder, a type of propellant which at the time was obsolete. However, it could be found in unused ammunition that was at least ten years old. Churchill also gave expert evidence that the scratch marks on the fatal bullets and on a shell case found in a stolen car were identical to those made by test firing the guns which had been confiscated from the suspects. To establish the uniqueness of the evidence, specialists at Woolwich Arsenal fired 1,300 similar pistols. They were able to establish that not one of them produced the same breach markings as those on the bullets used in the killing of the constable. This evidence helped to convict the suspects.[15]

Experts wishing to compare the striations on a bullet taken from a victim with those on a bullet fired from the gun of the suspect, will usually fire a bullet from the gun to be tested into a tank of water. The water stops the bullet without damaging it so that the markings on the bullet are not distorted or altered in any way. They then use a special instrument called a **comparison microscope**. This instrument has two inspection lenses which feed into the same field of view. The two images created by the device are presented side-by-side to the person looking down the microscope. In Figure 5.11 a comparison-microscope image of two bullets is shown.[16]

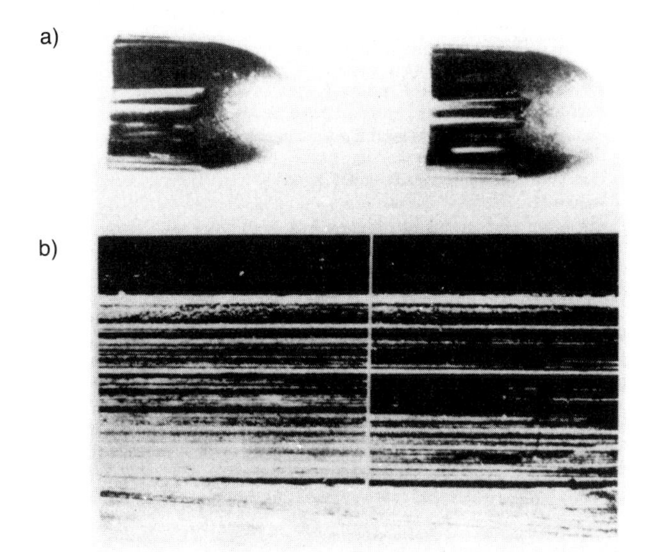

Figure 5.11. The "scratch" lines on a bullet are a unique stochastic signature identifying the gun from which the bullet was fired.[16] a) Two bullets to be compared. On the left is a bullet recovered from a crime scene, on the right, a bullet fired from the suspect's weapon. b) When viewed through a comparison microscope, it can be seen that the striations on the two bullets match exactly.

Not only do the striations on the bullet constitute a stochastic signature that enables the expert to deduce which weapon fired a bullet, but the marks on the ends of a cartridge created by the firing pin, and the mechanism which ejects the bullet from the gun are quite useful. This fact is illustrated by Figure 5.12. Knight also tells us that an important piece of evidence, which can be studied by the ballistic expert, is the wad made of felt, cardboard or plastic that forms part of the filling of the propellant part of the **round of ammunition**. (The entire assembly of detonator, propellant, and bullet constitutes the round of ammunition.) Knight states that expert forensic examination of wads can help to identify both the type and brand of ammunition used. In addition, the range of discharge can be determined, since the wad travels a much shorter distance than the bullet or shot discharged from the gun.

Figure 5.12. Firing pin marks and marks made by the ejector mechanism are additional stochastic signatures which can be used to identify a weapon. The ends of two cartridge cases of bullets fired from the same gun are shown. Marks from the firing pin and those from the breech and ejector mechanism are visible and match exactly.[3]

Another important problem tackled by the ballistic expert is to determine how far away the victim was from the gun firing the bullet. The series of pictures shown in Figure 5.13 are useful for this purpose. They show a study on the appearance of damage inflicted by a 0.32 calibre bullet fired from a Colt revolver at specified distances. When one gives a number such as 0.32 in association with a bullet, one is stating the diameter of the bullet in fractions of an inch. The size of the bullet is referred to as the **calibre** of the weapon. The term **revolver** indicates that the chambers of the gun revolves to deliver a series of bullets to be fired without having to stop and reload the gun. Occasionally, in western movies, the hero counts the number of shots fired by the villain and then emerges triumphantly to arrest the villain when he has run out of bullets. In other situations, the guns of the cowboys

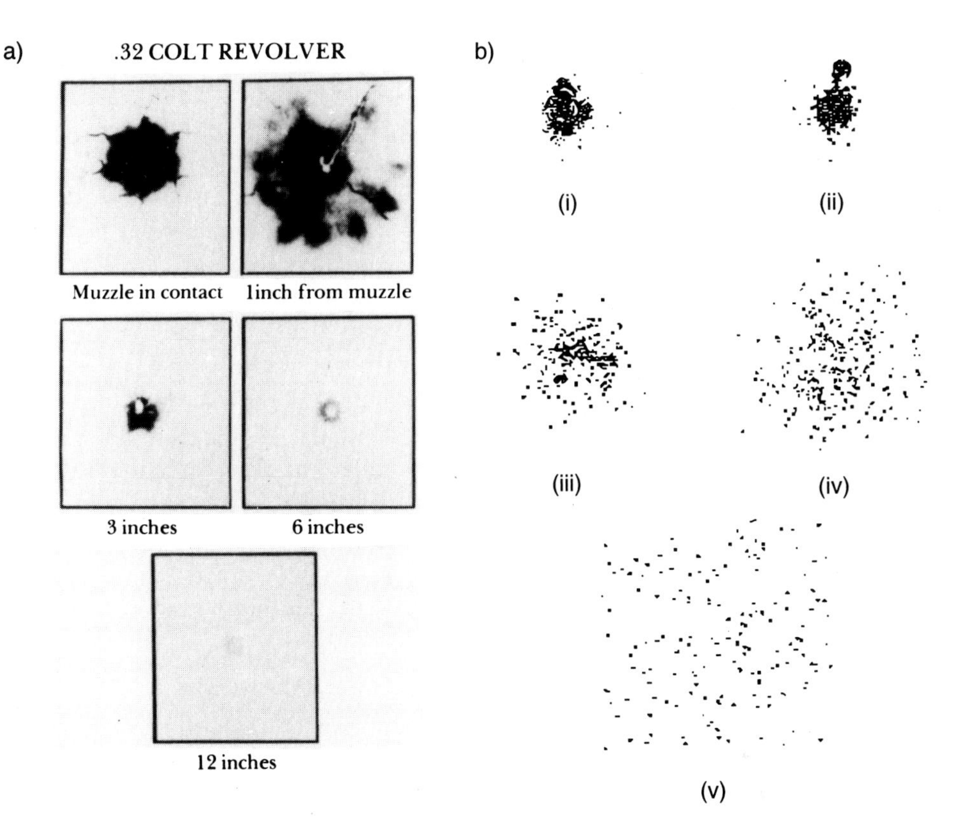

Figure 5.13. Determining the distance between the victim and the gun is an important aspect of the work of the ballistics expert. a) Test cards showing the effect of distance on the structure of the wound from a 0.32 calibre revolver (1 in = 2.54 cm).[3] b) The effect of distance on the dispersal of 12 gauge shotgun pellets. The shots were fired from (i) 30 cm (ii) 2 m (iii) 4 m (iv) 6 m (v) 13 m. (Reproduced with permission of Butterworth and Company Publishers Limited.)

appear to have an infinite number of rounds, defying the known structure of weapons used in hand to hand combat.

In Figure 5.13 (b) the way in which the pellets from a shotgun spread out in space is illustrated by a series of diagrams. Commenting on the difference in the effect of bullets and shotgun wounds, H. J. Walls tells us:

> *A bullet is as fatal at many hundred yards as at just a few feet. Even a 0.22 caliber bullet may penetrate flesh at over a mile. On the other hand, whereas a shotgun at close range is a terrible weapon blasting a ragged hole big enough to put a fist in or blowing half a skull away, once the shot charge is well spread out (some tens of yards), its effect, with small shot sizes, is more likely to be painful rather than lethal.*[17]

When discussing the way in which the effect of the shotgun diminishes with distance, Walls reviews the following case which is worth quoting at some length. Walls calls it the case of the mad farmer:

> *The farmer fired his shotgun at his wife as she was fleeing down the lane. It was of some importance to know how far from him she had been at the time of firing. If the distance was a few yards, then the farmer being familiar with guns would not have been too mad to know that a hit might well have been fatal, and he would have been charged with attempted murder; if on the other hand, the distance was say 40 or 50 yards he would equally have known that the hit was not likely to have been dangerous and a lesser charge would have been preferred.*[17]

Walls tells us that the only fact that he had to go upon was a medical report showing that the wife had 13 lead pellets in her legs and buttocks. Walls, therefore

> *prepared on large sheets of brown paper a number of identical projection outlines of a pair of legs and buttocks based on the dimension of the wife and fired at these from varying distances with the gun and type of cartridge in question. Finally, counting the shot holes within the outline at each distance.*

From this experiment Walls found that, on the average, the gun had to be fired at a distance of 40 yards before the number of the pellets expected in the target area fell to 13. Therefore, the farmer was charged with the lesser offence. Walls tells us that at the trial he was asked to produce the outline charts he had used and he tells us: "I well remember how manfully the judge and council tried to wipe the grins from their faces."

Walls also reports an investigation in which the distance of the gun from the victim was an important piece of evidence. The case involved a young lady who kept ponies and a young man who's attentions were not welcomed by the woman. The young lady frequently complained that people left the gate of the field open allowing her ponies to stray. Walls tells us that one evening the young man came to the police with a complaint. He said he had seen two men tampering with the gate as he was passing. He shouted to them, whereupon they ran away and one of them turned and fired a gun. In support of this odd story, he produced an undoubted bullet wound in his arm and a corresponding hole in the sleeve of his jacket. Examination of the sleeve showed an appearance identical with that made by a 0.22 weapon with the muzzle pressed right against the fabric. The young man possessed an 0.22 target practice pistol, and in the words of Walls, "he appeared to have indulged in a rather painful method of courtship!"

5.4 Body Armor

Great advances have been made in recent years to protect police officers and others who are potentially the targets of assassin's bullets. These developments are based upon the availability of a new plastic material called Kevlar, which is a registered trade mark of the DuPont Corporation.[18] This compound was developed in 1965 for making automobile tires. Earlier, in 1935, the DuPont corporation had invented nylon which had proven very strong and suitable for light weight body armor. (It is estimated that nylon-fiber flack jackets cut casualties amongst bomber crews by 60 % in the later years of the Second Word War[19]). Kevlar has 2.5 times the tensile strength of nylon. That is, it can withstand 2.5 times the stress that a nylon fiber can withstand. In Figure 5.14, the chemical constitution of some of the more

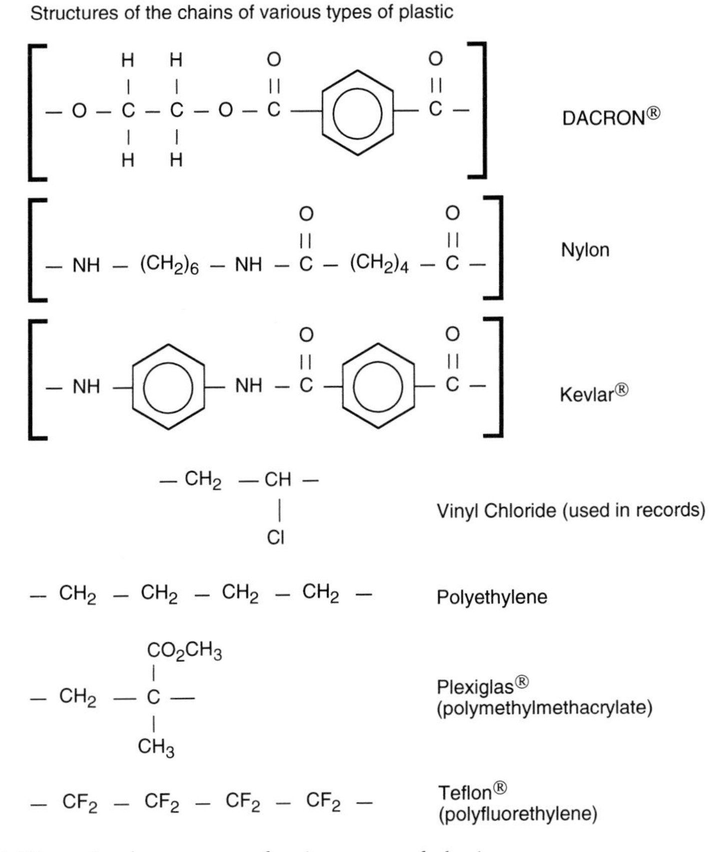

Figure 5.14. The molecular structure of various types of plastic.

familiar plastics are shown. Technically, Kevlar is a polymer cousin of nylon. Scientists know that one of the factors that determines the effectiveness with which a material dissipates the energy of a speeding bullet is the speed of sound in that material. (As will be discussed in greater detail in the next chapter, a sound wave is made up of a sequence of compression waves moving through a material, such as air or water.) J. E. Gordon commenting on the ability of objects to absorb energy states:

> *The speed of sound in steel, aluminum, and glass is approximately 4,800 meters per second (11,000 miles per hour). This is much faster than the speed of sound in air. Such speeds are far faster than any hammer blow and considerably faster than the flight of bullets. The result is that a hammer or a bullet is pressing against its target for a period, perhaps of about 100th of a second, which is very long compared with the time to conduct the energy away from the point of impact in the form of compression waves which are in fact sound waves.*[20]

In an article on Kevlar, McKean points out that because sound travels three times as fast through Kevlar as it does through nylon, the stress of the bullet hitting a vest made out of Kevlar is more rapidly distributed over the entire vest.[21] The effectiveness of body armor made out of Kevlar is illustrated in Figure 5.15. Because lead is a relatively soft metal, it is deformed by a collision with the body armor (the **penetration pressure** is the force exerted by the bullet over the area of contact. The deformation of the bullet by the Kevlar is helpful in preventing penetration of the armor, since the bullet exerts less pressure on the potential point of penetration as it deforms on contact with the armor. Armor piercing bullets used by the army are made of steel, which does not deform on contact, and, therefore, is more effective in penetrating any armor that it encounters. It should be noted that the absorption of the overall energy of a bullet, while preventing bullet wounds cannot protect the wearer against bruising.

Electron microscopy studies of different fibers used in making body armor, carried out by the United States Army's Nantick Laboratories in Massachusetts show that nylon fibers melt when a bullet strikes them. Glass-like fibers that do not melt are subject to transverse fractures which reduce their effectiveness in body armor. Part of the effectiveness of Kevlar armor is that the threads of Kevlar pull into long slender fibres absorbing energy before they break.[19]

Pieces of plastic of the kinds shown in Figure 5.14 are made up of long chains of the molecules shown. One of these is the plastic known as Teflon, which is famous for its non-stick ability. (Teflon is also a registered trade mark of the DuPont Corporation.) The atoms inside the Teflon molecule are all bonded to each other so well that they have no residual attraction to any other molecules. For this reason, Teflon is a very good lubricant used to coat frying pans to prevent food from sticking during the cooking process. The

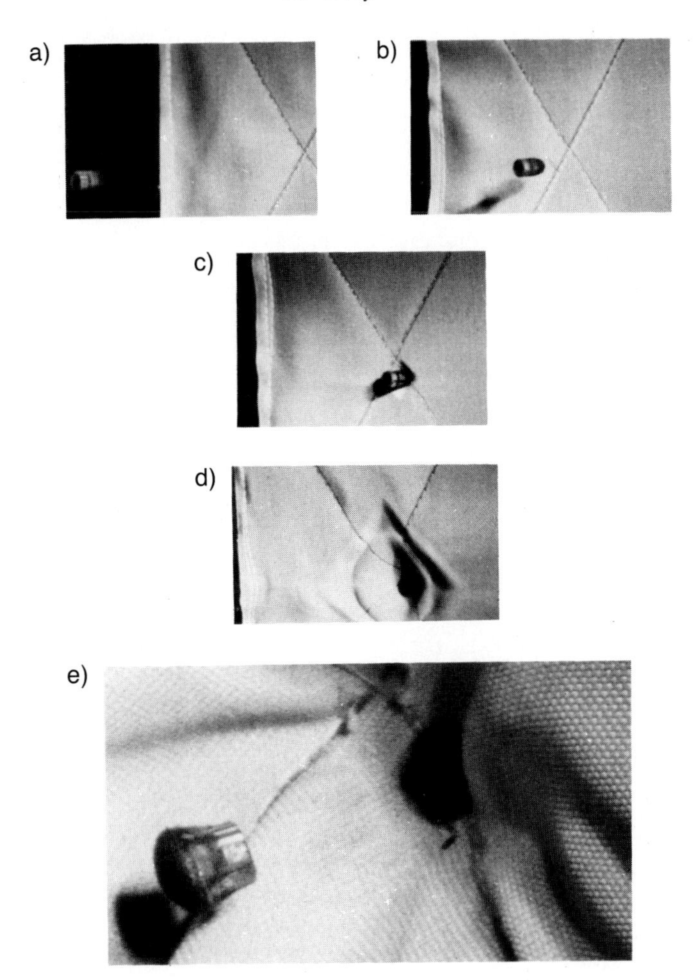

Figure 5.15. The protective power of body armor made from Kevlar is illustrated by the sequence of photos above. The sequence shows a 0.38 calibre bullet traveling at 244 m/s (800 ft/s) stopped in flight and blunted on impact by seven layers of Kevlar fabric. Each picture has an exposure time of less than 1 microsecond. (© E. I. Du Pont de Nemours and Company.[18])

manufacture of Teflon-coated bullets is forbidden by law in many western countries, because such bullets penetrate the body very effectively. Their path through any body or object is lubricated by the outer coating of Teflon. Indeed, Teflon-coated bullets were once known as "cop killers." The search for stronger and stronger fibers to build better armor is an ongoing task, and recently scientists have announced that they may be within reach of designing a fibre three times as strong as Kevlar.[22]

A popular crime fighter seen in children's television cartoons is Spiderman, a super hero able to spin webs to capture evil characters. In real life, the spider may be able to teach humans how to spin super-light-weight armor that is even better than the Kevlar armor used by today's law enforcement officers. It has recently been reported that scientists are collecting spider threads to study their structure in the hope of evolving lighter and stronger body armor.[22] The technique being used by the scientists to collect special silk from a spider is shown in Figure 5.16. The work is being carried out by a team of engineers and molecular biologists at the Natick Research Center in Massachusetts. Apparently, a spider such as the golden orb weaver (nephila clavipes) produces seven different kinds of silk from seven different glands on its body. Some of these silks are used for wrapping prey and coating the web, others are used for attaching web strands to trees.

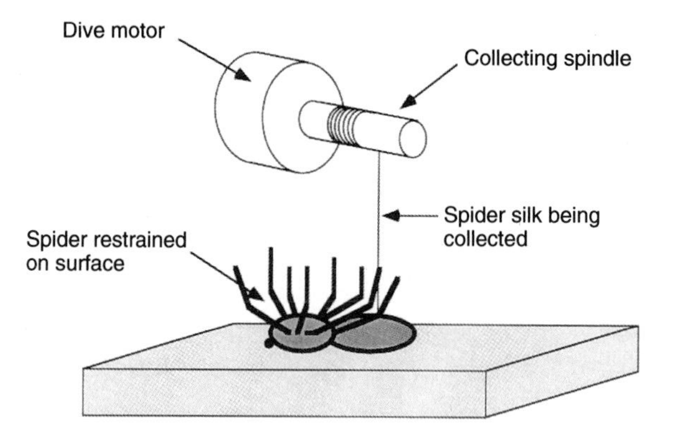

Figure 5.16. Scientists are trying to make synthetic spider silk so that they can manufacture super light-weight armour. In experiments, spider silk is collected from a golden orb weaver by gently reeling up the silk released by a restrained spider using an electric motor.

The golden orb weaver comes from Panama. It was chosen because it is big and produces a lot of strong, golden silk. The particular type of thread that scientists are interested in is the silk that the spider uses for the drag lines it drops into space to wait for a victim. This is also the thread it uses to make the long segments of the web, which are attached, for instance, to trees. When an insect flies into a web, the fibers stretch and absorb the energy created by the impact in the same way that body armor must absorb the energy of a bullet plowing into it. The scientists obtain the thread by first taping the spider upside-down onto a table. They then grasp the end of the fiber protruding from the gland with forceps and wrap it around a spindle driven by an electric motor. Stephen Fossey, one of the researchers recounts that:

we can get three to five milligrams of silk at one session which is a strand about 320 meters long. The obtaining of the silk does not hurt the spider which is ready to be treated the next day.[22]

Fossey tells us that Kevlar can elongate up to four percent before breaking, whereas the spider silk can stretch by as much as 15 % before breaking.[22] Scientists predict that the spider silk is probably at least three times as efficient as Kevlar in stopping bullets. They are attempting to understand the chemical structure of the silk so that they can synthesize it in a laboratory. Synthetic spider silk would then be used to make super light-weight body armor.

In November 1992, it was announced that bullet-proof vests are now available to protect police dogs against bullets. The equipment was first shown at the 99th Annual Association of Police Chiefs in Detroit. The vests, which cost about 750 U. S. dollars, are designed by the American company, "Second Chance." As the news story pointed out, police departments invest a lot of money in the training of dogs and want to be sure that they are protected. Some police forces, including the New York City police, say that death contracts have been issued by organized crime on their dogs, especially those involved in drug detection.[23]

References

1. Photograph by W. G. Hyzer, Industrial Research.
2. For a discussion of why the spinning of a bullet stabilizes its path in flight, see the discussion of bullet kinetics in B. H. Kaye, *Golf Balls and Other Interesting Missiles*, to be published by VCH Verlagsgesellschaft, Weinheim.
3. F. Smyth, *A Cause of Death. the Story of Forensic Science*, Van Nostrand Reinhold Co., New York, 1980.
4. O. Johari, I. Corvin, "Scanning Electron Microscopy and the Law: A Review," *Canadian Research 24* (1976), 24.
5. L. C. Nickolls, *The Scientific Investigation Of Crime* Butterworth, London, 1956.
6. M. S. Germani, "Application of Automated Electron Microscopy to Individual Particle Analysis," *American Laboratory*, April 1993, 17.
7. J. A. Cooper, "The Criminal Leaves His Card," *Industrial Research*, 15 November 1977, 22.
8. L. Taylor, "Fluid Extraction and Analysis of Aged, Single Base Propellants," *American Laboratory*, May 1993, 22.
9. Waters Associates Inc. Maple St. Milford Massachusetts 01757, U. S. A. is a major company manufacturing chromatographic equipment.
10. The information of Figure 5.10 is taken from the technical literature of Bio-analytical Systems Inc. Purdue Research Park, 1205 Kent Ave. West Lafayette, IN 47906, U. S. A.
11. L. Cherry, "Their Blood Cried out for Vengeance," *Science Digest*, May 1981, 60.
12. "Smoking Gun," *Discover*, November 1990, 22.
13. M. Hamer, "Forensic Science Goes On Trial," *New Scientist*, 9 November 1991, 30.
14. J. Lloyd, "Can We Balance Science And Justice," *New Scientist*, 21 September 1991, 12.
15. B. Knight, "Murder in the Laboratory," *New Scientist*, 25 December 1986, 59.
16. J. Broad, *Science and Criminal Detection*, Macmillan, London, 1988.

17. H. J. Walls, *Forensic Science. An Introduction To The Science of Crime Detection*, Frederick A. Praeger, New York, 1968.
18. Kevlar is manufactured by DuPont Nemours and Company, Inc. Textile fibers department, Kevlar special products, Centre Rd. Building, Wilmington Delaware.
19. For a discussion of the history of armor and a more comprehensive view of the use of Kevlar see discussion of armor in reference 1.
20. J. E. Gordon, *The New Science of Strong Materials*, Second Edition, Princeton Paperback Printing, Princeton, 1986.
21. K. McKean, "Longer Living Through Chemistry," *Discover*, July 1981, 80.
22. J. Beard, "Warding Off Bullets By A Spiders Thread," *New Scientist*, 14 November 1992, 18.
23. "Putting on the Dog" (news story) *Toronto Star*, November 22, 1992, A2.

Chapter 6

Electronic Ears and Eyes

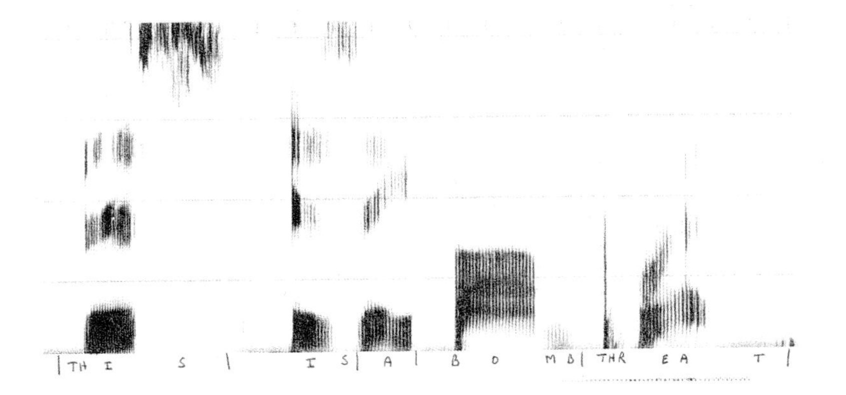

Chapter 6

Electronic Ears and Eyes

6.1 Is That You Charlie?

Surveillance is a major part of both modern detective work and the activities of security forces. Those concerned with this kind of work have turned to the scientist for help. One of the important problems facing security forces is establishing the authenticity of spoken messages such as those accompanying a bomb threat or a demand for ransom. In the early 1990s there was a great deal of controversy in Great Britain over the authenticity of recordings claimed to have been made of Prince Charles talking to a girlfriend on a cellular phone. Later on there was some controversy about the authenticity of a recording of a discussion between Princess Diana and Charles again over cellular phones. Some people claimed that the recordings were false and that they were fed to the media by British Intelligence officers who wished to cause a scandal. Again in the early 1990s considerable discussion was generated by an analysis of the famous Churchill war-time speeches. Based on speech analysis, some claimed that the speeches were read by a professional actor who was imitating Churchill. Speech analysis is still considered a controversial area of forensic science. All we can do in this section is look at the techniques used by the scientist to analyze speech messages so that the reader can follow and understand the discussions about speech analysis and recognition in forensic science.

The basic technique for analyzing speech relies on speech spectrograms, which display the information carried in spoken words. The title page of this chapter shows a spectrogram of the phrase, "this is a bomb threat."[1] To understand the structure of a spectrogram, we need to have a basic knowledge of how speech is generated and recorded. In Figure 6.1 the essential steps involved in generating a speech spectrogram are illustrated. **Sound waves** are a series of redundant compressions created by a vibrating object. The simplest sound that the scientist can work with is the note made by a tuning fork. The **tuning fork** is a U shaped piece of metal mounted in the middle as illustrated in Figure 6.1. The frequency with which the compression waves arrive at the human ear determines the sensation of pitch or height of a note. Thus, if the notes arrive at the human ear with a frequency of 440 cycles a second, the person hears the note "middle A" as played on a piano.

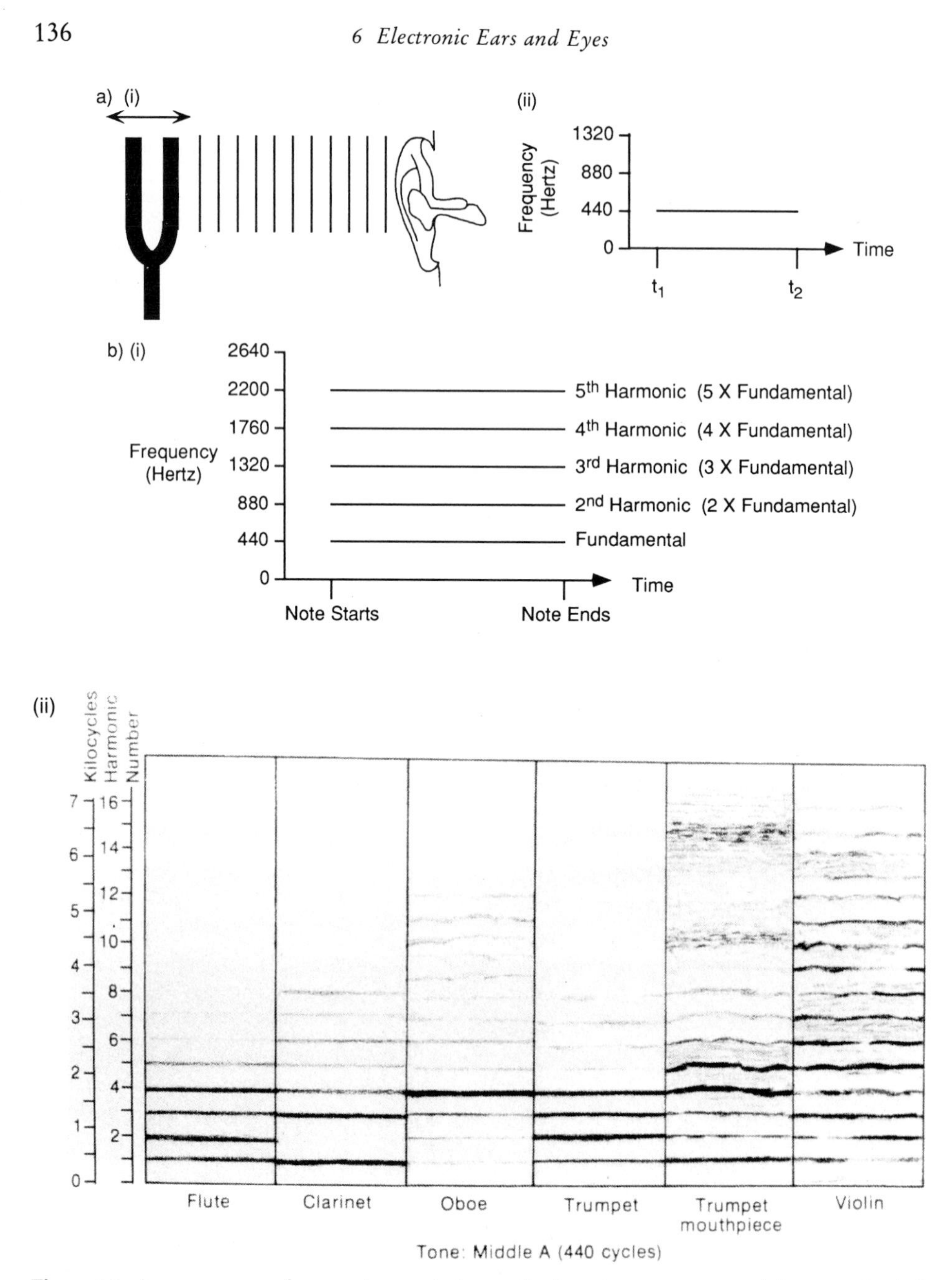

Figure 6.1. A spectrogram of a complex musical note displays the components of the note clearly.[2] a) (i) As the arms of a tuning fork vibrate it sends out a series of compression waves the frequency of which determine the pitch. (ii) Spectrum of a tuning fork at "middle A". b) (i) A flute creates a "pure" sounding note with a fundamental frequency and several harmonics.
(ii) Spectrograms of the sounds of some common instruments. Depending on their shape, musical instruments emphasize certain harmonics. (From *Man and His Technology*, 1973, reproduced with permission of McGraw-Hill).

An average bass singer can form musical notes having frequencies in the range of 75 to 300 cycles per second. The musical notes formed by a soprano voice range from 230 to 1400 cycles per second. The lowest tone that can be heard by the human ear is around 16 cycles per second; this is the lowest note played on a wind organ. However, spoken human words, and musical notes in general, have more than one frequency in the complete wave generated by a voice. The note "A" as played on different musical instruments has many different frequencies in the composite sound that we hear. To represent the various frequencies present in a musical note, scientists created what is known as a **frequency spectrum**. For the simple note made by a tuning fork, the frequency spectrum over the period of time $t_1 \rightarrow t_2$ is a very simple line, as shown in part of Figure 6.1 a)(ii).

When discussing the structure of a musical note, one often talks of the **harmonics** of a note. The harmonics are multiple frequencies of the basic frequency or fundamental tone. Of the musical instruments that humans have created, the instrument that produces the most pure tone, with only a small number of harmonic frequencies present in each musical note, is the flute. The components of the a middle A played on a flute and as analyzed by scientific instruments is shown in Figure 6.1 b) (i). In the language of sound physics, a subject more commonly known as acoustics, it is said that a note played on a flute has the first five harmonics present in its **acoustic spectrogram**. Acoustic spectrograms show the relative energy present in the various harmonics of a note by recording the intensity of the line representing the harmonic in the spectrograph. The acoustic spectrograms for a middle A played by different musical instruments is shown in Figure 6.1 b) (ii). When one looks at the sound spectrograph of the trumpet mouth piece, the fact that it is wavy as one goes from left to right indicates that, during the period that the note is being blown by the trumpeter, there is some variation in the exact frequency and strength of the harmonics. Note that the fundamental tone is pretty steady over this period and that it is the harmonics that wobble. This gives the effect of "vibrancy" to the note. If one compares the acoustic spectra for the trumpet and the trumpet mouth piece, one can see that the higher harmonics are suppressed in the sound of the trumpet This is due to the structure of the trumpet. It can be seen that the musical note with the highest number of harmonics present in the note, up to 16, is the violin. This is why one may think that the violin is sounding a higher note than the basic note. The strengths of the various harmonics in the trumpet mouth piece and in the violin are indicated by the densities of the line. If one looks at the note for the oboe, one notices that the fourth harmonic is stronger than the fundamental tone.

Before compact discs took over the market, people had phonographs (record players) which would play vinyl records at different speeds. If one inadvertently put a recording of a bass singer, which should have been played at 33 rpm (revolutions per minute), on at 78 rpm the singer's voice would be changed into a high soprano. This is because the notes were coming out of the phonograph at about double the original frequency. Today, one can buy a telephone accessory that will decrease the frequency of the speaker so that a woman will sound like a man with a husky voice. This is useful if one suspects that a thief is making random calls to homes to check who is in. One can also make a child sound like the same husky voiced male. These types of telephones are available from major suppliers of telephone equipment. The chipmunks of Walt Disney are recordings of human voices played at higher speeds to make them into squeaky sopranos.

After this discussion of the information summarized in a sonic or acoustic spectrogram, we can now look at the structure of spectrograms generated by human voices. In Figure 6.2(a) the spectrograms of three different people saying, "man shot Bill," are shown. When the computer starts to process such information, one of the first problems that it has to deal with is the fact that people speak at different speeds. The top two spectrograms show the appearance of the phrase when it is spoken by two different male voices. As can be seen, the first man spoke more quickly than the second. When the woman says the same phrase, it should be noted that the 's' of the shot is more emphatic in the woman's voice, which results in a higher pitched sound.[2] For technical reasons, telephone systems often suppress high frequencies present in a voice. Thus 'top' and 'shot' can sometimes sound alike over the phone, because the high frequencies in the 's' are suppressed. Similarly, when one suppresses the high frequencies in "p" as compared to "t" the two letters can sound alike. Another problem when working with spectrograms is that words that sound alike look alike on their spectrograms, as illustrated by Figure 6.2(b)[3]. The way in which a word is spoken effects the spectrogram. Thus, shouting and whispering the same word (see Figure 6.3 (a)) can result in quite different structures. Very often when a person whispers he also lowers the frequency of the voice. This is why one can have a frustrating experience communicating with someone who is hearing impaired. Even though the person appears to not understand what you are saying in normal speech, when you attempt to communicate in a whisper to someone else in the room, the hearing impaired person can hear what you are saying. In a case like this, you have dropped the frequency of your voice into a region where his hearing is not effected.

Some numbers that can be confused in spectrograms are illustrated in Figure 6.3 b). In general, female voices are higher than male voices. What we recognize as the quality of the voice and regional accents, generate features of the voice spectrum, as illustrated by the information summarized in Figure 6.3 c). Although regional accents are a problem from the point of view of speech recognition by electronic ears, they are also a feature that can uniquely identify a speaker. For example, I am well aware that the way that I say two words very clearly demonstrates the fact that I grew up in the Northeast of England: these words are road and raining. I pronounce the

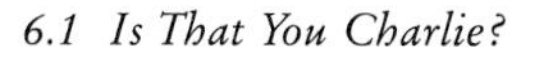

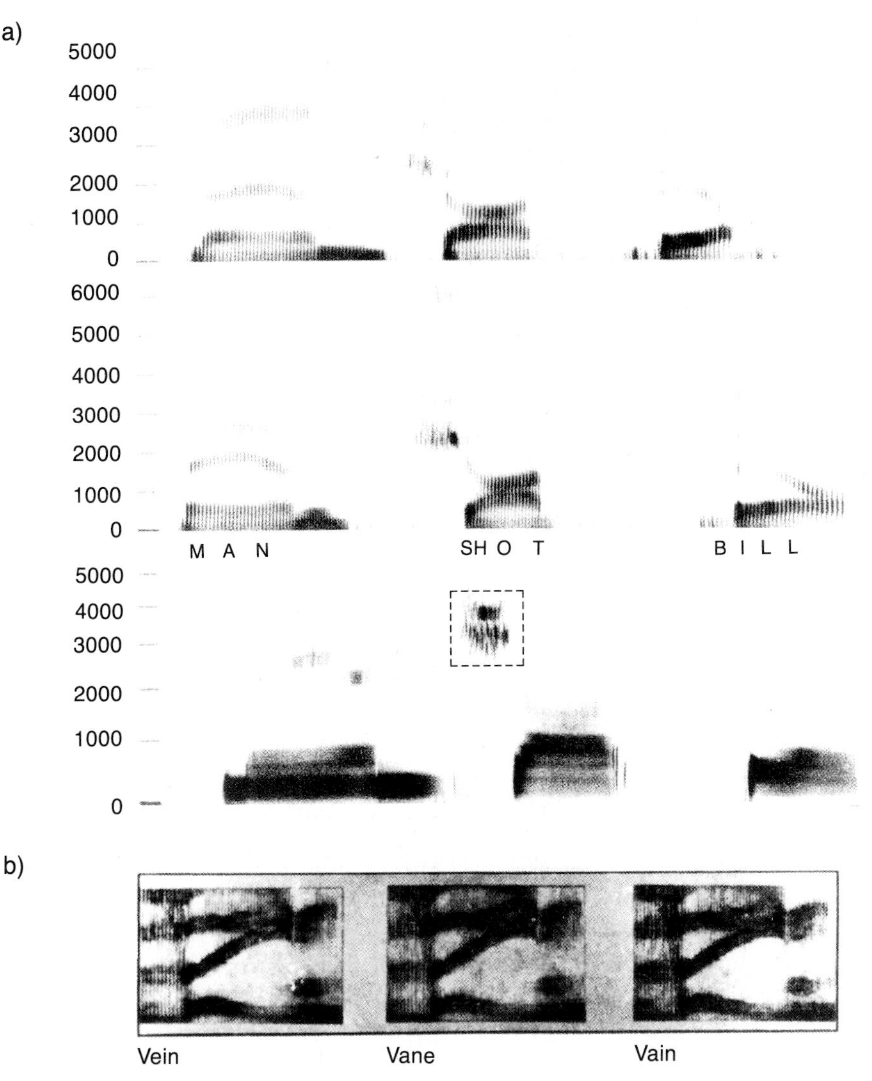

Figure 6.2. The basic speech spectrogram has various features attributable to the speaker's vocal anatomy and speech habits. a) Spectrograms for three different people saying "man shot Bill." The first two are males and the third is a woman. Notice that the female voice has an accentuated "S" as shown by the dashed box. (From *Man and His Technology*, 1973, reproduced with permission of McGraw-Hill). b) Words that sound alike have similar spectrograms.

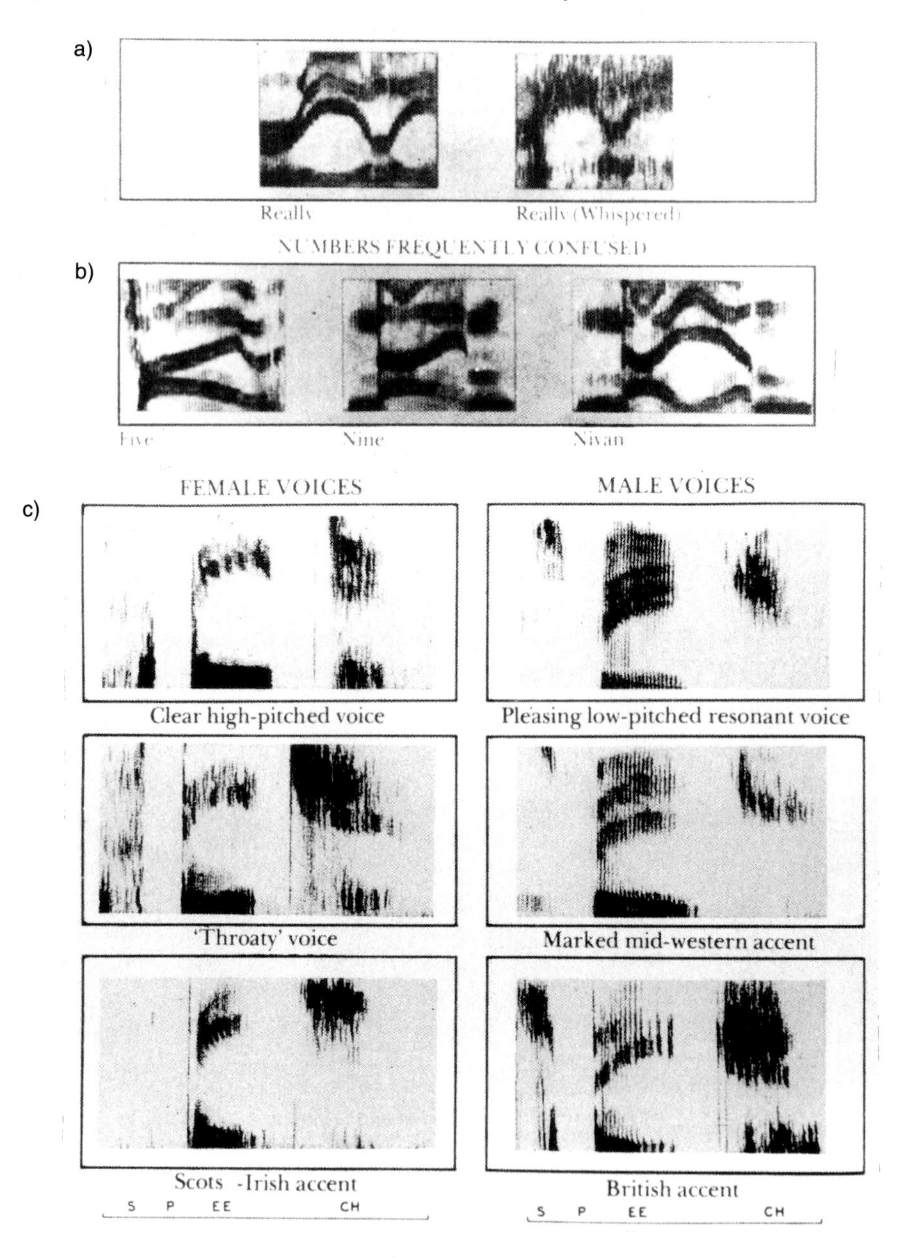

Figure 6.3. In the analysis of speech the way in which a word is spoken, similarity of sounds, and regional accents can cause problems of identification; on the other hand, a dialect can be a useful feature establishing identity. (Reproduced from *Science Spectrum.*) a) The way a word is said influences the structure of it's spectrogram. b) Numbers are frequently confused. c) A spectrogram is influenced by the sex of the speaker and regional accent.

word road with a very long 'o', and I pronounce the word raining as if it were spelled reigning. Thus, you cannot tell if I am discussing the weather or the reigning of Queen Elizabeth II. Again, in my dialect I do not distinguish between the sound used to reproduce the words fare, fair, and fur. In spite of these difficulties of interpreting speech, major research efforts are being carried out to develop automatic speech recognition.[4,5] This is because the communication industry would like to be able to develop computers that can accept spoken orders, as well as intelligent typewriters that will process speech.[6]

The speed with which speech-recognition equipment is coming into use for forensic purposes was recently described by Elizabeth Geake. In a review article she described an experiment proceeding in Scotland at the Center for Speech Technology Research at the University of Edinburgh.[7]

The system is being developed in co-operation with Cairntech, an electronic company, and the Fife Constabulary (Fife is an area of Scotland.) The speech-recognition system being used will enable police officers on the beat to check car number plates by talking to a computer, which has the ability to speak back. The police officer will pick up a microphone and say a word which allows the receiving station to know that he wants a voice-oriented national computer check. A voice synthesizer will say, "please go ahead." The officer will then say his or her police number, the license plate number, color and make of the car. The computer will speak back the name and address of the owner. Further developments of the technique will involve scramblers to encode messages for security reasons. It is expected that a commercial version of the system will cost $1,600 per patrol car. Speech recognizers work by breaking words down into their **phonemes**, that is, their constituent sounds, and comparing these with a library of standard spectrograms, known as speech **templates**, stored in the computer. Each speech template is an average of words spoken by many different people. The Center for Speech Technology has built up a database from the voices of Fife police officers who presumably have a distinctive Scottish accent. The Center recognizes that they eventually may need to have a set of stored words for each officer likely to use the system. The licence plate speech recognizer only has to cope with a vocabulary of about 50 words, and since the police already follow fixed routines for requesting information, the system will know in advance which sequence of words to expect.

In many high-technology industries in the United States, voice recognition is already the basis of allowing access to personnel working in a given location.

A famous political incident in which voice recognition was involved occurred in 1967. In the summer of 1967, Israel and several Arab nations were involved in a war. In the midst of the conflict, Israeli officials held a press conference to release a tape recording which they claimed came from a conversation between President Nasser of Egypt and King Hussein of Jordan.

The alleged conversation was said to have taken place between the two heads of state and concerned the possibility of blaming the United Kingdom or the United States or both for the destruction which had been inflicted on the Arab Nations by the Israeli Air Force. The translation of the alleged conversation reads:

> NASSER: *Will we say that the United States and England or just the United States?*
> HUSSEIN: *The United States and England*
> NASSER: *By God, I will make an announcement, and you will make an announcement, and we will see to it that the Syrians make an announcement that American and British aeroplanes are taking part against us from aircraft carriers.*
> HUSSEIN: *Good, alright.*

Immediately, the authenticity of this conversation was challenged by reporters. To decide whether or not the conversation was real, or fabricated by the Israeli intelligence forces, copies of the tape were sent to Lawrence J. Kersta, a New Jersey communications engineer who specialized in voice identification. Kersta compared the speech spectrograms of the alleged voice of Nasser with certain basic sounds and words from a similar analysis of Nasser's speeches. Kersta was convinced that the voice alleged, to be Nasser's, was indeed that of the President of Egypt.

Thus far in this discussion, we have looked at the processing of speech by computers, but it should not be overlooked that the human ear is often a very good analyzer and recognizer of speech. Like all other human skills, the ability to recognize a voice varies from person to person, just as musical ability varies from person to person. However, voice recognition has been used in Canadian courts. In a procedure called a **voice lineup,** victims of certain crimes listen to voice recordings of different people. The technique was developed by Calgary Police staff sergeant Dwight Mayer in 1981, when the police department in Lethbridge (200 km south of Calgary) invited him to investigate the case of a family of four that had been terrorized and held captive for several hours. Mayer arranged to have a suspect's voice recorded without his knowledge. Mayer then had six other people with similar speech patterns, including actors and other experienced public speakers, read the same words into a tape recorder. Each family member then listened to the tape separately, and all four identified the voice of the suspect as that of the masked man who had held them captive. He was later convicted and sentenced to 26 years in prison.[8] Mayer said that victims have been able to identify the suspect positively in 74% of the cases where they heard a voice lineup within twelve months of the crime. With witnesses who were not victims, the success rate is lower, about 33%. The technique has not been used widely but has been successful in Alberta.

6.2 Should We Eat More Carrots?

The title of this section will be mysterious to those of you who have never heard the myth that if one eats more carrots one can see in the dark. Stories concerning the origin of this myth vary, but one is that it was promoted by the British Government to explain the mystery of how British pilots were apparently seeing German aircraft at night. The reason for British pilots' high rate of success at fighting the German bombers at night was that the British were the first to use radar. The story about the carrots was spread to mislead the German scientists trying to work out the reasons for the success of the British. Nobody seriously believed the story, but it serves as an introduction to the idea that, with special help, we can see at night. The title page of this chapter carries a picture taken at night of an arrest being made by a police officer. In Figure 6.4(a) another picture taken at night is shown. In Figure 6.4(b) the soldier equipped with night-vision goggles is shown carrying out a task. The night goggles and the infrared cameras used to take pictures at night operate with radiation of wavelength of 8–10 μm, which is

a)

Figure 6.4.

b)

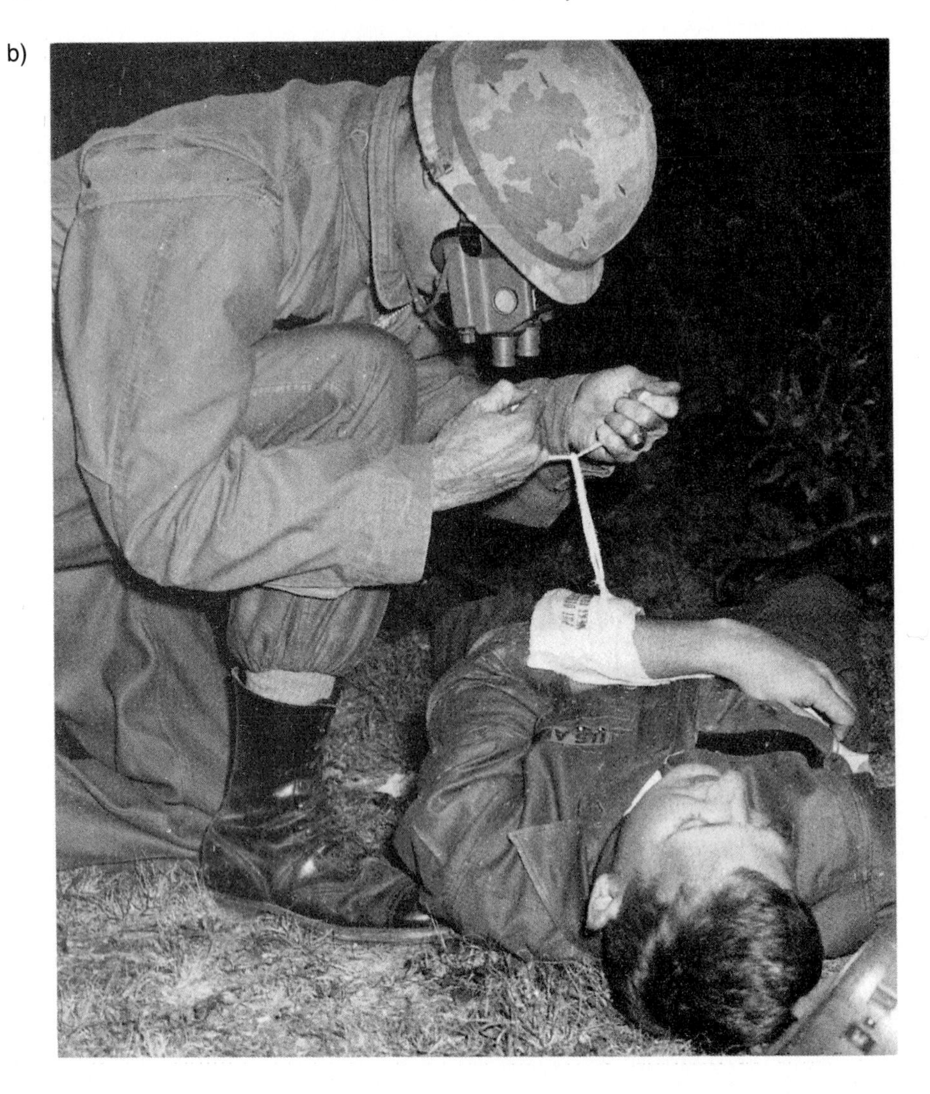

in the infrared range of the electromagnetic spectrum. All objects emit radiation. (For a discussion of the radiation given out by different bodies at different temperatures, see discussion of black body radiation and the greenhouse effect in reference 9.)

It is not possible to characterize a given object with a radiation profile containing all infrared wavelengths, because infrared radiation of specific wavelengths is absorbed by water vapor and carbon dioxide in the air. The absorption bands of such infrared radiation are shown in Figure 6.4(c). Because of this absorption phenomena, infrared cameras operate over the wavelength range shown in Figure 6.4 (c)(iii). Glass absorbs radiation of this

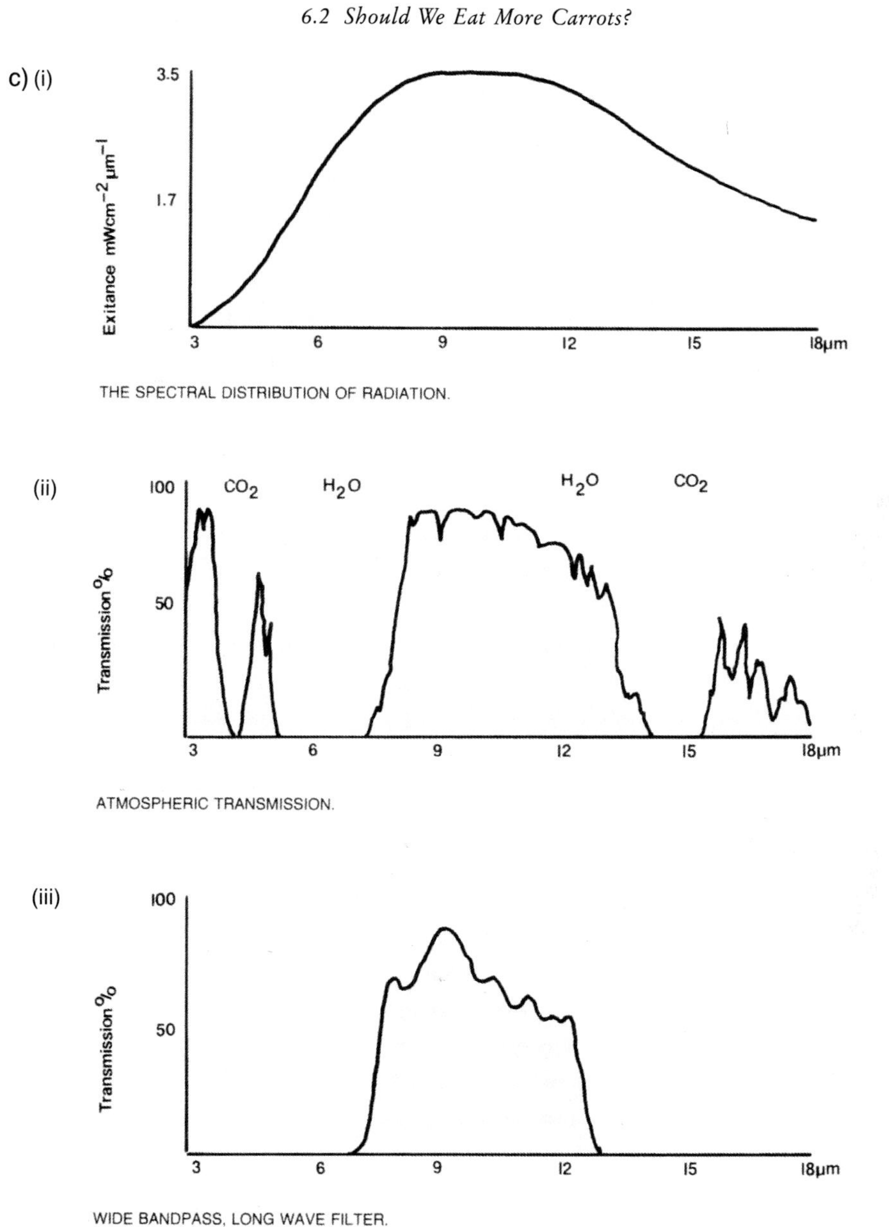

Figure 6.4. Infrared cameras enable security forces to carry out surveillance tasks at night in total darkness. a) (page 143) Picture taken in the dark by an imaging system sensitive to infrared radiation, which detects differences in temperature of objects.[15] b) Persons equipped with night vision equipment can carry out tasks at night. c) Infrared cameras and night vision goggles operate over an infrared wavelength range of 8–13 μm. (i) Typical radiation spectrum of an object a few degrees above zero. (ii) The effect of absorption by water vapor and carbon dioxide gas on the spectrum of (i). (iii) The operational range of infrared cameras and goggles.

wavelength. This is why green houses work; they trap re-emitted radiation from warm soil.[9] One of the substances that is transparent to infrared radiation over the 8–13 micrometer range is germanium. It has a refractive index of 4, this makes the construction of fairly simple lenses to focus the radiation possible.[10] Inside the infrared camera, the energy received through the germanium lense is processed by an image intensifier. Essentially, this creates a flow of electrons, the intensity of which varies with the energy received at the camera. These create a visible image on a phosphor screen, as illustrated by the sketch in Figure 6.5.[11] Infrared cameras and night-vision goggles can see a human face at a range of several hundred meters. In the daytime, infrared cameras can see through the camouflaged pattern to pick up the presence of hot objects, such as a gun that has been recently fired, and a human face against a background of tree foliage.[12–16] Modern infrared cameras can take pictures of a body and draw contours of different temperature regions. These cameras are widely used in medicine to detect such things as cancerous tumors and blood clots blocking arteries. (Cancerous tumors are rich in blood and look warmer than the rest of the body, and a blood starved area of the brain will be cooler than adjacent regions.)

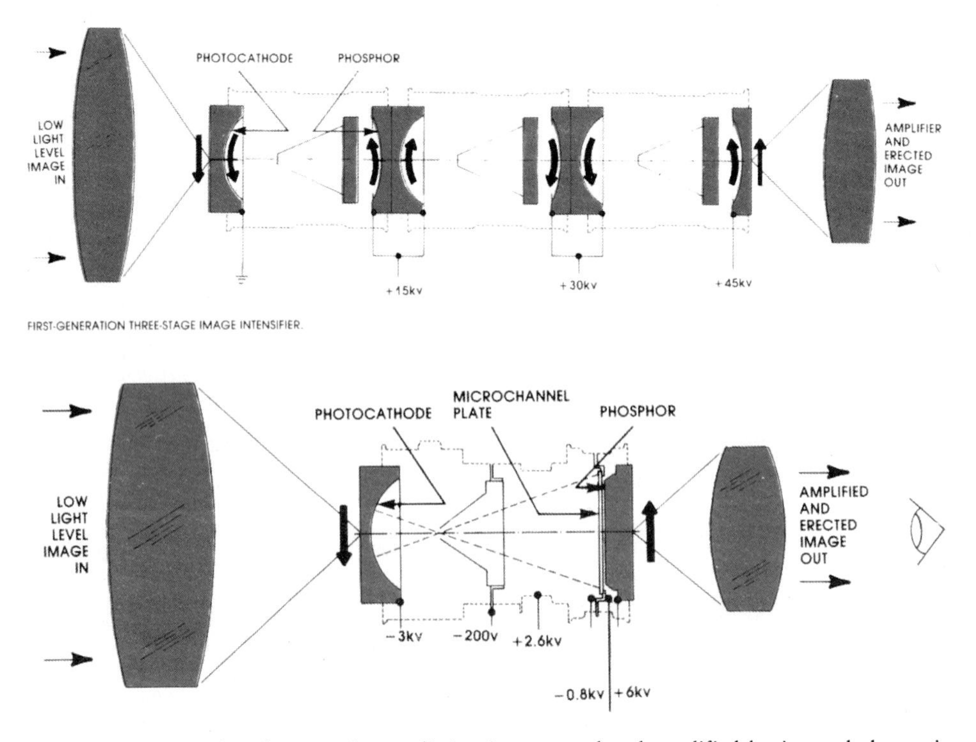

Figure 6.5. In an infrared camera heat radiation is converted and amplified by internal electronic components into a visible image on a phosphorescent screen.

Infrared cameras operating from helicopters can be used to search for bodies buried in shallow graves. Heat produced by the decomposition of the body and its interaction with the earth causes a hot spot which is picked up by the camera.[17]

6.3 Speed Reading

No, this section is not going to be a lesson in how to improve your reading speed; it is going to discuss the way in which robots can detect how fast a car is moving and automatically issue a speeding ticket. There is some confusion in the public mind over the techniques that are used to monitor the speed of automobiles. A modern radar trap does not use radar to monitor the speed of a car. It uses a device which should be more properly called a **Doppler gun** to measure speed. The term **radar** came into use during the second world war. It is a word made up from the phrase "**ra**dio **d**etection **a**nd **r**anging." In the original use of radar, a small pulse of radiation was directed towards the suspected location of an aircraft. If some of the radiation was returned, then something must have reflected it. The large metal mesh bowls rotating on the top of the towers of airports are the receivers for pulses of radiation sent out to detect the movement of aircraft near the airport. The speed of the radio waves is identical to that of light waves. From a knowledge of the time at which the pulse was sent out to search for an aircraft and at which it returned to the location of the receiver, one can deduce the distance of the aircraft from the radar installation. To work out the speed of the aircraft, one sends a second pulse after a measured time interval, and from the change in distance of the aircraft in the time interval between pulses, one can determine the speed of the aircraft. Early radar devices for detecting the speed of cars on highways used essentially the same technology to measure the movement of the car between two pulses of radiation sent to and reflected from the moving car.

Modern speed detectors are not radar devices; they operate on what is known as the **Doppler effect**. The Doppler effect is named in honor of the Austrian mathematician and physicist Christian Johann Doppler (1803–1853). We are not sure how Doppler came to investigate the phenomenon now carrying his name, but it may have been because he noticed that the sounds made by a steam engine changed as is went past a stationary observer. This phenomenon is obvious to anyone who has listened to a siren on an ambulance or a fire engine moving past them on the street. As the siren goes past, there is an apparent drop in the sound of the siren. This drop in the observed frequency is due to the speed of the moving sound emitter. The effect is illustrated schematically in Figure 6.6. Scientists measure fre-

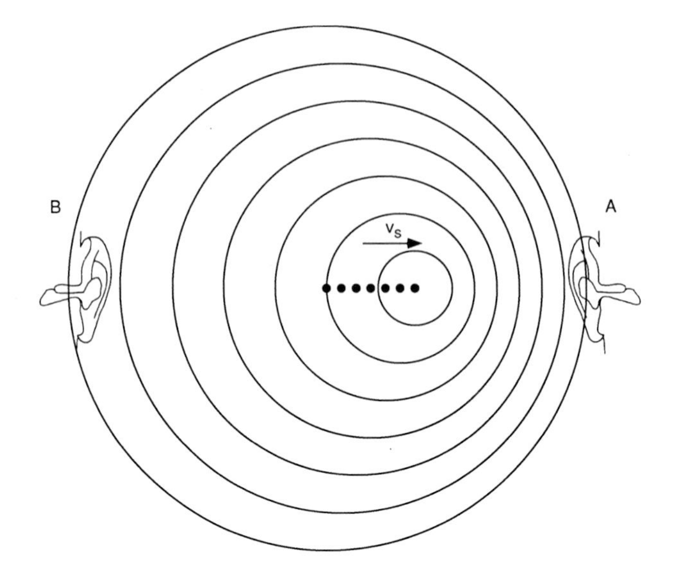

Figure 6.6. As the source of a sound moves toward the listener at A, the movement of the source causes the compression waves to arrive more frequently than the source is actually producing them. In a similar manner the compression waves arrive less frequently at B as the source moves away.

quency per second in a unit known as hertz (Hz). This unit comes from the name of a German physicists Heinrich Hertz (1857–1894) who studied the fundamentals of sound and light waves. When an ambulance speeds down a street at 60 mph, with a siren, which gives out a steady sound of 440 cycles per second (or in modern scientific terminology, 440 hertz), the movement of the ambulance towards the observer results in the siren sounding as if it had a frequency of 479.1 Hz. When the ambulance is moving away from the observer, the frequency of the sound drops to 406.8 Hz. For people familiar with the musical keyboard, this is equivalent to the sound shifting from an A sharp to an A flat. The Doppler shift is also effective for detecting microwaves bouncing off objects such as moving cars. There is a double Doppler shift involved when monitoring a moving automobile. As the automobile receives the microwaves, they are shifted in frequency by an amount calculated to be 1.28×10^{-7}. When the automobile reflects the waves, there is a second Doppler shift of the same magnitude, so, for a car moving at 85 mph, there is a shift of 614.4 Hz in the microwave frequencies. This can easily be measured by a technique known as **beat frequency**. This phenomenon is also used in tuning a guitar. If you have an “A” on one string and then you press a fret on the guitar to have the other strings sound in the same frequency, if the guitar is out of tune and the second string gives out a note of

445 Hz, then the human ear will hear a wowing sound, at a frequency of five cycles per second for this example, when the notes from the two strings combine to create what is known as a beat frequency. This beat frequency occurs because of the interference of the sound energy from the two slightly different notes. A person tuning a guitar hears the beat frequency and then alters the tension in the second string so that the beat frequency disappears. In the modern Doppler-speed gun, the reflected microwave is mixed with the original frequency to produce a beat frequency which is transformed into a speed estimate that is directly displayed on the panel of the instrument. The speed of microwaves is so high that if the police officer's signal hits the car when it is 100 meters from him, it takes only 0.66 microseconds for the reflected microwaves to return to the car. During this time interval a vehicle moving at 85 mph has only moved a distance of 0.025 millimeters. Fantastic, but true!

There is a widely-circulated anecdote about a famous physicist and professor at Johns Hopkins University, R. W. Woods, involving the Doppler effect. He is reputed to have gone through a red light. He, however, claimed that the Doppler effect made the red light look green. If we carry out the necessary calculations, we can show that Professor Wood would have had to be travelling at a speed which could have amounted to billion dollar fine for speeding. However, it is reputed that because of Wood's bamboozling of the Judge with the Doppler effect, he was acquitted!

Many motorists have invested in devices, which can detect microwaves being beamed at a car or being broadcast from a source some distance away. These devices are known as "**fuzz busters**" or "**anti-radar**" devices. People investing in such devices should be aware that in many parts of the world the use of anti-speedtrap electronic devices is illegal. The presence of such a device in the car can lead to confiscation of the instrument and the issuing of another ticket.

In November 1992 I was a passenger in a taxi in Newcastle, England. As the taxi passed through a traffic light, another car sped past the taxi, obviously trying to beat a yellow light. As he moved past, there was a bright flash of light, which the taxi driver advised me was a robot taking a picture of the speeding car. He told me that the robot would issue a traffic ticket to the speeding car along with an electronic print-out of the speed of the vehicle. Police forces in many parts of the Western world are now introducing this system. For example, in October 1992, in London, England, police installed devices along a 40 mph zone Near west London. In three weeks, 22,000 motorists were clocked doing more than 60 mph. The problem now is that the robotic system for issuing traffic tickets is bogged down by all the work it has to do! There is some controversy over this type of police monitoring, but it is very probable that the system will become widespread since it takes pressure off the police, freeing them to fight crime instead of quarreling with angry motorists.

In Great Britain, automobile insurance companies are pressing the government to adopt a different type of surveillance system, one that would be fitted to every private car.[18] This type of device is known as a recording tachometer. It is similar to the **black box** carried on all commercial aircraft that records the details of the flight of the aircraft. The term tachometer comes from the Greek word for speed. The recording tachometer fitted to a car could be examined by the police to discover the exact time of the crash, the speed of the car at that instant, if the car was accelerating or decelerating, if the brakes had been applied, and if the lights were on. The association of British insurers say that its members pay out more than 8 billion dollars a year for crash damage and they claim that fitting tachometers to cars would cost about $200 a car. Scientists who have carried out tests with recording tachometers say that fitting the device definitely changes driver behavior. Already, in Great Britain, commercial truck drivers must have their vehicles fitted with the equipment. Fincham, commenting on this fact, pointed to the important human factor when he reported that drivers worry about being found out. He says that some truck drivers whose vehicles were equipped with the devices have been found at the scene of accidents trying to eat tapes from the recording instruments in their trucks. Again, as in the case of automatic traffic ticketing robots, a major problem in accepting the electronic surveillance by automated tachometer would be public resistance to what they regard as an invasion of privacy. It is amazing how much disaster the public will accept in defence of their freedom. They would rather have the freedom to be killed by a drunken driver than allow an electronic monitor to record the pre-crash movements of a deadly automotive missile.

6.4 What About Those Watching Electronic Eyes?

The robot traffic cop issuing tickets is only the latest development in the movement toward using television cameras to constantly survey all areas of modern life. As the cost of television cameras and computer systems fall, we can expect more and more pressure to install surveillance systems to prevent crime. In Great Britain, there were two spectacular cases involving the use of video surveillance cameras in everyday life. In February 1993, a two-year-old boy, James Bulger, was abducted from his mother in a shopping center in Liverpool. The surveillance cameras at the shopping mall had recorded a picture of the small child being led away by two youths. Later, the body of the child was found; he was murdered. These pictures were used to track down and arrest the two suspects. In March of the same year, two men were recorded on security cameras at Harrod's (a famous London store) and then again, in Victoria Station in London shortly before two bombs went off.

Images of the suspects were broadcast to the public over television, and in just six hours after the pictures were shown, two men were arrested.

Currently, the quality of cameras used in video surveillance is relatively poor, although specialists predict that by 1998 new lower cost television cameras will be available to generate high-quality pictures. These pictures can be used to carry out computer recognition procedures by comparing pictures from security cameras with a library of potential trouble-makers. In the case of the video of the two youths abducting the small child, the expert who processed the pictures tells us that:

> *The camera system was not designed to record enough detail to identify individuals. The lense covering was dirty and the video tape had been reused many times. This combination is not unusual.*

The picture was enhanced by electronic processing at the IBM Scientific Center in Winchester, England. Rick Turner, a member of the IBM team that enhanced that particular picture, estimates that in Great Britain the police request 20,000 video pictures be enhanced every year. Some enhancements are used for intelligence rather than for evidence in court. The requests are so plentiful that in England IBM has set up a special bureau offering this service. Only certain types of enhancements are permitted for use in court, and any change must be reversible in front of court witnesses to show that information has not been added to the original picture.[19,20] Alan Hewitt indicates that surveillance operators are currently prepared to go with low-quality cameras because in 95 % of all cases, "would be" criminals are deterred by the fact of knowing that they are being watched. Only in five percent of cases proceeding to a prosecution is a good quality picture required. However, this situation is likely to change, because the quality and price of television cameras for remote surveillance is dropping rapidly in price.

One of the technological innovations which is making it possible to install more video cameras for surveillance is the development of fiber-optic cables. Transmitting a video picture over a telephone line requires a great deal of signal capacity. However, a thin fiber of glass the thickness of a hair can carry as much information as a set of copper cables eight inches in diameter. (In the early 1990s, a major source of copper for industry was the telephone cables being replaced with glass fibers in cities!) The cost of fibre-optic links between cameras and surveillance centers is falling rapidly, due to developments in the telephone and television industries. Figure 6.7 shows recent techniques for enhancing video camera pictures, which also help illustrate some of the problems of transmitting pictures over a telephone cable. The first in the series of pictures has been artificially degenerated to a lower resolution. Each of the visible squares in this degenerated picture is described as a **pixel**, which is the short word for picture element. To transmit a picture over a telephone line, one must send the address of each pixel and then send information about the color of that particular pixel. You may like to test for yourself the human eye's amazing ability to integrate a relatively coarse picture, such as the middle one of the three pictures shown on the right-hand side of Figure 6.7. If you hold the picture at a distance of about a 1.5 meters and squint, you can actually see the face in the degraded picture better than if you look with wide open eyes and hold the page 15 cm away. If you oscillate the picture a small amount at a distance of about 30 cm, the integration performed by the eye is even better.

The pictures shown in Figure 6.7 are of interest because they demonstrate a very advanced concept for recognizing elements of the picture, which has a fractal structure. In 1977 a new branch of mathematics was invented by Benoit Mandelbrot called **fractal geometry**. The ordinary geometry that we all learned at school is known as **Euclidean geometry**, after the Greek mathematician who invented it. Fractal geometry deals with the structure and abilities of ragged lines to fill space.[21]

The basic idea involved in the concepts of fractal geometry can be appreciated from the information summarized in Figure 6.8. In classic science, a line has a dimension of 1 and an area has a dimension of 2, whereas a cube

Figure 6.7. Computers can be used to improve the quality of a picture from a surveillance television camera. Shown above is a simulated representation the results of a computer enhancement of a poor quality surveillance tape image.

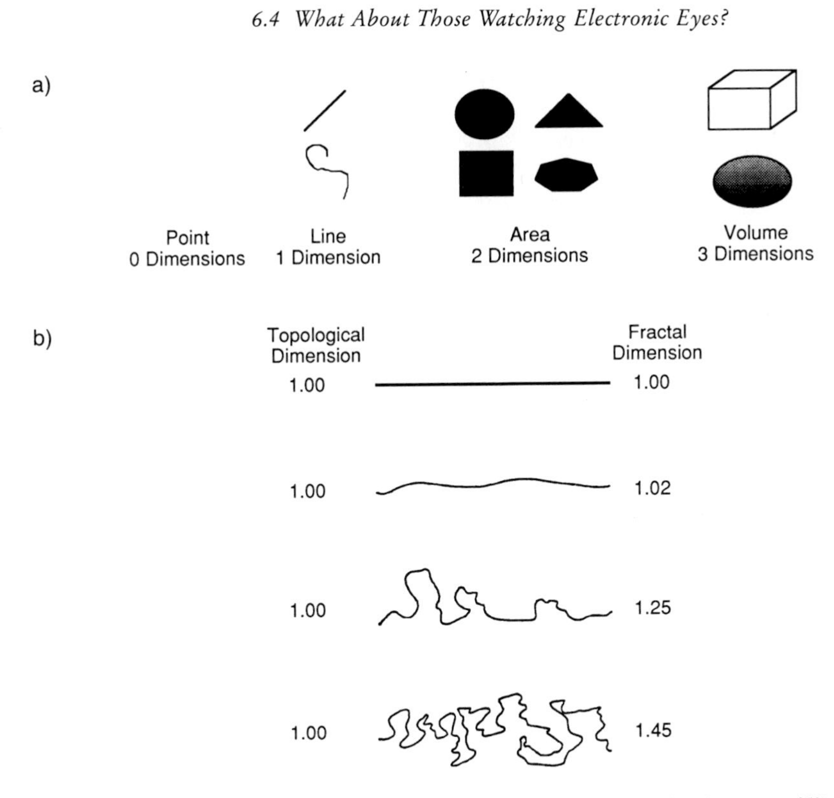

Figure 6.8. The concepts of fractal geometry can be used to describe the space filling ability and rugged structure of various objects. a) Objects are traditionally viewed as having one, two or three dimensions. b) All lines have the topological dimension of 1 but ragged lines fill space more efficiently and can be assigned a "fractal dimension."

has a dimension of 3. Mandelbrot suggested that if we wish to describe the raggedness of a line, drawn in two-dimensional space, we can attach a fractional number to the basic dimension of the line to describe its ability to fill space. The combination of the dimension of the object and the fractional number is known as the **fractal dimension** of the line.

A piece of paper has a dimension of two. As you crumple it, it occupies space more efficiently, eventually assuming a dimension of three, if you can compress it to a high density. En route to this stage, a fractional number is added to the number two to create a fractal dimension. In a fractal dimension such as 2.4, the 2 tells you about the nature of the surface of the paper, and the fractional number tells you how tightly it has been crumpled.

Thus, the lines of Figure 6.8(b) have the fractal dimensions shown. It is much easier to transmit the fractal dimension of a ragged line for computer processing, than it is to send the image of the line as a pattern of pixels over a telephone connection.[21–23]

Thus, a spokesperson for a company called Origin Redhill estimates that by using fractal logic to compress the information in an image, one can reduce the number of signals from 750,000 to 5,000. Comparing two pictures, each of which requires 5,000 pieces of information, can be carried out on a personal computer in a fraction of a second. Barnsley, one of the developers of fractal compression technology, calls the technique a **fractal transform**. The older, pixilated transmission is called **bit mapping**.[23] Before fractal transform compression, one would have needed a large central computer to carry out the comparison of two pictures based on 750,000 pieces of information. The company Origin Redhill has already sold their system to the police. It is anticipated that in a few years, a police officer could use a mobile phone to transmit a picture of a person caught committing a crime. Within minutes, the computer would be able to return the criminal record of the person from its memory bank, if the person already has a police record. Again, forensic scientists can expect to have much better equipment in the future for these types of tasks, because of the need to develop security systems for airports and to protect politicians. In 1992, specialists at the University of Wollongon in Australia began developing an automatic face recognition system for installation at Sydney's International Airport. It was estimated that the system would be complete in 1993, and installed at the airport by 1997. The Australian system picks out people's facial characteristics, such as eyes, nose, mouth, chin and hair. For each characteristic, it recognizes variations in the size and shape, giving more than half a million different faces. Using a picture of a face from a standard security camera, the system uses what is known as a **neural network** to compare it with a data base of up to two thousand suspects' faces. Neural network is the general name for a relatively new computing device, which learns to recognize patterns even without being given a set of instructions on how to recognize them. It is an intelligent machine that learns from a training set.[24]

It is difficult to explain in simple terms the operation of such devices, but these intelligent computer systems are finding applications in several areas of forensic science, as we will learn when we look at the problems of fraud, connected with such diverse products as pure olive oil and real champagne. The way in which one uses a neural network to carry out a recognition task can be appreciated by considering a relatively simple pattern recognition problem. Such a recognition task is being studied by my group at Laurentian University in co-operation with Dr. Bonifazi at the University of Rome. Carbon-black pigments are substances used to reinforce rubber tires and they are present as pigments in paint, ink, and many other products. The pigments are made by burning oil with a shortage of oxygen. The resulting black fumes are collected and constitute the carbon-black pigment. In Figure 6.9(a), a set of profiles from two different grades of carbon black made by a pigment company are shown. The scientific problem Dr. Bonifazi and I are tackling with the aid of neural networks is the quantization of the differ-

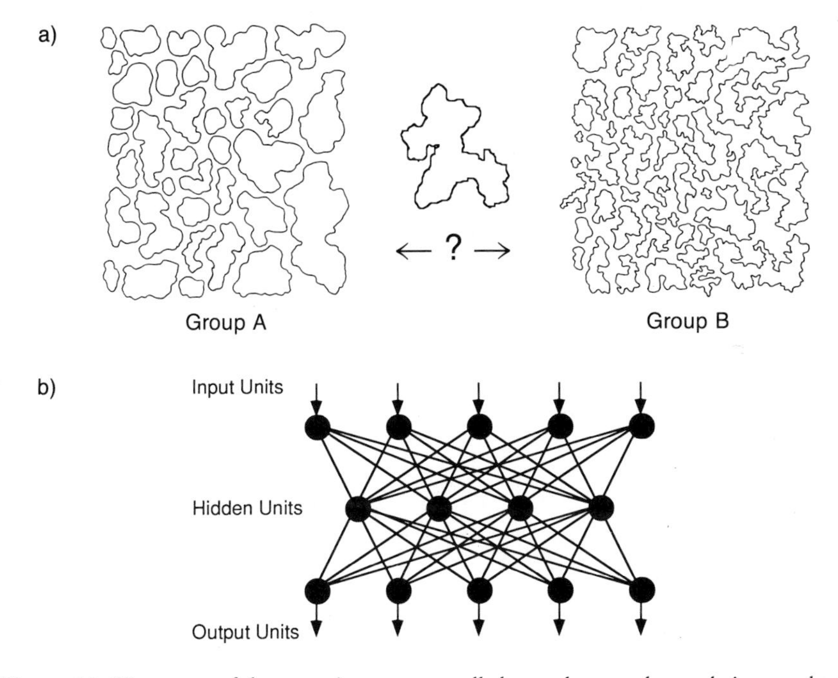

Figure 6.9. New, powerful computing systems called neural networks are being taught to recognize complex patterns such as faces and speech. a) In recent work by Bonifazi and Kaye, a computer was trained to recognize two different grades of carbon black and to decide which grade should be assigned to a new particle. b) The basic structure of a neural network.

ences between the two sets of profiles. In a classical analytical procedure, one would measure the profiles directly, determining, for instance, the fractal dimension (raggedness), area, and major dimensions. If one uses a neural network, the procedure is very different. The neural network has three main elements: an input array of sensors, a set of randomly connected units behind the input, and the output array (see Figure 6.9 (b)). In a training session, all of the profiles of population A, and then all of the populations of B would be given as input. Then the outputs from all of the different training profiles would be stored in the computer memory. If one now was to show to the computer a profile such as the one shown with a question mark, the computer would look through its memory and match the new profile to all of the existing memory patterns. It would then decide whether the new profile belonged to the pigment group A or B. Neural networks are called parallel processing units (see discussion earlier of the techniques for fingerprint recognition), and because they are becoming very important in industry, the cost of such units is falling very rapidly. It is likely that neural networks will be used in surveillance technology and in new procedures for speech recognition.[25]

Another surveillance technique for use at the airport and other public places is being developed by Iterated Systems of Atlanta, Georgia. Its Matchmaker program uses fractal mathematics to match faces. In a presentation of the power of the program, Iterated Systems demonstrated, in March 1993, how the computer system was able to recognize Bill Clinton in a photograph in which he was surrounded by his staff. Earlier, the system had been taught to recognize Clinton from a picture in Newsweek magazine. Dennis Colomb, Vice President of Iterated Systems said that the computer found Clinton based on the structure of his eye brows, forehead, and hair. In a review article of the power of these new face recognition systems, Elizabeth Geake makes the following statement:

> *in restricted areas, employees will no longer need identity passes. Their faces will be recognized as they pass through the entrance to their work place. The movements of confused or dangerous people in hospitals or prison could be monitored automatically. Credit and bank card fraud will shrink when a person's face can be scanned and checked at the teller's work place or at a cash dispenser. Wanted criminals could be intercepted at airports before they have even opened their passports.*[20]

One area of surveillance where the falling cost and better performance of cameras lead to widespread application is in stores, where they can be used to prevent shop lifting. Figures given out by the retail council of Canada show that 2.7 billion dollars worth of goods were stolen from Canadian stores in 1992. An average of 7 million dollars per day in merchandise is stolen by customers and employees. Much wider use of surveillance cameras in stores can be realized, as the price of television cameras drop. In a recent review of surveillance cameras, Keape pointed out that costs in the United States have already dropped to the point where parents are using their own small cameras to watch children in another room, while they go about their household work. It is predicted that, in a very short time, home owners will be able to video-phone home and see that everything is in place. A small town in Great Britain, King's Lynn, Norfolk, has installed a network of 45 cameras, which monitor a leisure center, 17 car parks, an industrial estate, and the streets of a housing estate. Signals from all of these cameras are fed to a central control room equipped with a hot line to the police station. These pictures are fed through optical fibers. Operators hired from a private security firm monitor 22 screens, 24 hours a day. Normally, pictures from all of the cameras are recorded by time lapse recorders, which take just one frame of each of the cameras in turn. But if the operator sees anything suspicious, one camera can be switched to continuous recording to capture all the action in more detail. Since the system has been installed, theft from cars fell from 207, the year the cameras were installed, to seven, a year later. The police also made 139 other arrests for offenses such as drug pushing and defacing property. Many specialists argue that the public in general is not willing to accept such surveillance of their lifestyle. However, in the town of King's Lynn only one person out of 2,000 interviewed commented that he feels the quality of life is negatively affected by knowing that he is being watched. 53 % of the respondents felt that the closed-circuit television viewing made them feel safe; 62 % thought it would deter crime, 74 % thought it would detect crime, and 80 % welcomed the installation of video cameras in public places.

Scientists are aware that human beings are not very good at watching an unchanging scene for a long period of time. They become bored when given a task such as that of monitoring all the computer screens in an array. Therefore, computer specialists are starting to develop robotic monitors that can alert a human when something changes on a particular region of a video screen. Thus, a system known as Sentinel – being developed by EDS Scion of Camberley in Surrey, detects objects of a particular size, stopping in particular regions of the screen, a task which is important when monitoring a car park. Sentinel can also be set to record everyone who leaves a building through a certain door and to note which direction they head in or if anyone is moving around after working hours.

In this chapter, we have considered the development of electronic ears and eyes. In the next chapter, we look at the possibility of developing synthetic noses for sniffing out drugs and explosives in security work.

References

1. The speech spectrum on the cover of this chapter is taken from L. Cherry, "Their Blood Cried out for Vengeance," *Science Digest*, May 1981, 60. The photograph is used with permission from *Science Spectrum*.[10]
2. The graphs of Figure 6.1 are taken from E. E. David Jr., J. G. Truxal (Eds.), *Man and his Technology*, McGraw Hill, New York, 1973.
3. F. Smythe, *Cause of Death: the Story of Forensic Science*, Van Nostrand Reinhold Co., New York, 1980.
4. E. Joseph Simmons Jr., "Speech Recognition Technology," *Computer Design*, June 1979, 95.
5. R. H. Bolt, F. S. Cooper, E. E. David Jr., P. B. Denes, J. M. Pickett, K. N. Stevens, "Identification of a Speaker by Speech Spectrograms," *Science 166* (1969), 338.
6. R. B. Cox, T. B. Martin, "Speak and the Machines Obey," *Industrial Research*, 15 November 1975, 16.
7. E. Geake, "Police Get Direct Line To Computer," *New Scientist*, 16 January 1993, 20.
8. J. Stuart, "The Sound of Assailants" *MacLeans*, 30 July 1984.
9. B. H. Kaye, "Golf Balls and Other Missiles," to be published by VCH Verlagsgesellschaft, Weinheim.
10. J. A. Clarke, H. Hawden, "Lenses to See in the Dark," *Science Spectrum Notes 134* (1975), 2.
11. J. B. Waugh, "Progress in Night Vision," *Optical Spectrum*, November 1980, 56.
12. B. Goldberg, "Night Vision Systems," *Electrical Optical Systems Design*, January 1974, 18.
13. See trade literature of AGA Thermovision, 550 County Ave. Secaucus, New Jersey, 07094.
14. In 1993, a night vision camera could be purchased from the Edmund's Scientific Company for just under $ 4,000.00.

15. Night vision goggles are available from the optics division of Bell and Howell, 7100 McCormick Road, Chicago Illinois, 60645.
16. D. M. Szles, " A Design Study of Intrusion Protection," *Optical Spectrum*, July 1978, 41.
17. This aspect of forensic science is discussed in Denis Shaw, "Physics in the Prevention and Detection of Crime," *Contemporary Physics 17* (1976), 307.
18. E. Geake, "Black Box could Put Brakes on Bad Driving," *New Scientist*, 9 November 1991, 26.
19. E. Geake, "Tiny Brother is Watching You," *New Scientist*, 8 May 1993, 21.
20. E. Geake, "The Electronic Arm of The Law," *New Scientist*, 8 May 1993, 19.
21. Fractal Geometry was invented by B. Mandelbrot and is described in B. Mandelbrot, *Fractals: Form, Chance and Dimension*, Freeman, San Francisco, 1977.
22. The data of figure 6.8 are taken from B. H. Kaye, *A Random Walk Through Fractal Dimensions*, Second Edition, VCH Verlagsgesellschaft, Weinheim, 1994.
23. M. F. Barnsley, *Fractals Everywhere*, Academic Press, San Diego, 1988.
24. "Neural Networks Become More Human," *R & D Magazine*, May 1993, 71.
25. The basic operation of a neural networks and the processing of speech are explained in D. Shanks, "Breaking Chomsky's Rules," *New Scientist*, 30 January 1993, 26.

Chapter 7

Bloodhounds Real and Synthetic

Chapter 7

Bloodhounds Real and Synthetic

7.1 Dogged Detective Work

The title page of this chapter shows one of the stars of the detective world. In real life, Sergeant, a frisky black labrador from Quebec, is used to track down drugs.[1] Dogs are also used to enforce laws associated with hunting and fishing in the Sudbury area.[2] Agriculture Canada uses beagles to track down smugglers who break the law by bringing meat and other food products into Canada.[3] In 1993 it was estimated that a good tracker dog used in law enforcement costs approximately $70,000.

The use of dogs to track down criminals and political opponents is an old technology, as witnessed by the meaning of the word **sleuth**. In medieval English, sleuth meant the track of an animal or person. It is related to the old Norse word sloth. By the 1400s, the word was only used in compound nouns such as sleuth-dog and sleuth-hound. The term sleuth-hound was used in Scotland specifically for a kind of bloodhound used to hunt game or to track down fugitives. Thus, in the history and chronicles of Scotland written in Latin in 1536 we are warned that:

> *He that denies entrance to the sleuth hound shall be considered a participant in the crime and theft committed.*[4]

In North American English the words sleuth-hound came to be used to describe a detective; a word which was quickly shortened to sleuth. Anyone who has watched the work of a bloodhound in a movie knows that the way to avoid leaving a trail that a bloodhound can follow is to find a stream and move along through it for some distance before emerging to make a getaway. One of the most pleasant experiences of my career was to work with Dr. Andrew Dravnieks, one of the worlds leading experts on odors, during my stay at the Illinois Institute of Technology Research in Chicago (1963–1968). I asked Dr. Dravnieks why the bloodhound was so good at following a trail when human beings did not appear to be able to track a suspect by means of odor. Dr. Dravnieks advised me that if I was willing to run along a suspect's trail with my nose two millimeters above the ground, I may have some success in the task! In order to understand why a bloodhound needs to keep his nose to the ground and to be able to assess the advantages and problems of using real dogs in detective work, we first of all need to understand the molecular nature of odor and how human beings and animals detect odor. These subjects are discussed in the next section.

7.2 Why do Onions Make You cry?

We will start our exploration of the molecular structure of odors by looking at why your eyes water when you cut into an onion. In Figure 7.1 the chemistry of various things that happen during this task are illustrated. The chemistry of the molecules given off by an onion has been investigated in some detail. Pharmaceutical companies are beginning to discover that some of the claims made about onions by folk medicine may indeed have some

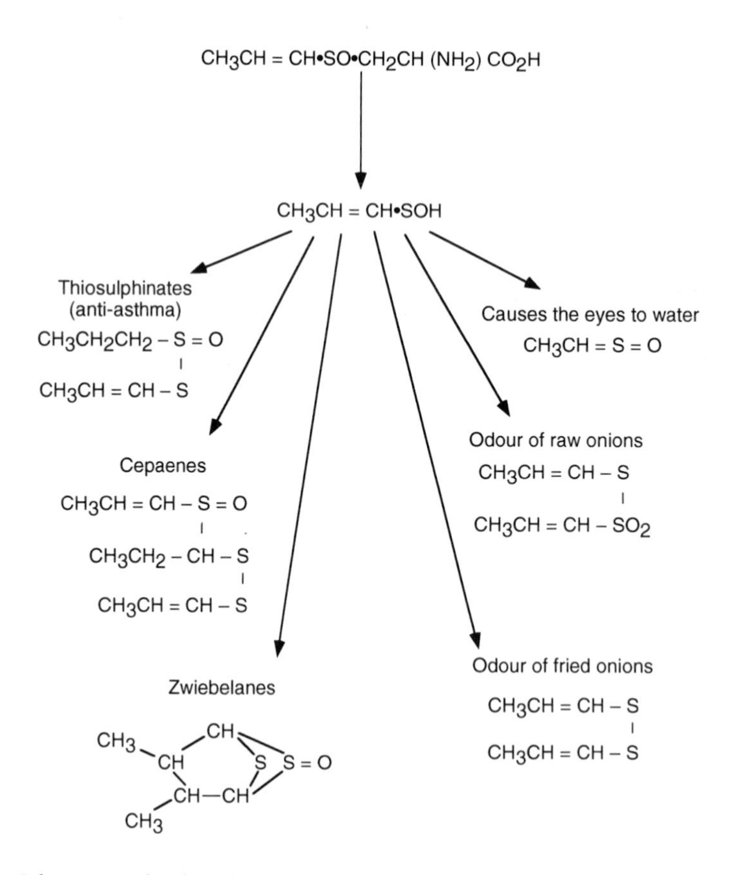

Figure 7.1. Odors are molecules which, when drawn into the nose, stimulate receptors and cause the sensation of smell.

substance. Thus, two Japanese chemists have found that compounds from an onion slow down the rate at which blood platelets clump together during clotting, which might explain why onions are reputed to be good for the heart.[5]

Other scientists have found that onion extracts can stop asthma attacks in guinea pigs. As illustrated in Figure 7.1, when you first cut into an onion, a compound that is given off reacts in the air to form another compound, which is known as propenesulphenic acid. It should be noted that many odors are acids. This is why in household cleaning an effective absorber of odors is baking soda; it is mildly **alkaline**. Alkaline substances react with acids to neutralize them. Therefore, the powdered baking soda not only offers a very high surface area for the adsorption of odors but the odors react chemically with the powder and are neutralized. The same strategy is adopted in the making of powders for cat litter boxes.

The molecules given off on cutting an onion can rearrange themselves quickly to form a compound known as propanethial oxide which is very **volatile**. (The Latin word *volare* means "to fly." A substance is said to be volatile when the molecules can easily escape from the liquid as a vapor.) When propanethial oxide dissolves, it produces sulfuric acid, which irritates the eye and initiates weeping. Two traditional ways of avoiding the effect of onions can be understood from the physical properties of the chemical responsible for this irritation. The first is that the molecules, which cause weeping, are water soluble; so you can avoid the weeping problem by peeling the onions under water. Molecules responsible for leaving an odor trail are dissolved in water and washed away from the immediate point of release. (Fugitives adopt the same strategy when they try to avoid detection by walking in a stream.) Another technique for avoiding having your eyes water while chopping onions is to chill the onions before chopping them. This lowers the volatility of the molecules escaping from the onion. We shall discover later in this chapter that if we want to capture an odor for study with scientific instruments, a primary method of odor collection involves lowering the temperature of the molecules to the temperature of liquid air (approximately −180 °C). Under these conditions the energy of the molecules is lowered to the point where it is easy to capture them on an oiled surface. In continuing our saga of the onion, I would like to quote the words of John Emsley:

> *frying onions pushes the reaction along another route to give the sweet smell of bispropanesulfide, otherwise known as fried onion smell.*[6]

In Figure 7.2 the structure of the human olfactory receptor is shown. To be detected by the brain, odor molecules must land on a thin layer of mucus on the olfactory membrane. These molecules are then detected by the nerve endings in this membrane, which lead to the brain. The nerve cells in the olfactory membrane have many **cilia** (thin hairs), which can be attacked and

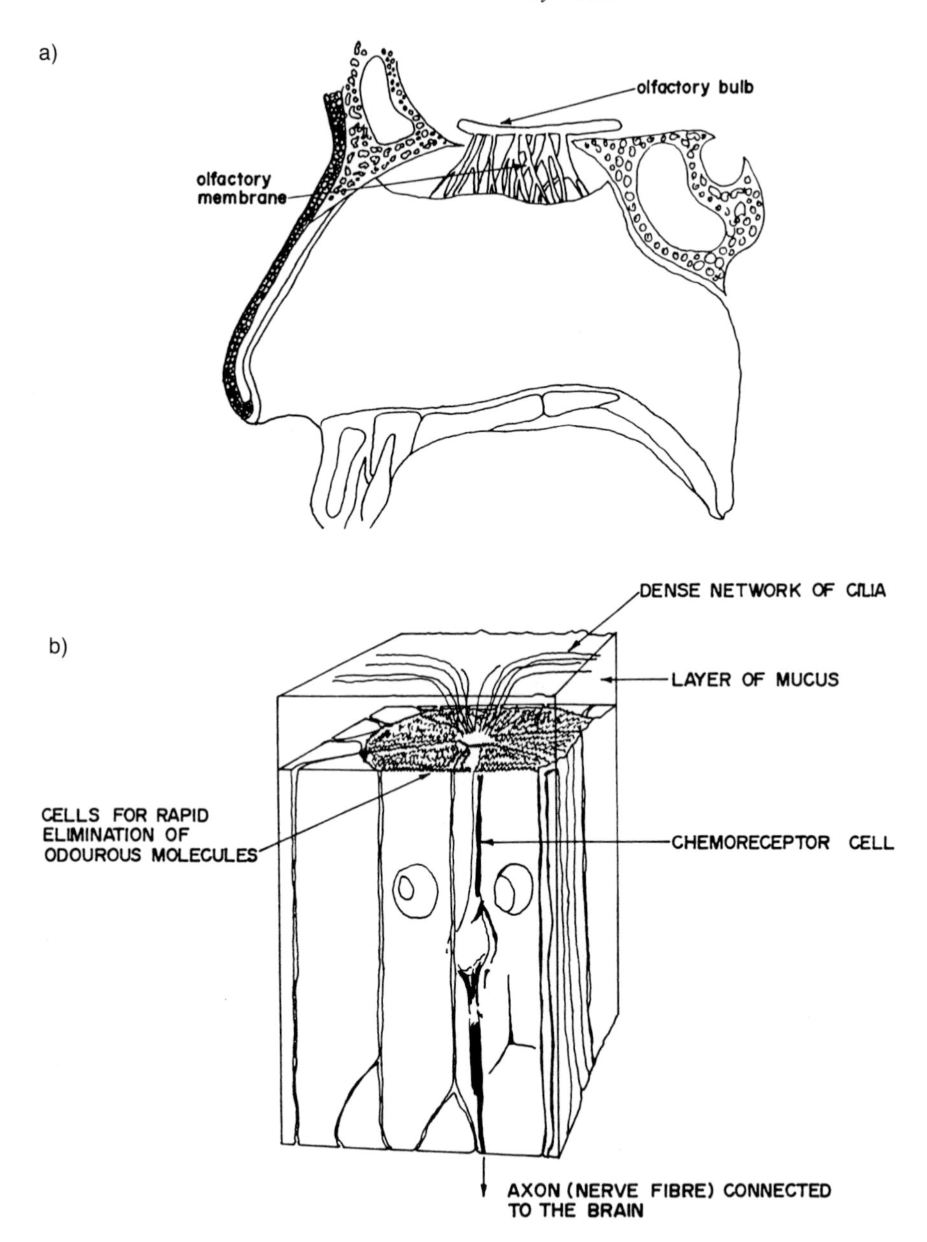

Figure 7.2. The sense of smell arises when odor molecules are captured by the olfactory membrane. a) Structure of the nose. b) The basic odor receptor in the human nose.

paralyzed by nicotine. This is why smokers have a very poor sense of smell. Smell and taste are closely linked and people with damaged cilia also lose the ability to taste finer aspects of flavor.[7–11]

It should be stressed that the sensation experienced as odor occurs as the brain interprets what is going on in the olfactory membrane. Because it is a brain-organized reaction, persistent odors are blanked out in the sensory-reception area of the brain. A persistent odor "fades out" after 45 seconds. Furthermore, extensive exposure to an odor can fatigue the nose so that it no longer responds to a low level of that odor. Close to Sudbury there is a paper mill town called Espanola. Visitors to Espanola often notice a persistent odor of boiled cabbage (see Figure 7.3), whereas the residents of the town are relatively unaware of the odor.

If an investigating officer arrives at the scene of a crime and is aware of a distinctive odor, it may be worthwhile recommending that air samples be taken and analyzed, as described later in this chapter. However, it should be noted that the investigating officer may be distracted from the odor and, therefore, its significance, because his brain blanks it out after 45 seconds. When we sniff at an object, we momentarily increase the air flow over the olfactory membrane to increase the amount of odor molecules being captured by the membrane and detected by the brain. Many animals appear to have larger olfactory membranes than humans, which may be one of the reasons why they are more effective hunters. Some dogs have been selectively bred to enhance their odor tracking abilities. Breeds such as bloodhounds, beagles, spaniels, and labradors may have bigger olfactory membranes, which enable them to be more effective at capturing an odor that they are trying to follow.

Dogs use the same sniffing action to enhance their odor capturing ability. Linking the molecular structure of a substance to the sensation that a molecule creates in the brain is a difficult subject. Thus, as shown in Figure 7.3(b), some very different molecules can create the sensation of bitter almonds in the brain. This is particularly important when it comes to detecting poisons such as hydrogen cyanide. It is not always appreciated that hydrogen sulfide, which smells of rotten eggs, is a compound that belongs to a family of chemicals known as mercaptans, which are also very poisonous. The fact that it smells like rotten eggs sometimes deludes people into thinking it is a relatively safe compound. Sudbury, Ontario, is the home of a major nickel-smelting operation, and some days when the weather conditions are right, one can smell the hydrogen sulfide being given off from the smelter process. The nuisance aspect of odors is an important branch of civil litigation that is too large a subject to be covered in this book.

Closely related to the smell of rotten eggs is the odor given off by a skunk when it uses chemical warfare against its enemies. Apparently, a similar odor is given off by beer which has spoiled in the can. For this reason, spoiled beer is known locally as skunky beer.

a)

H – S – H
Hydrogen Sulphide (Rotten eggs)

$CH_3 – S – H$
Methyl Mercaptan (Boiled cabbage and paper mills)

$CH_3 – CH_2 – CH_2 – CH_2 – S – H$
N-Butyl mercaptan (Skunk)

b)

HC ≡ N
Hydrogen Cyanide

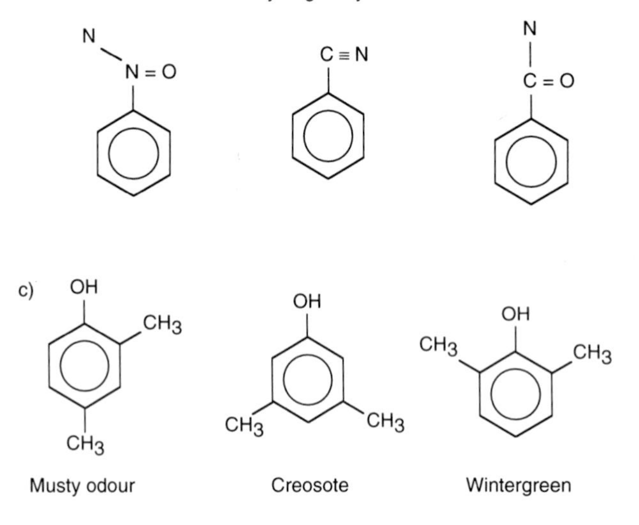

Figure 7.3. Linking odor sensations to molecular structure is not an easy task. a) Rotten eggs and skunk smell belong to a family of chemicals known as mercaptans. b) Some different molecules can give rise to the sensation of bitter almond. c) The chemical structures of some easily recognized odors.

In Figure 7.3 (c) the chemical structure of some familiar odors is given. The reason why a tracker dog keeps his nose close to the ground is that immediately above the surface of the ground there is hardly any wind, particularly if there is grass or other plant cover. As we will discover later when looking at the problems of sniffing out dynamite from a cardboard box, any fibrous material will tend to absorb an odor and then give it off slowly. This is why fabrics retain odor molecules.

The fine particles of soil and earth can absorb large quantities of odorous molecules. A colleague at the university, Dr. Scott Fairgrieve, was involved in the investigation of a murder case in which the victim's body was disposed of by burning it in an area of a garden. As he researched the earth, he noticed a smell of gasoline and took samples for analysis, since it appeared likely that gasoline had been used as an accelerant in the burning. (Accelerant in this context means something to get the fire going). In many arson cases, accelerant may still be present in material near the point where a fire was started, and odor analysis of fumes from an object contaminated by the accelerant may generate useful evidence (see discussion of the characterization of gasoline later in this chapter).

A rather unique crime occurred in Sudbury on Saturday, July 12, 1993, when a person was charged with mischief under $1,000 for throwing a container of foul smelling mining products at a house in the city. (Assault with a deadly odor?) In fact the man had thrown a cylinder of mercaptan, which is used in the mining industry as a safety measure. When mining machinery fails, often there is no means of communicating with the miners underground except through the ventilation system. And if the situation is life threatening, the mining company will release mercaptan into the ventilation system of the mine. When the miners smell the foul smelling compound, they know that safety precautions must be initiated. All miners employed at hard-rock mines in Canada are trained to recognize the foul odor of mercaptans so that they will be aware of attempts to warn them of danger. An assault like this one gives new meaning to the colloquial expression "creating a stink about your grievance!"

Natural gas, which is an explosion hazard and can suffocate people if it fills a house, is also an odorless hazard. It is common practice in many countries to add a foul smelling compound to natural gas before it is piped into homes. This means that if there is a gas leak in the house, the owner can detect the smell. Again, part of the problem is that if you enter a house and smell such an odor, you must act quickly, otherwise you will be lulled into accepting the odor because your brain will fade out the signal very quickly.

7.3 Dog Detective Performances

Now that we understand the molecular origin of odors, we can review the successes of dogs as detectives, and we can also find out some of the problems that dogs may face in this kind of work.

The rate of movement of molecules in the air is amazingly high. For instance, oxygen molecules travel at an average speed of 0.46 kilometers a second in air at 0 °C. The reason why odor molecules do not leap from the surface where they are created and vanish in the distance is that, although they are moving around at high speed, they also frequently collide with the air molecules. This means that their path of dispersal is a random set of staggers back and forth, and the overall **diffusion** (the technical term for spreading) of an odor from a point is much slower than you would estimate based on a knowledge of its raw speed between collisions. However, molecules can travel very great distances. Thus, a male Gypsy moth can smell a female Gypsy moth at a distance of several miles and actually locate the female giving out the odor over such a distance. An odor that influences behavior is known as a **pheromone**. Musk, a major ingredient in expensive perfumes, is believed to be a pheromone that stimulates sexual behavior in humans.

In an earlier section, the value of a tracker dog in 1993 in Canada was estimated as being of the order of 70 thousand dollars. The major reason for this sum is the expense of training the dog. The way in which Agriculture Canada trains one of its dogs to detect smuggled goods in the baggage of airline passengers is described in some detail in an article in Reader's Digest. Digger, the dog in question, was rescued from a Vancouver Dog Shelter operated by the S. P. C. A. (Society for the Prevention of Cruelty to Animals). The person in charge of selecting dogs, Jane Boulton, took Digger out on a trial run, and she recognized that the dog was a natural for the task. Boulton states:

> *Digger, hot on the tarmac, nose to the ground, was undeterred by roaring planes. Inside the airport, sniffing happily among a sea of legs and baggage carts, he delighted in greeting passengers with a friendly wag of his tail.*[2]

Digger remained at the airport for six months so that he could feel completely at ease in the facility. Scent training began February 1990, in a large empty hall free of sound, movement, odors and glaring light. Strewn about the floor were a variety of bags and suitcases, some empty, some containing easy to detect foods. When Digger lingered at a bag with food in it, Boulton made him sit, praised him profusely, and gave him a biscuit. Later, she withheld a reward until he sat down on his own at the bag containing food and pawed at it. As the training proceeded, Digger was gradually introduced to items more difficult to detect under normal noisy working conditions. After four months of training, Digger moved to Toronto where he worked with his newly appointed handler.

A typical day in the life of Digger is described in the article, and interestingly, it is noted that the identity of Digger's hometown near the airport is a secret, because, as his handler explains, Digger makes enemies and some of them might try to harm him. Digger's sense of smell is estimated to be five hundred times more powerful than that of a human and he can pinpoint minute quantities of about 75 dairy, plant, and animal scents concealed in suitcases, wrapped in paper, cloth, or plastic. Just as Digger is trained to detect food products, other dogs are given different training to enable them to track down specific drugs. When we discuss the building of robotic sniffers to replace dogs, we will look at the odor signatures of some drug compounds. You will recall that Sergeant, a black labrador, was used quite successfully in Quebec to track down narcotics.[1] In Northern Ontario, conservation officers need to check on the amount of fish and game taken by fisherman and hunters and prevent the poaching of bear, moose, and deer.[3] In 1993, the Ontario Provincial Police had 15 general service dogs: 4 that sniffed out explosives, 2 for narcotics, and the others were general purpose tracking dogs. Rocky, a German Shepherd (a breed of dog called an Alsatian in Europe for the region from which they were originally trained as sheep dogs), have been used to sniff out leaks of insulating fluid from electric cables buried underground. In Alberta, Duke, Ponch, Searge, Buster, Kitty, Ginger, Snipper, Zoey, Pepper, and Squirt are a group of labradors trained to find leaks in natural-gas pipelines. The gas is laced with a compound known as Tekscent R, a secret mix of odoriferous chemicals, which the dogs can detect as it leaks from holes in the pipelines and up through the soil.[12,13]

Quafe, who developed Tekscent, said that developing the compound took a lot of chemical tricks. To be effective, the chemical must mix totally with the fluid in the pipeline, so that no matter where the hole is, top or bottom, the chemical will leak out. At the same time, it must remain completely insoluble in the fluid so that when it escapes from the pipe it vaporizes and percolates through the soil to the surface. Quafe settled on a chemical with an odor reminiscence of rotting vegetation. Other chemicals were added to give the right physical properties. Quafe has found that the dogs are capable of picking up concentrations a billion times smaller than the best machines available can. It takes about 14 weeks to train the dogs, and up until July 92, they had tackled 78 leak jobs and had a success rate of 97%. One of their toughest assignments was a pipeline in the alligator swamps of Louisiana, where a pipeline was buried in about 1.5 meters of mud with 1.5 meters of water on top of it. Quafe tells us that in this case the dogs were trained to work from a shallow draft swamp boat, and the dogs found the leak.

Every day, the British parliament goes to the dogs at 1:30 p.m. In the words of Tom Dalyell, a regular correspondent to the New Scientist and a member of the British Parliament:

> *Every day in the British Parliament, beautiful labradors and spaniels whisk around parliament to see that no morning visitor has left a time bomb of explosives under a seat to blow us all up after business starts at 2:30 p.m. No visitor can go round the chamber after the dogs have been round.*[14]

Tracking a suspect is one thing, but identifying a suspect by letting the dog pick out the suspect from the equivalent of a "scent lineup" is more controversial. Some recent work by Brisbine and Austaead has suggested that dogs may not be as good at picking out people as they are at tracking them down. However, Sergeant Ian De Bruin, of the dog section of Rotterdam's police in the Netherlands, carries out three hundred identification tests each year and claims a 99% success rate. De Bruin has accumulated a scent bank that contains more than a hundred samples taken from the scenes of major crimes. When put in sealed containers, he claims that scents can last for at least three years.[15] The Dutch supreme court has ruled that canine scent identification is allowable as evidence in court, if corroborated by the knowledge of an experienced dog handler. A similar precedent was established in Scotland when, in 1960 in the Scottish high court, the appeal of a man who had been convicted of breaking into a house with the intent of stealing was turned down. He had been convicted on the evidence that a police dog had tracked him from the scene of the crime to his home and later selected his shoes from an array of shoes. In a unanimous decision the Scottish judges agreed that the evidence was allowable and sufficient for a conviction. The handler had worked with the dog for more than 18 months and attested in court that the animal was thoroughly reliable.

7.4 Red Herrings and Work Fatigue

There is no doubt, however, that one of the problems associated with the use of dogs to track down drugs and other smuggled materials is that they do become fatigued and they can be distracted. This aspect of tracker dog performance has become a part of everyday speech. It is where the term **"red herring"** comes from; it describes an argument put forth in a debate; the main intent of the argument is to distract from the main topic of the discussion. The term arises from the well-known fact that one can distract a hunting dog, such as those used in fox hunting, by dragging a piece of strong smelling fish over the scent trail made by the fox. The favorite type of fish used for this diversionary practice is smoked, cured herring, which is red in color. Another name for the same type of cured fish is **kipper**. Unfortunately, fugitives rarely have friends ready to drag kippers across their trail, but in England, where some oppose fox hunting as a cruel sport, groups who oppose hunting use kippers to confuse the hounds. Using a kipper to distract fox hounds is using what is known as a masking odor.

In the area of smuggling drugs, criminals have been known to wrap the drugs with smelly substances such as mothballs to try to divert the sniffer dogs. Scented sprays sold to make the air of a house "fresh" are also masking odors. One of the main advantages of a synthetic bloodhound, of the type to be described in the next section, is that they he is not overwhelmed by masking odors.

In the article on the dog used by Agriculture Canada, the handler describes Digger's off days. Occasionally, "Digger will sniff at bags and just go through the motions of working but actually do very little." These lapses of attention are dealt with by confining Digger to a Kennel in the office of the handler for an hour or so. If the dereliction of duty persists, Digger is temporarily relieved of his duties and checked into a commercial kennel for a few days until in the words of his handler, "bored and aware that he is being rebuked, Digger welcomes getting back to work, which he attacks with renewed gusto."

If science can develop synthetic bloodhounds, another advantage will be that they can work all day and never get tired. Furthermore, they do not have to be exercised and fed. Perhaps, however, the greatest disadvantage of the real sniffer hound is that eventually it will grow old and must be retired. In the next section, we will look at the physical basis of the design of a synthetic bloodhound which is already finding applications in forensic science.

7.5 K.9 Synthetic Bloodhounds

Devotees of the science fiction television series "Dr. Who," will recognize the pun in the title of this section. A robotic dog called K.9 featured in many of the Dr. Who episodes (pronounced canine, which is the scientific name for a dog). The first step in developing a synthetic bloodhound is to collect samples of odor-bearing air. Odor molecules will usually dissolve in fat and oil, and the earliest method for collecting fragrances – the first step in the manufacture of perfumes – was to place paper smeared with oil close to the flower giving off the fragrance of interest. The fragrant molecules dissolved in the oil and were then driven out of the oil by heating the paper. One of the reasons paper can be used in this technique is that the mat of fibers constituting the paper has a very high surface area. Today many papers are loaded with clay powders to give them opacity. Dried ink is also a powder that has a very high surface area, and, therefore, ink pigments will collect a large number of odor molecules. In Figure 7.4 highly magnified pictures of the structure of good writing paper and newsprint are shown. Before cigarette smoking was banned from most work places, I used to notice that when I opened up the envelope containing the cancelled checks returned from my bank, the paper of the checks reeked of tobacco smoke.

Any paper at the scene of the crime should be examined and tested to see if it contains odor molecules that could be a clue to the occupation or whereabouts of the criminal. For instance, air purging of the paper of a ransom note into the electronic nose of a modern synthetic bloodhound may indicate the location where the note was written.

Modern instrumentation for collecting a sample of air to look for odor molecules is shown in Figure 7.5. This was the type of equipment (called fluidized bed odor sampling equipment) used by my colleagues in the mid 1960s, when designing a dynamite sniffer for use in checking if explosives had been placed on commercial aircraft. A fluidized bed is constructed by

Figure 7.4. To the naked eye, paper appears as a smooth continuous surface. These pictures, taken by an SEM at various magnifications, illustrate the fibrous nature of paper. Because of the high surface area of the paper/filler system, it is very good at absorbing odor molecules. a) Top side of good quality, clay-filled writing paper, showing clay filler fine particles. b) Bottom side of the clay-filled paper shown in (a). c) A photomicrograph of an inked area of newsprint, showing pigment particles attached to the fibres. d) A low-magnification image of newsprint. (Images courtesy of Paprican.)

placing powder on top of a grid through which air or other gases are forced. As the air starts to move up the column at a certain velocity, the powder grains are lifted up by the gas flow and swirl and bubble like a liquid. This agitated bed of airborne fine particles is known as a **fluidized bed**. Modern chemical engineering processes often use a fluidized bed to encourage reactions or adsorption phenomena between gases and solids. For the experiments that my colleagues were carrying out, a very fine gold powder coated with an oil known as an apiezon oil was used, which has a low volatility. (Chemistry students often come across this type of oil when they are working with vacuum equipment in the laboratory.) As mentioned earlier, one can slow down molecules of odor, making them easier to capture, by lowering the temperature of the air in which they are moving around. My colleagues were carrying out the fluidized bed sampling on equipment that was chilled to the temperature of liquid air (for reference purposes, the temperatures of some liquified gases are given in the legend of Figure 7.5). To operate the sampling equipment, the air under test would be used as the fluidizing gas for oil-coated gold fine particles.

The second part of modern odor analysis is the use of appropriate chromatographic or other equipment to analyze the captured odor molecules. In the case of my colleague's "explosives sniffer," after the air under test had been passed through the sampler for a given period of time, the powder containing the captured odor molecules was heated in another chamber to drive off the odor molecules into a gas chromatograph. The output was then calibrated to show the various types of molecules present in the test air. Although nitroglycerin is a major component of dynamite, the equipment used for testing aircraft air could not look for this particular component of dynamite, since anyone travelling on the aircraft that has heart problems would likely have nitroglycerine pills in their possession. The people who are developing explosive sniffers do not publicize which chemical they look for, or the techniques that are being used to build chromatography equipment. However, there are indications that a major effort is going into the development of miniature equipment to make the synthetic bloodhounds of tomorrow tiny portable devices, rather than large, cumbersome pieces of equipment. In an article published in 1990, it was claimed that scientists had built a high-pressure liquid chromatograph with a two micron diameter column, which performed better than conventional laboratory equipment.[16] We can expect the rapid development of miniaturized equipment because it can be used in military environments to enhance the ability of security forces to find hidden bombs.

In the 1960s it was widely rumored that the U. S. military were developing portable sampling chromatographic equipment to detect ammonia. People who are afraid apparently give off ammonia. This is said to be the reason that dogs become agitated when certain people are near them. The ammonia given off because of fear is sensed by the dog and it becomes agitated by the

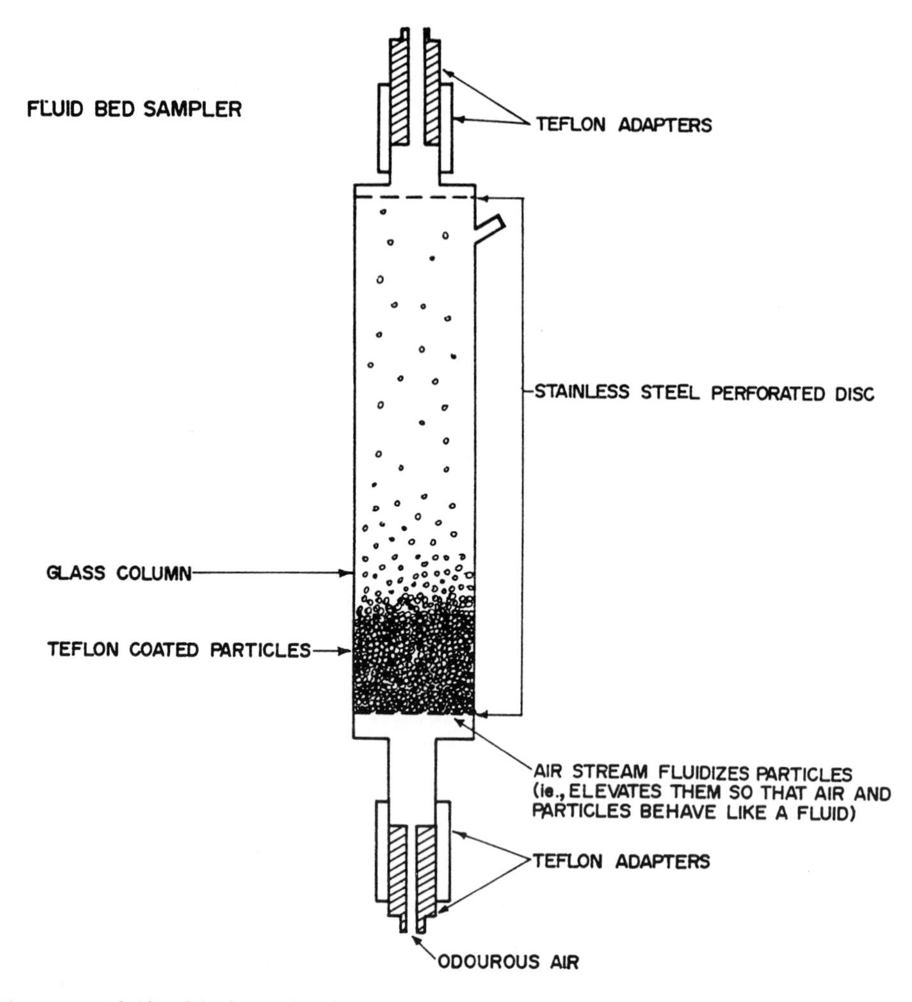

Figure 7.5. Fluidized bed samplers have an enormous surface area for odor capture. To improve their efficiency at capturing the molecules the fluidization chamber is chilled to the temperature of liquid air. The following are the temperatures at which some gases become liquid: oxygen −183 °C; nitrogen −196 °C; hydrogen −253 °C; helium −268.7 °C; ammonia −33.5 °C; sulfur dioxide −8 °C; carbon dioxide −78.2 °C; methane −164 °C.

odor. The owner will say that the dog can smell that the person is afraid. Again, the rumor is that when the U. S. helicopters started to fly over the jungle to sniff out hidden Viet Cong with the new equipment, the Viet Cong developed a strategy of hanging open bags of stale urine, which also smells of ammonia, to draw the helicopters fire.

One major interest of forensic specialists, apart from detecting bomb threats and tracing explosives used in terrorist attacks or bank robberies, is the possibility of refining the chromatographic analysis of odors to help identify people. This topic will be discussed in the next section.

Another major analytical procedure that can be used to detect very low levels of odor molecules in a sample of air combines chemical ionization of odor molecules with mass spectrometry. In his book on forensic science, Broad describes the instruments used as awesomely complex and overwhelming for the layman.[17] This is somewhat of an exaggeration, but it does indicate that the technique is complicated. Scientists in Canada have built a portable combination of these two instruments and mounted it in a mobile van. They call the combination of the two instruments **TAGA** (trace atmospheric gas analyzer). In this technique, the samples of air to be examined for their odor content or toxic chemicals, are first mixed with a carrier gas. They are then ionized by what is known as an ion–molecule reaction with a reactant ion.[18]

Ions are formed when a molecule is split into parts. The chemist uses the term **ion** to describe a part of a molecule with residual electric charge. Because they are charged, these fragments can be made to move by the use of electric voltage. A good example of ionization is the dissolution of sodium chloride (table salt) in water. The sodium chloride "splits" into two ions: a sodium ion with a positive charge and a chloride ion with a negative charge. These ions are free to wander about the solution, and that is what ion means; it comes from the Latin word *ire*, "to travel." This word has also given us exit, "to go out of" and circuit, "to travel in a circle." In the chemical ionization chamber, the molecules under test are changed into charged ions. In the mass spectrometer, the ions are characterized by the ratio of their mass to the charge. Using calibration odors and/or chemicals, extremely low concentrations of material can be detected and characterized. Thus, using the equipment more that 100 substances including acetone, ammonia, isopropyl alcohol, and nicotine have been detected directly in the exhaled breath of patients in hospital. The massive increase in the nicotine levels in the air of a room when someone lights a cigarette has been detected and measured. It has also been shown that non-smokers exposed to smoking in a confined space breathe in about 20% of the nicotine found in the breath of smokers. This type of information is becoming very important in civil-action suits in which people are claiming that their health was damaged by co-workers, and in law suits against the tobacco companies. These civil suits are a much bigger topic than can be coped with in this book.

Scientists working with TAGA equipped vans have made some dramatic measurements on cabin air from a D. C. 8 airliner in which a small quantity of dynamite had been placed. Using ion peaks specific to ethylene glycol dinitrate, a chemical found in dynamite, the presence of an explosive in the cabin was detected at a level of three parts per trillion. In reference 18 the analyisis of air inside a suitcase is described. To collect the air, the top of the suitcase was pressed down by hand, freeing a puff of air. This air was sam-

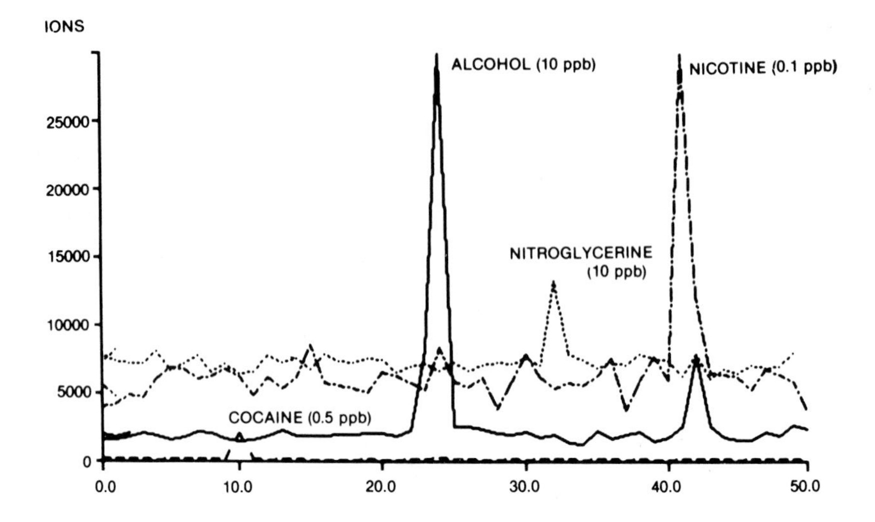

Figure 7.6. A combination of chemical ionization and mass spectrometry has proved to be very sensitive for characterizing odors (ppb: parts per billion). Data from the TAGA equipment for a test to determine the simultaneous presence of four compounds. (Reproduced from Canadian Research, April 1980.)

pled for 30 seconds. The gas sample was adsorbed onto a wire coated with a commercially secret film of an adsorbent called OV-117. The adsorbed odor was then **desorbed** (expelled) into the TAGA analyzer and a ten-second pulse of heat was applied to the wire coated with the adsorbing material. In this case, drugs placed inside the case were detected by the equipment (see Figure 7.6). We can expect rapid development of this type of equipment not only because of its security applications, but because of its use to track down pollution generating industries.[19] It will also be valuable in the quality control of food, where aroma is an important property.[18]

7.6 Olfactronic Signatures of Individuals, Locations and Compounds

In the children's story "Jack and the Beanstalk," the giant is reputed to have said, "Fee Fi Fo Fum I smell the blood of an Englishman." Perhaps the giant was an expert in recognizing odor signatures, the name given to the set of odors given off by an individual. If that odor is investigated using a combination of odor sampling, followed by gas chromatography, this is referred to as an olfactronic signature. The reason for this is that not all of the components in the odor signature are actually odoriferous (another term meaning odor bearing). In scientific studies of which components of an olfactronic signature are odoriferous, samples of each "peak" emerging from a gas chromatograph are diverted to sniffer ports where trained humans record their "smell reaction" to the different gases. Dr. Dravnieks, at the time that I was working in Chicago, was being kept very busy determining the effectiveness of anti-odorants and investigating odor nuisances in industrial and public buildings. To generate an olfactronic signature of a human, the individual under study has to be placed in a large glass bottle. Air is blown over the person and collected in an apparatus, which contains a high-surface-area powder coated with oil. The equipment used in this work is shown in Figure 7.7.

Figure 7.7. Photograph of the chamber used to isolate a person from odors in the ambient air in order to determine their personal olfactronic signature. Note that the open chamber is shown here. The lid is closed and sealed during the sampling procedure. (Used by permission of the Research Institute of the Illinois Institute of Technology.)

A typical olfactronic signature of a Caucasian male, as measured using the equipment of Figure 7.7, is shown in Figure 7.8(a). To store such signatures in a computer for recognition comparison purposes, the peaks of the olfactronic signature are represented by a series of bars, as shown in Figure 7.8(b). This bar code representation can now be stored in a computer and recognized with the type of machinery used to process postal codes and other bar code systems. Concerning the structure of human olfactronic signatures, Dravnieks has the following comments:

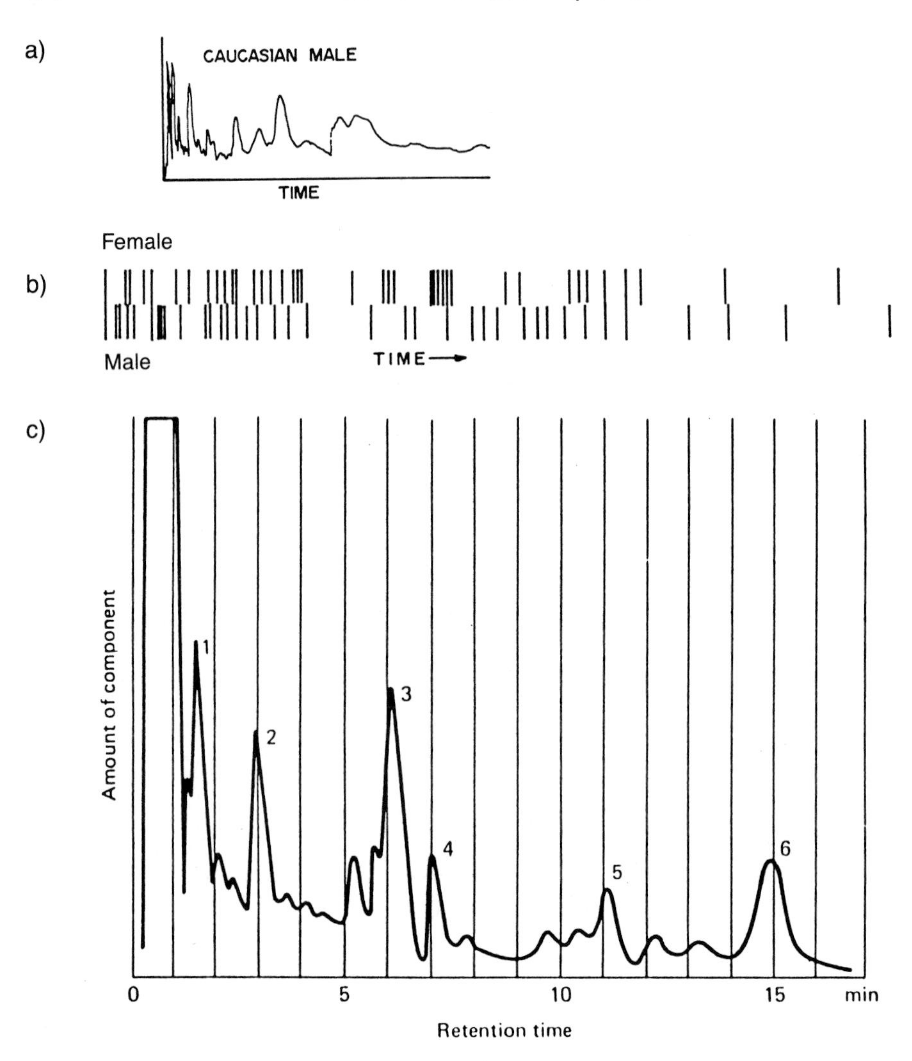

Figure 7.8. Olfactronic signatures can be stored in computers and recognized from memory using pattern recognition techniques. a) Olfactronic signature of a male subject. b) Comparison of simplified barcode olfactronic signatures of a male and a female subject. c) Olfactronic signature of a commercial gasoline. (Used by permission of the Research Institute of the Illinois Institute of Technology.)

> *An individual's vapor composition varies from day to day. Sixty to seventy percent of the features remain the same but the rest of the features reflect the food eaten, differences of activities on different days etc.*[20]

The olfactronic signature is sensitive to subtle changes in behavior and metabolism of humans, a characteristic that can be used as a powerful diagnostic tool in the study of drug addictions. Dr. Dravnieks and his colleagues have studied the detection of drugs using their odor signature. Some of these results will be discussed in the next chapter.

In Figure 7.8(c), the olfactronic signature of a commercial gasoline is shown. The gasolines made by various manufacturers are so specific in their constitution that an olfactronic signature will often be able to indicate where the gasoline was bought and when it was made in the refinery. This type of information is obviously important when investigating an arson incident. Recently, it was announced that tracker dogs are also being used to look for accelerants in arson cases.[21]

The usefulness of synthetic sniffing equipment for detecting explosives is illustrated by the data of Figure 7.9 reported by Dravnieks. Describing the experiment carried out to generate the data of Figure 7.9, Dravnieks states that the air was sampled from a cardboard box in which a stick of dynamite was stored for several days, and then removed. Ethylene glycol dinitrate (EGDN) vapor emitted by dynamite, not medical nitroglycerine tablets, is characteristic of dynamite. The vertical axis measures the amount of EGDN vapor in the air in the box. After the stick of dynamite was removed, the box was opened and repeatedly vented by blowing in air through a hose; this procedure was later repeated at irregular intervals. Even after several weeks, the fact that dynamite had been stored in the box was evident from air analysis.

People unfamiliar with the fibrous structure of cardboard, which is very similar to the various systems shown in Figure 7.4, are surprised by data such as that of Figure 7.9. Many people would fail to test the box after sev-

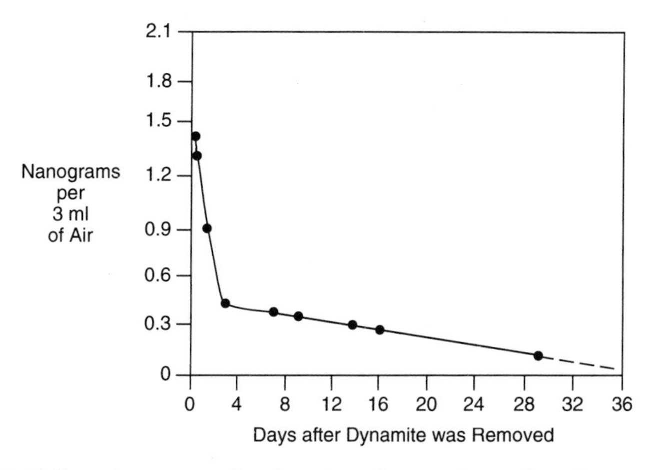

Figure 7.9. Delicate instruments for detecting odors can be used to detect containers which have held dynamite long after the dynamite has been removed. This could be useful in identifying persons plotting to use an explosive device in a terrorist action. Development of this kind of instrumentation is also useful in detecting bombs placed in aircraft. (Used by permission of the Research Institute of the Illinois Institute of Technology.)

eral weeks, suspecting that any odor present would have long since vanished. Dravnieks describes another experiment, which demonstrates the amazing ability of odors to linger around a person.

In another experiment with dynamite, the stick was handled with the hands. About half an hour later, despite hand washing in between, EGDN vapor could be detected in an air sample drawn from the semi folded palm.[20]

Perhaps we can summarize the material presented in this chapter by again quoting Dr. Dravnieks:

a dynamite sniffer is typical of an artificial nose, technologically bred to achieve a certain purpose. The device is transportable and reduction of it to a portable size is feasible. The olfactronic information in material left in hands and on surfaces can in principle give clues of the recent history of an individual's hands, his living and personal habit, and his occupation.[20]

References

1. "Lab Retriever is one of Canada's top NARC's," (news story), *Sudbury Star*, 5 October 1993, A1.
2. J. Brown, "Conservation Officer Says They Can Do Better Job With More Canine Helpers," (news story) *Northern Life*, 21 February 1993, 1.
3. S. Katz, "Digger The Super Detective," *Readers Digest*, February 1993, 32.
4. See the word Sleuth in: F. C. Mich (Ed.), *Mirriam Webster Book of Word Histories*, Mirriam-Webster, Springfield MA, 1991.
5. S. Kawakishi, Y. Morimitsu, *The Lancet 2* (1988), 330.
6. J. Emsley, "Onions Run Rings Around Chemist," *New Scientist*, 30 September 1989, 32.
7. D. G. R. Ottson, "How We Recognize Odors," *New Scientist*, 15 October 1970, 114.
8. R. Freedman, "How Do You Recognize That Smell?," *New Scientist*, 18 October 1973, 190.
9. J. Barynin, "Measuring Odor Pollution," *New Scientist*, 15 October 1970, 116.
10. S. K. Freeman, "Odor," *International Science and Technology*, September 1967, 70.
11. A. Dravnieks "Odors as Signatures," *New Scientist*, 15 September 1966, 622.
12. W. Thomas, "Canine Detectives," *Imperial Oil Rev.*, Winter 1991, 28.
13. P. Park, "Labradors on the Scent of Pipeline Leaks," *New Scientist*, 23 July, 18.
14. T. Dalyell, "Salter's Chance to Rule the Waves," *New Scientist*, 30 November 1991, 62.
15. C. Puttnam. "Can Police Dogs Really Sniff Out Criminals?," *New Scientist*, 14 September 1991, 24.
16. A. Birchall, "A Whiff of Happiness," *New Scientist*, 25 August 1990, 44.
17. J. Broad, *Science and Criminal Detection*, Macmillan, London, 1988.
18. J. N. Labows, B. Shushan, "Direct Analysis of Food Aromas," *American Laboratory*, March 1983, 56. This article describes the use of the TAGA equipment to look at the odor of fruit and meat samples such as Knackwurst.
19. F. W. Karasek, "Instrumented Van Helps Profile Spread of Toxic Chemicals During Mississauga Train Crash Crisis," *Canadian Research*, April 1980, 30.
20. B. K. Krotoszynski, J. M. Mullaly, A. Dravnieks, "Olfactronic Detection of Narcotics and Other Controlled Drugs," *Police*, January/February 1969.
21. D. St. Pierre "Fire Search Goes to the Dogs," (news story) *Sudbury Star*, June 18 1993, B1.

Chapter 8

Drunken Drivers, Drugged Individuals and Distracted Horses

How much is too much?

The amount of alcohol in your bloodstream is determined by the number of drinks you've had, your weight and the number of hours since your first drink. The following charts help you approximate blood alcohol levels. This is a only rough a guide — numbers will vary from person to person.

1. Count the number of drinks you've had (1 drink equals 1 1/2 oz. spirits, 5 oz. table wine, 3 oz. fortified wine, 12 oz. (1 pint) regular beer.
2. Refer to the blood alcohol chart (chart A). Find the number of drinks and your body weight and then find the corresponding blood alcohol number.
3. Find the number of hours since your first drink (chart B). Subtract the corresponding mg% number (bottom line, chart B) from the blood alcohol number from step 2.

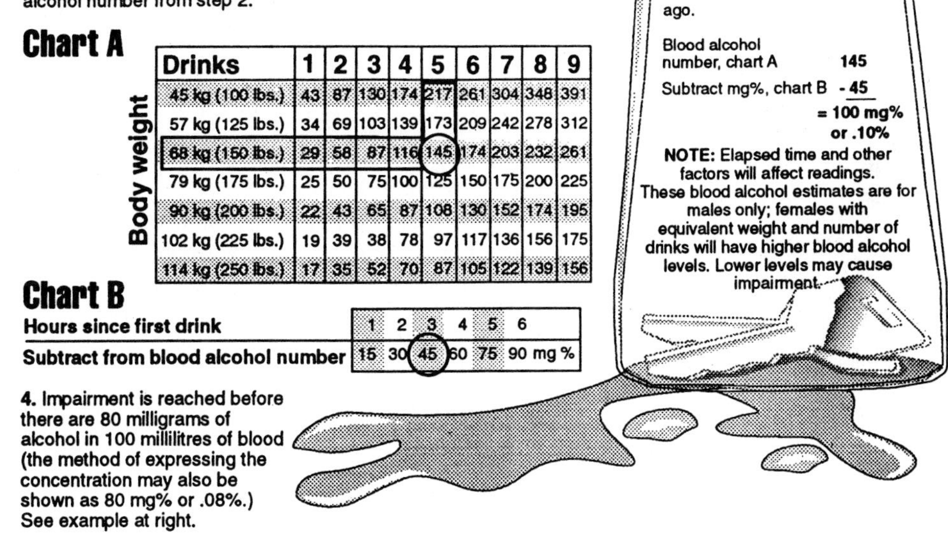

Chart A

Body weight / Drinks	1	2	3	4	5	6	7	8	9
45 kg (100 lbs.)	43	87	130	174	217	261	304	348	391
57 kg (125 lbs.)	34	69	103	139	173	209	242	278	312
68 kg (150 lbs.)	29	58	87	116	145	174	203	232	261
79 kg (175 lbs.)	25	50	75	100	125	150	175	200	225
90 kg (200 lbs.)	22	43	65	87	108	130	152	174	195
102 kg (225 lbs.)	19	39	38	78	97	117	136	156	175
114 kg (250 lbs.)	17	35	52	70	87	105	122	139	156

Chart B

Hours since first drink	1	2	3	4	5	6
Subtract from blood alcohol number	15	30	45	60	75	90 mg %

4. Impairment is reached before there are 80 milligrams of alcohol in 100 millilitres of blood (the method of expressing the concentration may also be shown as 80 mg% or .08%.) See example at right.

Chapter 8

Drunken Drivers, Drugged Individuals and Distracted Horses

8.1 Alcohol Abuse

The reader may be surprised to discover alcohol in a chapter which deals with the problem of drugs. Usually, the citizen thinks of alcohol as a stimulant, whereas in fact it is a drug. In his book on forensic science H. G. Walls states:

> *Contrary to popular belief, it is not a stimulant; it is a narcotic...The apparently stimulating affect of drink is due to the fact that drinking is usually done in stimulating company and circumstances and that alcohol effects first those parts of the brain which govern the inhibitions. Its slightly anesthetic actions makes the drinker less aware of depression or fatigue. Hence, the euphoria felt after a drink or two. The fine edge of neuromuscular coordination is blunted; hence the slurred consonants and, at a more advanced stage, the staggering gate. At still higher blood concentrations, the respiratory center is affected; hence the stertorous breathing associated with "sleeping it off," and hence eventually, death from respiratory paralysis caused by acute alcoholic intoxication.*[1]

The major interest of the forensic scientist in the effects of alcohol on individuals arises from the fact that many road accidents involving death and destruction are facilitated by drivers that have taken alcohol. Perhaps one of the most unfortunate aspects of taking alcohol is that it decreases motor skills such as those required in the driving of a car at the same time as it increases confidence – a combination of effects that often leads to disaster. Unfortunately, many individuals are unaware of the real physical and psychological effects of drinking and boast about being better drivers when they have had a few drinks.

Nickols presents a chart showing how, as the level of alcohol in the blood increases, the chances of being involved in a road accident also increase. Thus, a person who has a hundred milligrams per hundred milliliters of alcohol in his blood is seven times more likely to be involved in a road accident than someone who is sober.[2] Professor Cohen of Manchester University in Great Britain, has made a scientific study of drivers' willingness to take risks after drinking alcohol.

In his experiment Professor Cohen used as test subjects a panel of 59 skilled bus drivers. The control group received no alcohol; the others were given either small or moderate doses of whiskey, producing blood alcohol levels up to nearly 60 milligrams per hundred milliliters (60 mg/100 ml). They were then asked to estimate the minimum width of a gap, as marked by a moveable pair of vertical posts, through which they thought they could just drive their buses. The results showed that, even at these relatively low blood-alcohol levels, the drivers were more rash than when quite sober and some were ready to attempt with apparent confidence to drive their buses through gaps that were in fact narrower than the vehicles.[1]

Walls points out that detectable deterioration in the ability to drive a car begins to show at 30–50 mg/100 ml and that obvious effects begin to appear at blood levels of alcohol in the range of 60–80 mg/100 ml. He states, however, that the non drinker or occasional driver will probably begin to feel and show the effects of alcohol at blood concentrations lower than those given above. On the other hand, the regular or problem drinker may well be able to carry around 150–200 mg/100 ml without apparent discomfort. There have been numerous cases recorded of true alcoholics who still appeared passably sober at 300 mg/100 ml. In the words of Walls:

> *any concentration above 400 milligrams per hundred milliliters is very dangerous and could be fatal.*[1]

The actual level of alcohol permitted in a driver's blood varies from country to country. Currently in Canada and England, it is 80 mg/100 ml, but certainly in England there is a move to lower the level to 50 mg/100 ml, which is the level enforced in Sweden. It is not appropriate in this textbook to discuss the level at which these figures are set. Our mandate is to look at the way in which scientists measure the amount of alcohol in the bloodstream, so that the various regulations in different countries can be enforced by the appropriate authorities.

To scientists, the word alcohol means a group of chemicals with the structure OH (oxygen, hydrogen) at the end of a chain of atoms. The component of liquor and beer, which gives it its strength is a chemical known as **ethyl alcohol** or ethanol. A related compound, **methyl alcohol** is a potent poison that can cause blindness and other major problems for anyone who ingests it. In North America, illegally produced liquor is described as moonshine. Moonshine liquor, unfortunately, can sometimes contain methyl alcohol, because amateurs making the brew do not fully understand the chemistry of producing alcohol. Chemically, ethyl alcohol is a liquid which mixes freely with water, and the blood of the body is 70 % water. Alcohol taken into the mouth is transported rapidly into the bloodstream. This means that ethyl alcohol can be detected in the bloodstream within a couple of minutes of taking a drink. When the distribution of alcohol in the body has reached an equilibrium, the alcohol content of any tissue fluid, blood, lymph, cerebral spinal fluid, saliva, gastric juice or urine will be proportional to its water content.

In Figure 8.1 the type of information given out by the local police force on what is likely to be a blood content of alcohol after a number of drinks is given. It should be noted that a German scientist, Otis, who carried out a detailed statistical analysis of the available data, concluded that the probability of accidents starts to increase with any blood alcohol concentration above zero and, thereafter, increases exponentially.[1] One of the important points to appreciate when discussing how the body deals with alcohol is to realize that the liver processes alcohol at a steady rate and does not increase its speed of

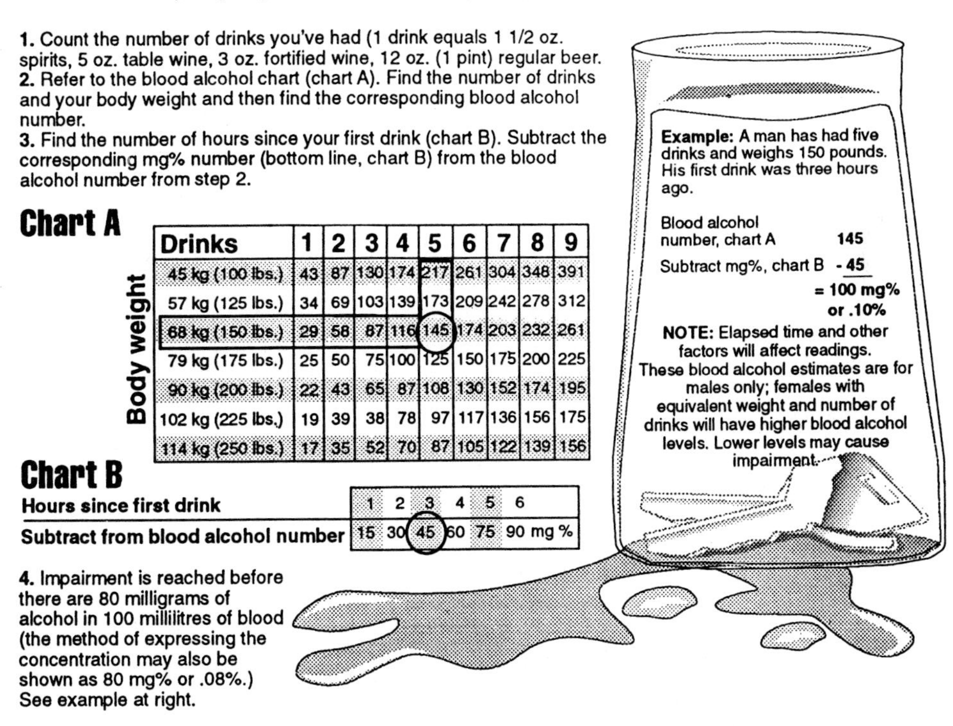

Drinks	1	2	3	4	5	6	7	8	9
45 kg (100 lbs.)	43	87	130	174	217	261	304	348	391
57 kg (125 lbs.)	34	69	103	139	173	209	242	278	312
68 kg (150 lbs.)	29	58	87	116	145	174	203	232	261
79 kg (175 lbs.)	25	50	75	100	125	150	175	200	225
90 kg (200 lbs.)	22	43	65	87	108	130	152	174	195
102 kg (225 lbs.)	19	39	38	78	97	117	136	156	175
114 kg (250 lbs.)	17	35	52	70	87	105	122	139	156

Hours since first drink	1	2	3	4	5	6
Subtract from blood alcohol number	15	30	45	60	75	90 mg %

Figure 8.1. The liver gets rid of alcohol (C_2H_5OH) from the body at a steady rate. The alcohol remaining in the bloodstream depends upon the number of drinks, the body weight of the drinker and the period over which the drinking occurred.

processing if the amount of alcohol in the body increases. This fact is shown by the data summarized in Figure 8.2 (a). The fact that the decay lines defining alcohol levels in the blood over a period of time are parallel, irrespective of the starting level of the blood alcohol concentration, indicates that the rate of elimination is independent of the actual alcohol level in the blood.

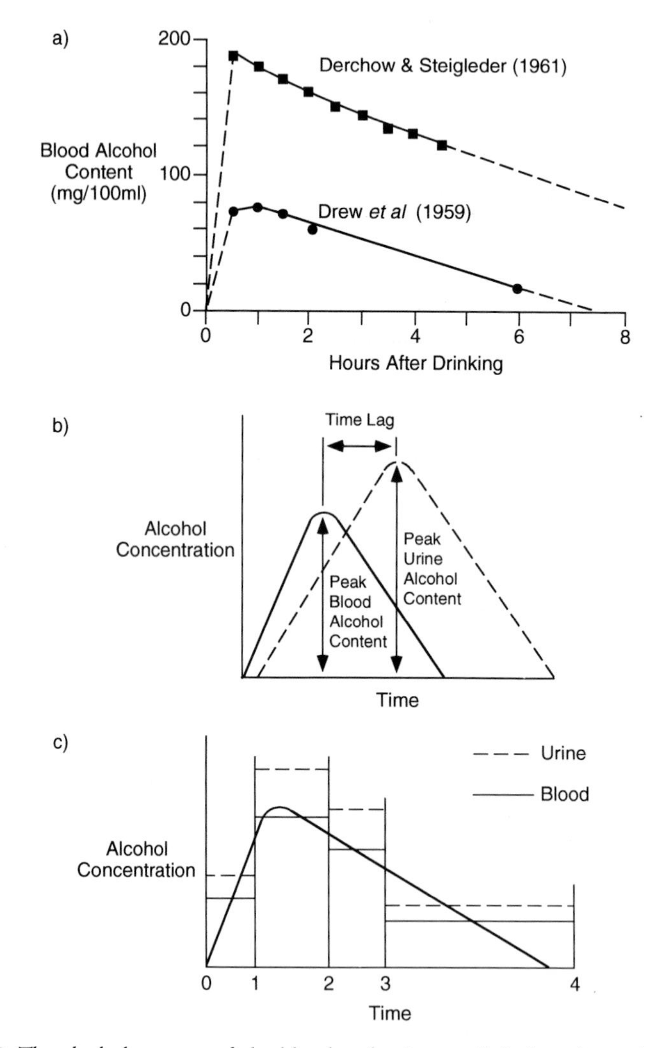

Figure 8.2. The alcohol content of the blood and urine are linked as shown by the data summarized in this set of diagrams.[1] a) Two blood alcohol curves for large and small amounts of alcohol. b) Relationship between blood and urine alcohol levels. The peak urine alcohol concentration is higher than that in the blood but is reached later. c) Urine alcohol levels can be taken as often as the bladder is emptied. Samples taken at times 1–4 show that the alcohol level in urine is 4/3 of that in the bloodstream.

Because the methods of testing for alcohol vary from country to country, we will restrict the discussion given here to general principles of testing, rather than to any one particular instrument. Readers in Ontario can read about the rules and regulations with regard to taking samples from motorists suspected of having blood alcohol levels above the legal limit, and the testing procedure to be followed to ensure the legal acceptability of the evidence, in a booklet prepared by the Center for Forensic Sciences in Toronto.[3]

8.2 Testing for Alcohol

The physical basis for testing the breath of the driver for alcohol content is described in by H. G. Walls in his book on forensic science.[1]

> *Alcohol evaporates from the blood passing through the lungs into the air in the tiny sacks of the lungs called* ***alveoli****. The concentration of alcohol vapor in the alveolus depends upon the vapor pressure of alcohol at the temperature of the body and its concentration in the blood.*

The most commonly accepted ratio of the alcohol level in the blood related to the alcohol level in the alveoli is 2,100 to 1. That is, one unit volume of blood contains the same amount of alcohol as 2,100 unit volumes of air at the temperature of the body. Translated into the limit of 80 milligrams of alcohol in a hundred milliliters of blood (the legal limit in many countries at the current time), the air level is circa 35 micrograms per hundred milliliters of breath (35 mg/100 ml). Again in the words of Walls:

> *The figure found when investigating the breath will be too low if the breath tested is not true alveolar air, that is deep breath from the end of a strong expiration of breath. It will be much too high if any alcohol remains in the mouth from recent drinking.*[1]

In one type of breath analysis system the driver blows into a mouth piece, which contains yellow crystals of acidified potassium dichromate. The alcohol in the breath changes the dichromate to a green chromium sulfate. Using another device, a fixed volume of deep breath is passed into a standardized glass container of diluted chromic acid. The alcohol produces a color change which is measured using photocells and electronics. In a variation of this principle, the chromic acid to be mixed with the alcohol vapor in the breath is supported on a column of solid inert granules. When a fixed volume of breath containing alcohol is passed through it, the column changes color and the length of the column which changes color can be related to the alcohol vapor concentration.[1,4]

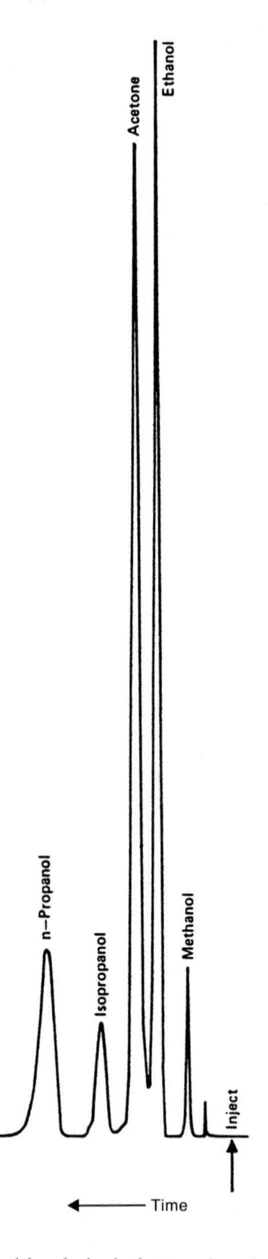

Figure 8.3. A routine chromatogram for blood alcohol. Running time 3 minutes.[5]

When testing blood for alcohol content, the vapor from a blood sample is passed through a **gas chromatograph**, and the resultant chromatogram output shows the various chemicals in the blood, along with their concentrations. Thus, in Figure 8.3, the chromatogram for a test carried out on a blood sample is shown.[5] Gas chromatographic equipment of the type used to generate the data shown in Figure 8.3 is calibrated using known amounts of pure ethyl alcohol. Urine samples are also tested by the same basic technique.

Pure ethyl alcohol is widely used in industry as both a raw material and a solvent. When I was a student in Great Britain, the use of ethanol in teaching laboratories was closely controlled. We had to sign a register for every spoonful of alcohol that we used in our experiments. However, inevitably some of it found its way into illegal use, and a co-student of mine used to make a very potent rhubarb wine laced with pure alcohol.

Manufacturers of illegal alcohol, known by the colorful name of bootleggers, have been known to high-jack tankers of pure ethanol to enrich their illegal liquor. (Apparently the term **bootlegger** comes from the fact that some of the earlier smugglers of spirits used to wear high boots to hide the containers of alcohol built into the leg of the boot.) An analytical technique, which can be used to measure alcohol in the blood, or any other liquid, is infrared spectrometry. It was used to convict some bootleggers who had stolen a cargo of ethanol. The empty truck, which had once contained the ethanol, was found abandoned on a back road. A little later, a bootlegging operation was raided in another part of the state and a rather large quantity of 95 % pure ethanol was discovered. The stolen ethanol was from a solvent recovery process used by a pharmaceutical manufacturer. It was being sent from the recovery unit to another nearby factory. It so happened that this ethanol, was contaminated with 0.4 % methyl ethyl ketone, a very unusual contaminant for ethanol. The scientists at Perkin-Elmer were asked to investigate the case. In Figure 8.4 the infrared absorption of methyl ethyl ketone, ethanol, and the ethanol mixed with methyl ethyl ketone are shown. In the words of the report on the investigation given in the Perkin-Elmer literature:

> *Comparison of the spectrum of the bootleggers alcohol with that of the batch from which the stolen alcohol was taken showed a ketone absorption in each of the same intensity. These two batches were identical in every other way and with this evidence the state police were certain of the source of the ethanol in the bootlegger's storage tank.*[6]

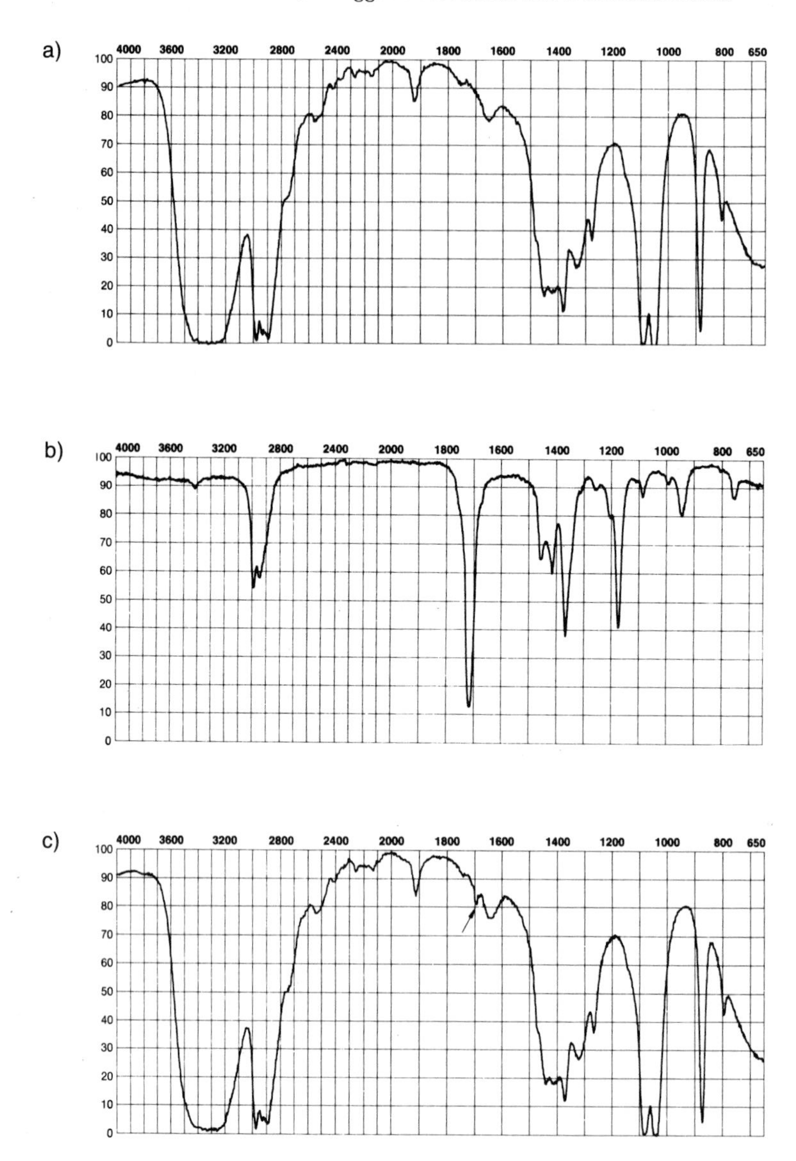

Figure 8.4. Infrared absorption spectra can be used to study the origin of ethanol and liquid mixtures containing ethanol.[6] (Reproduced by permission of the Perkin-Elmer Corporation.) a) Infrared absorption spectrum of commercial quality ethanol. b) Infrared absorption spectrum of methyl ethyl ketone. c) 95 % ethanol with 0.4 % methyl ethyl ketone found in bootleggers' vat and the solvent recovery unit. Note the small peak indicating the presence of a small quantity of methyl ethyl ketone.

8.3 Detecting and Analyzing Drugs

The word drug is used very loosely in everyday speech. A medical dictionary lists four different definitions of a drug. For our purposes, the appropriate definition is:

> *a substance other than food intended to affect the structure or function of the body.*[7]

Chemistry textbooks will describe drugs as:

> *substances not naturally associated with cell metabolism that have a profound effect on the metabolism, especially of the brain and nervous system. Common types of drugs include stimulants, pain killers, tranquillizers, antibiotics, anesthetics, hallucinogens, and nerve poisons. Just how these substances produce their effect is well-known for some and completely unknown for others.*

Testing for the presence of drugs in the body of the person taking the drugs is an important aspect of forensic work and could become a very important topic, if public employees are subject to mandatory testing for drug use because of public concern for safety. Whole books have been written on the subject of drug testing, and we will look at some of the highlights of the forensic involvement with drug testing. In a very readable account of the problems faced by law enforcement officers with regard to illegal drugs, the following story is told:

> *Two undercover Chicago police officers banged loudly on the scarred door of a seedy low-rental apartment in a run down section of the windy city. The pusher, suspicious of the insistent pounding, rushes to flush away the evidence. The police are armed with a search warrant and know that possession must be proved to make the search and arrest valid. They rush the door and catch the dealer flushing more than a pound of heroin, along with their case, into the Chicago sewers. Moving quickly, the policemen rip the porcelain from its floor fixtures and pour the liquid from the bowl into a gallon gas jar. "Take this to forensics they've got some newfangled equipment to find that heroin" demands one of the officers as the other turns to the suspected pusher and states "You're under arrest, You have the right..."*[8]

The "newfangled equipment" referred to was the instrument that we already mentioned, which is a combination of the gas chromatograph with a mass spectrometer (GC/MS). In Figure 8.5 the data from a study of both a standard sample of heroin and the material present in the water trap of the toilet, as analyzed by the Chicago police, is shown. The article by Boese on the use of this equipment to study drugs contains useful information on the

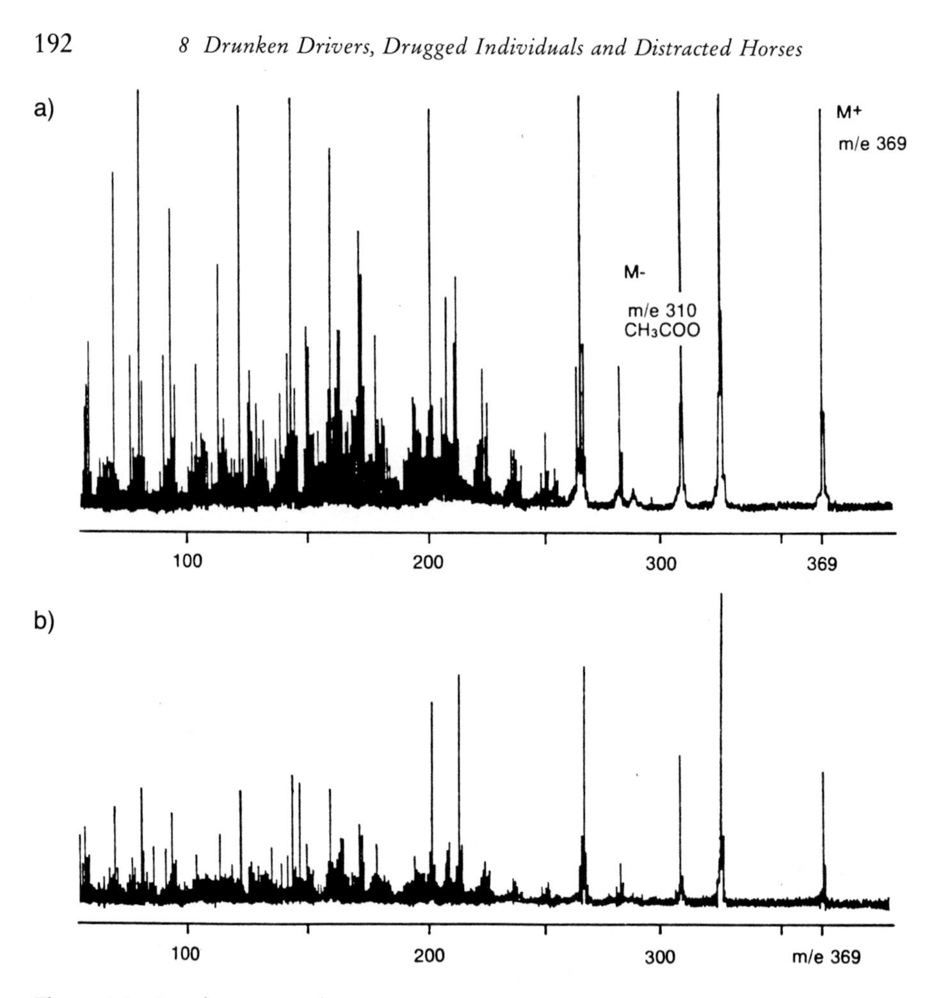

Figure 8.5. Gas chromatography/mass spectrometry(GC/MS) is a sensitive analytical technique for detecting and characterizing drugs.[8] a) spectrum of a standard sample of heroin. b) spectrum of water taken from the plumbing trap of a suspect's toilet.

problems faced by police when trying to enforce the appropriate laws on drug possession and sales.[8]

Heroin is defined in a medical dictionary as a strong physiologically addictive narcotic of the formula $C_{21}H_{23}NO_5$. It is made by acetylation of morphine, but it is more potent than morphine. It is prohibited for medical use in the United States, but is used illicitly for its euphoric effect.[7]

Morphine and heroin are derived from opium. **Opium** is described as the dried narcotic juice of a Eurasian (those that grow in Europe and Asia) poppy. To collect the opium, the grower cuts the seed pod of the poppy and the white juice that oozes out of the cut when dried, constitutes the opium. (The word opium comes from a Greek word meaning the sap or juice of a plant). **Morphine,** the compound used as a pain killer, is derived from opium. It's chemical formula is shown in Figure 8.6. Morphine takes it's name from Morpheus the Greek god of dreams. **Codeine** is, in turn, derived from morphine. The word codeine means "poppy head." If you examine the two formulas for these substances given in Figure 8.6, you will notice that one OH group has been removed from the morphine and substituted with the chemical group OCH_3. The status of codeine as a drug varies from country to country. Thus, in the United States, codeine is a controlled drug and is not available over the counter. It is, however, frequently used in cough mixtures. In Britain and Canada codeine is available over the counter but even then the pharmacist watches out for people who purchase too much.

In the mid 1960s when I was living in Chicago, I missed being able to use codeine-based headache tablets. When I visited England I went into the local pharmacy and tried to buy three bottles of these tablets. The pharmacist at first refused to sell me three bottles and looked at me with some suspicion as a possible addict to codeine. I produced my U.S. citizenship card and managed to persuade her that I was a visitor stocking up for future head-

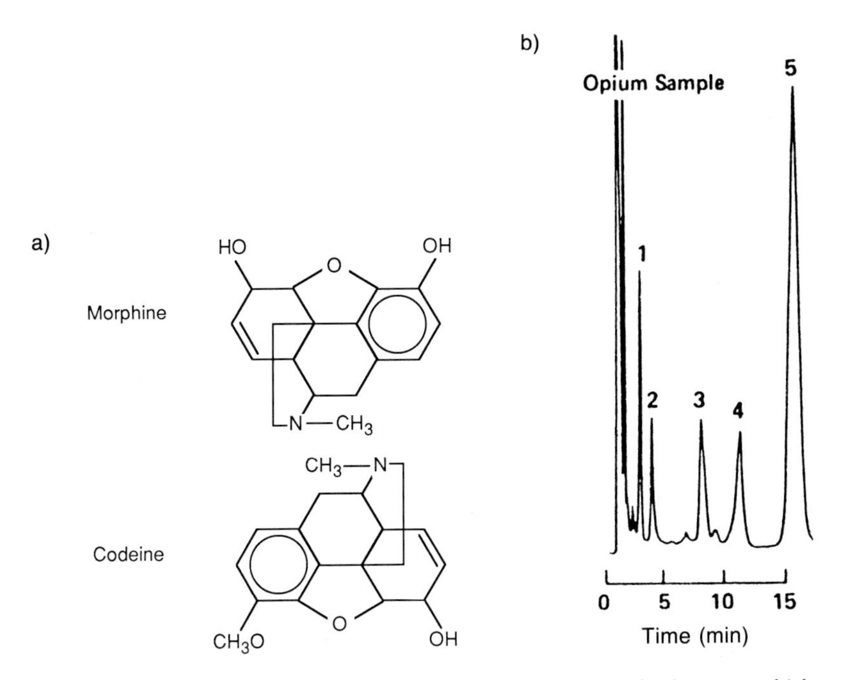

Figure 8.6. Drugs used by addicts are often complex mixtures of substances which can be studied by chromatographic techniques. a) Chemical structure of Morphine and Codeine. b) Chromatogram of an opium sample. (The peaks of the graph are: 1, morphine; 2, codeine; 3, thebaine; 4, narcotine; 5, papaverine) (Reproduced by permission of *American Laboratory.*)

aches. Also, I realized the possibility that if I had my luggage searched on the way back into the States, with this relatively large quantity of codeine-based headache tablets, this could have created problems with the customs officials. The fact that codeine is a controlled drug in the United States should always be kept in mind by travellers journeying there. The chemist describes compounds such as morphine and codeine as **alkaloids**. The medical dictionaries describe alkaloids as:

> *any of numerous, usually colorless, complex and bitter organic bases containing nitrogen and usually oxygen that occur especially in seed plants.*[7]

In chemistry a base is the opposite to an acid and is described as a alkaline compound. When we discuss the subject of poisons in the next chapter, we will discover that many plant alkaloids are extremely poisonous.

Because opium is a raw product from the juice of a poppy head, opium samples can contain many different compounds and alkaloids. Figure 8.6(b) shows the analysis of an opium sample, carried out using liquid chromatography. This is the usual format for reporting this type of analysis.[9]

Many drug substances give off enough vapor for them to be studied using odor analysis equipment of the type discussed in Chapter 7. Thus in Figure 8.7, the olfactronic signatures of samples of heroin, marijuana and LSD are shown.[10] It is because the drugs give off such odors that tracker dogs can be used to discover the presence of drugs at airports and in homes of the suspected addicts. It should be noticed that in Figure 8.7(a) it is stated that the heroin sample is 84 % pure. This qualifying statement is needed because most samples of heroin, as they are sold to drug addicts on the street, are adulterated with many substances to dilute the heroin. Thus, in the technical literature on drug analysis prepared by the Waters Corporation, the following statement is made:

> *Street heroin samples contain a variety of related compounds including monoacetylmorphine, morphine, acetylcodeine, opium alkaloids, and dilutants such as quinine, procaine, methapyriline and various sugars. A separation and identification of as many constituencies as possible is the goal in the fingerprinting of drug samples for comparison and identification purposes.*[11,12]

The various components of a sample of street heroin can be analyzed using liquid chromatography techniques, as illustrated by the data shown in Figure 8.8 and reported by Loveland and Williams.[13] An arrested drug dealer had samples of heroin, which were analyzed by capillary column gas chromatography. The results of this test are shown in Figure 8.9(a). In capillary column gas chromatography, narrow capillary columns (0.2 to 0.4 millimeters in diameter and 10 to 50 meters long) made from flexible fused silica

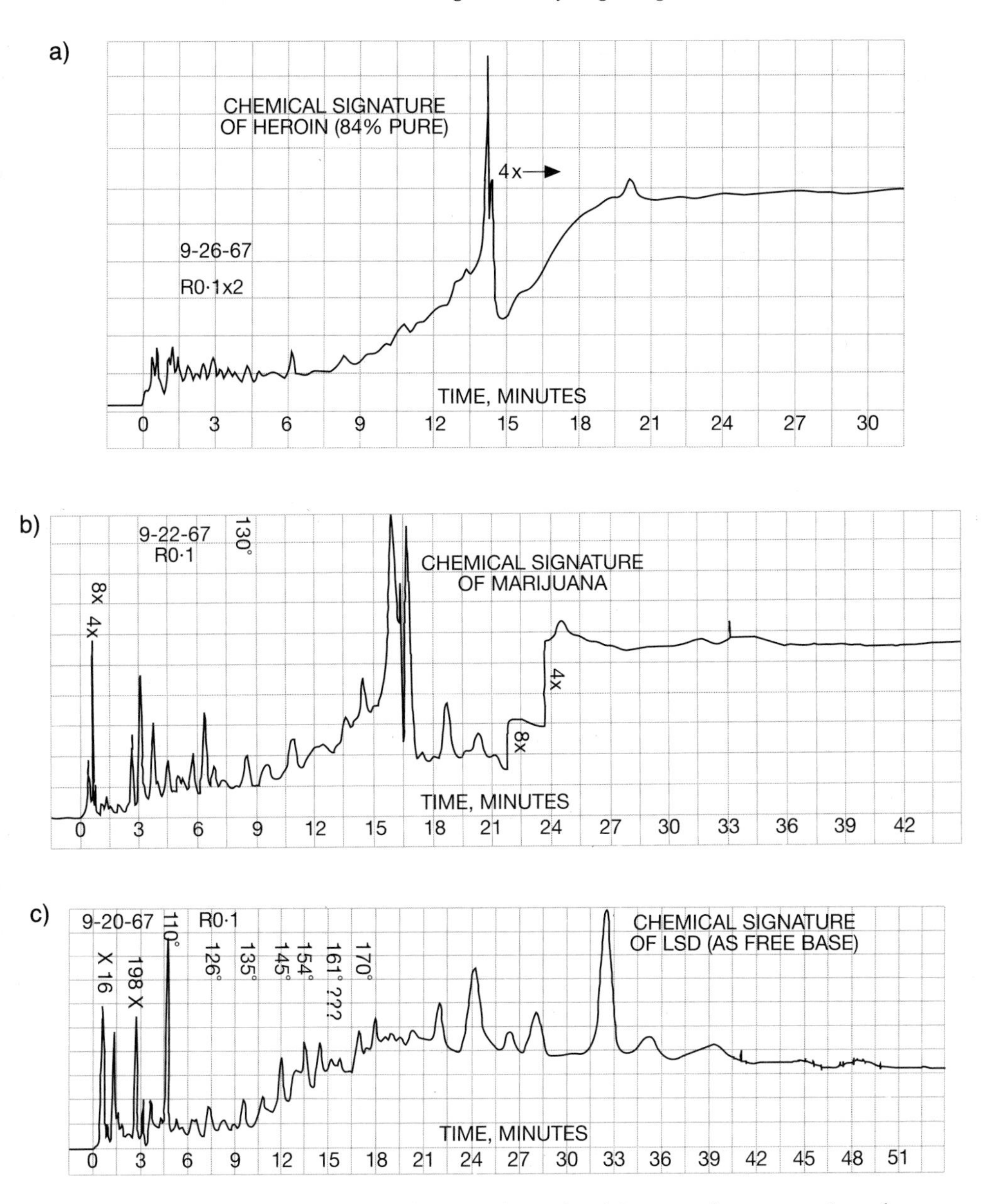

Figure 8.7. The vapor given off by a drug can be analyzed in a gas chromatography column to record the olfactronic signature of the drug.[10]

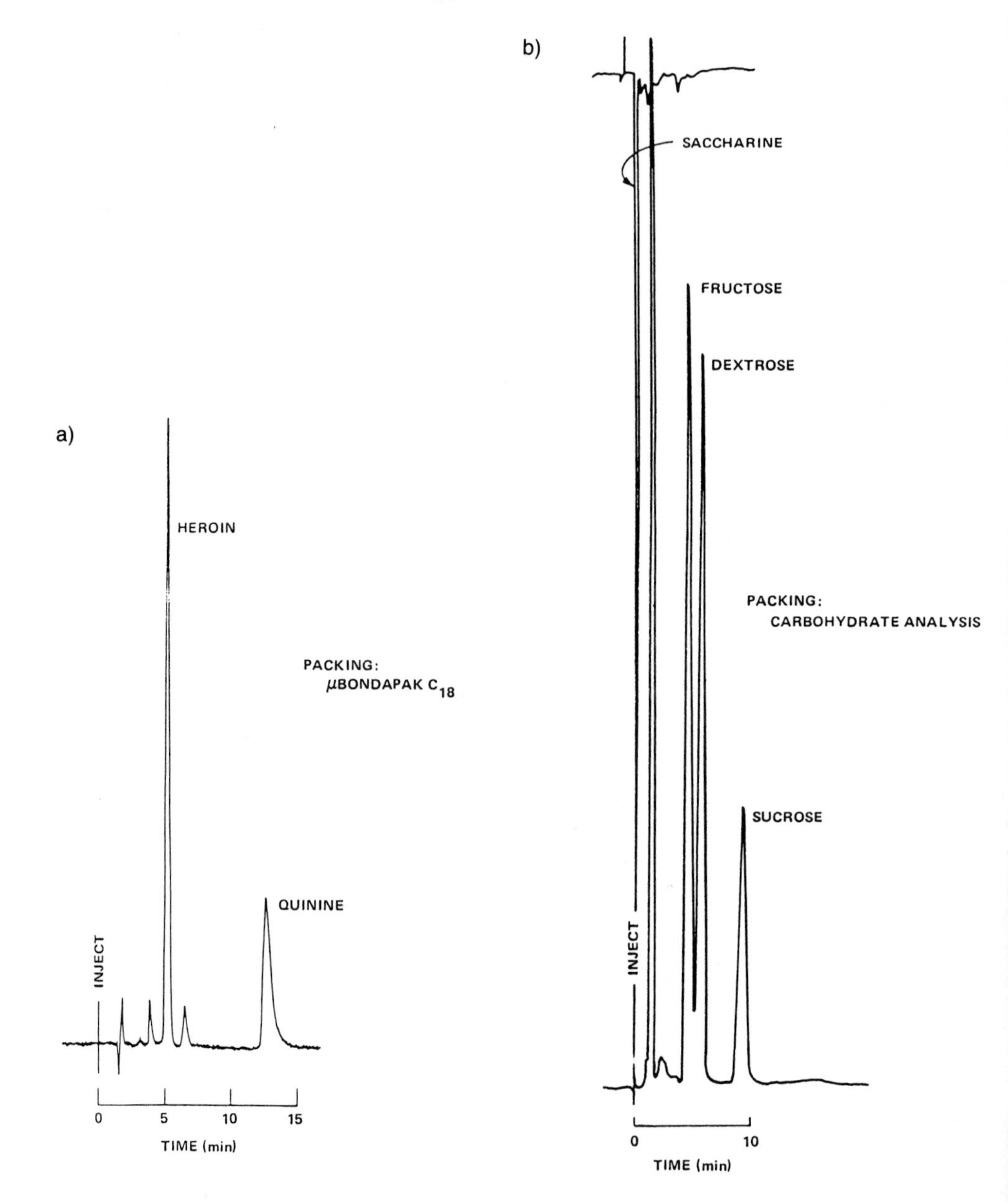

Figure 8.8. Identifying substances used to dilute heroin sold in the street is an important aspect of forensic work. (P. J. Twitchett, *J. Chromatography 104* (1975), used by permission of the Journal of Chromatography.) a) Liquid chromatography analysis of a sample of street heroin. b) Sugars are often added to street heroin to dilute the sample. Shown here is the chromatogram for a mixture of several sweeteners.

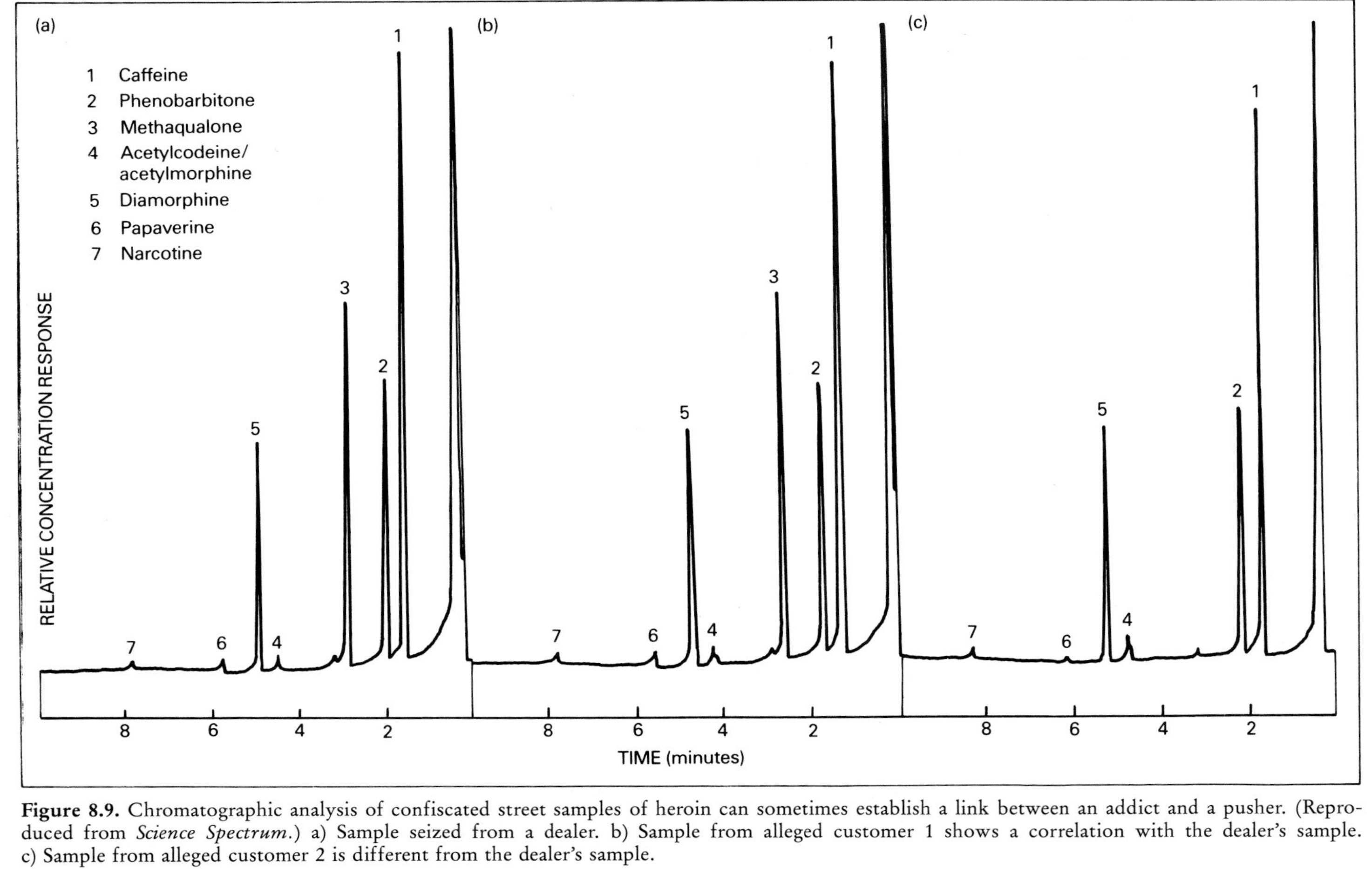

Figure 8.9. Chromatographic analysis of confiscated street samples of heroin can sometimes establish a link between an addict and a pusher. (Reproduced from *Science Spectrum.*) a) Sample seized from a dealer. b) Sample from alleged customer 1 shows a correlation with the dealer's sample. c) Sample from alleged customer 2 is different from the dealer's sample.

and treated internally with a non-extractable, chemically bonded stationary phase are used to give a highly resolved analysis of a compound. Samples seized from two alleged customers of the dealer were also analyzed. From the chromatograms of Figure 8.9(b) and (c), it can be seen that customer 1 had been dealing with the arrested dealer, whereas apparently, customer 2 had obtained the heroin sample from another source. The data of Figure 8.9 are taken from a review article by Michael Loveland and Ray Williams of the Metropolitan Police Laboratory, London, England.[13] In their article, dealing with the different methods of characterizing drugs, they describe an x-ray diffraction device for studying crystalline drugs. X-ray diffraction is similar to optical diffraction (discussed in Chapter 2), only instead of patterns of lines diffracting light, rows of atoms in a crystal diffract x-rays. In Figure 8.10 the x-ray diffraction system used by Loveland and Williams to study drug samples is shown. Commenting on their technique, they state that the same substance always gives the same pattern and in a mixture each component produces its own pattern independently of the others.

The library of chromatograms of heroin samples seized from many sources throughout the world is held by the laboratory of the government chemist in London, England. This laboratory carries out work on behalf of the British customs, which excises people and deals every year with many cases involving the importation of drugs. Such a library shows that there is a wide variation not only between heroin samples coming from India, Pakistan, and South East Asia, but also between different batches of heroin originating from the same regions within these countries. Such information, also obtainable from drugs such as amphetamines, enables forensic scientists to provide useful evidence on drug trafficking to authorities responsible for enforcing the law.[13]

One of the advantages of using liquid chromatography to look at the constituents of a drug mixture is that, as each peak indicates the arrival of a component at the exit of the chromatography column, the samples can be collected and inspected by other analytical methods. A very useful method for looking at the fractions collected from a chromatography column is infrared absorption spectroscopy. Commenting on this fact, Loveland and Williams state that:

> *the forensics scientists' best method for identifying most controlled drugs quickly and cheaply is the infrared adsorption spectrum characteristic of the drug, as produced after exposing it to a beam of infrared radiation. Confirmation can be achieved by comparing the spectrum of the unknown with spectra of authenticated drug samples in the same way that one would match a fingerprint. The task of confirmation is much easier nowadays because most modern infrared spectrometers are linked to microcomputers, which can be programmed to carry out the recognition task. The computer can also help with the identification of mixtures by subtracting from the*

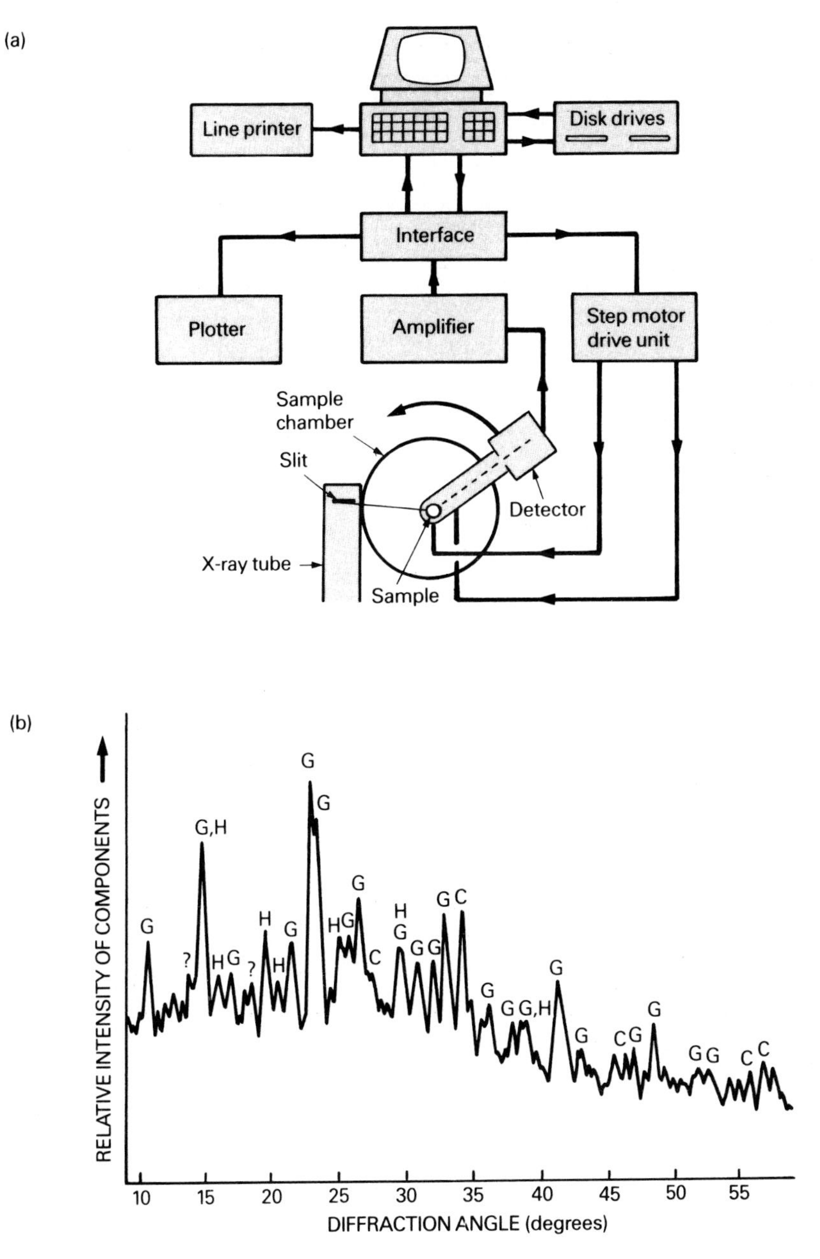

Figure 8.10. X-ray diffractometers can be used to identify drugs in confiscated material. (Reproduced from *Science Spectrum*.[13] a) Block diagram of an x-ray microdiffractometer system used to study crystalline drugs. b) Microdiffractometer plot of a typical street heroin sample. (On the graph, letters represent: H, heroin base (diamorphine); G, glucose monohydrate; C, calcium carbonate; ?, unidentified).

overall spectrum those portions due to an excipient in a powder, for example a sugar.[13]

The term **excipient** used in the above quotation is a term used in the pharmaceutical industry to describe something added to a mixture to make it a bulkier. Lactose (a milk sugar) is often used as an excipient to make tablets larger and easier to handle. In Figure 8.11, some infrared absorption spectrometers used in drug studies discussed by Swineheart and Gore are shown.[6] The first spectrum is that of a compound of heroin, of which only 1.1 micrograms was available for the investigation. The spectra of Figure 8.11(b) show the study of a sample of LSD extracted from a sugar cube, which was intended as a delivery system for the drug. The fact that the infrared spectrum of the LSD does not show any peaks associated with sucrose (see Figure 8.11 (b)(ii)), shows that the drug had been successfully separated from the sugar cube.

If one looks up the word **cannabis** in a medical dictionary, one finds two definitions. The first is a description of the plant family – a family of tall, rough annual herbs with waxed stems, leaves with three to seven leaflets, and pistol like flowers in spikes. The second definition is:

any of the preparations (as marijuana or hashish) of chemicals that are derived from the hemp plant and are psychoactive.

In many parts of the world smoking marijuana is not an offense, but it is a prohibited substance in North America and Great Britain. There is a story about a retired officer from the Indian army, living in a British town. Apparently he was walking down the street and was surprised to see the plant from which marijuana is derived growing in flower bed down the boulevard separating two streams of traffic. He found out that some of the gardeners, originally from India, had developed a profitable sideline associated with the job of growing flowers. The word **hashish** is the Arabic word used to describe the hemp plant. The leaves from the plant can be smoked or digested in various forms as an intoxicant. In the dictionary, we can discover that the word **assassin** was originally a word used to describe the fanatical Muslim sect who sanctioned the murder of prominent adversaries. It now means one who murders another either for hire or from fanatical motives. The word assassin is derived from the Arabic word meaning hashish eaters because the assassins drugged themselves with hashish in preparation for their crimes. Loveland and Williams comment that:

cannabis can be identified using relatively simple chemical tests and a microscope to confirm the presence of botanical features characteristics of the plant.[13]

Cannabis samples often contain contaminants such as fragments of an insect, which an expert biologist can identify as coming from a certain region and thus identify the source of the cannabis. Thin-layer chromatography can also be used to identify cannabis, along with many other drugs.[14]

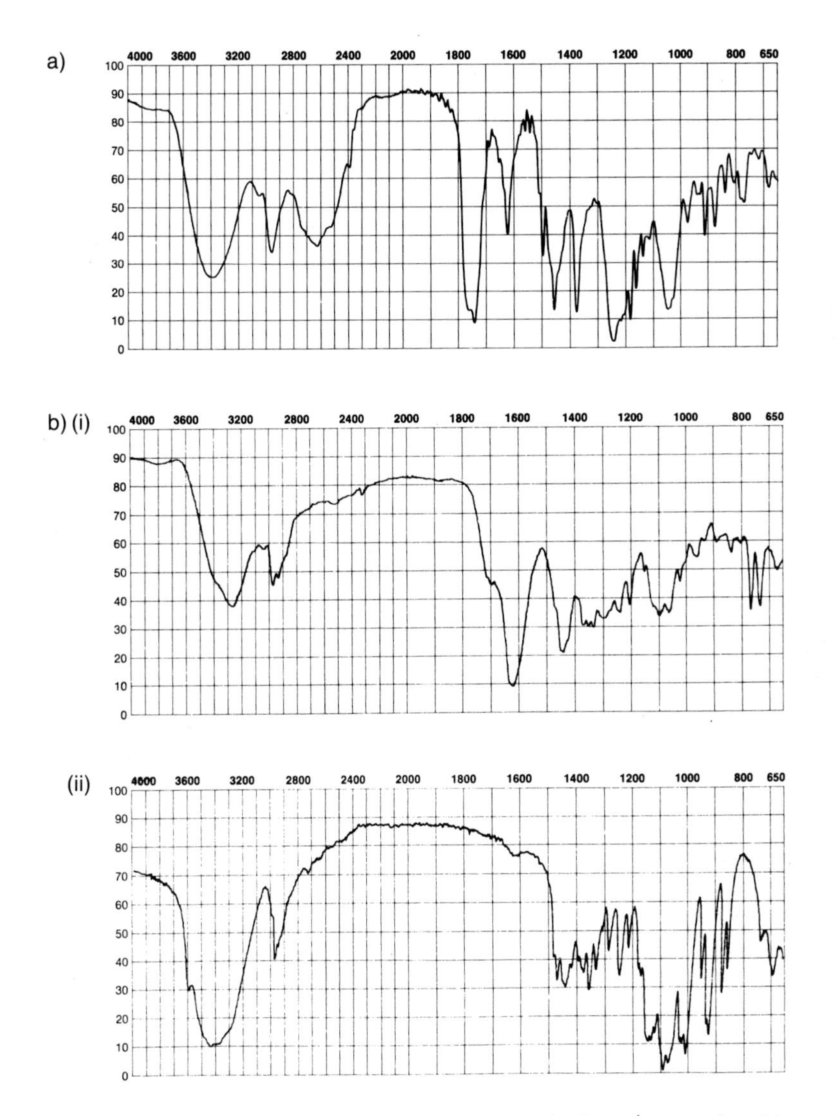

Figure 8.11. Infrared absorption spectra can be used to study drugs.[6] (Reproduced by permission of the Perkin-Elmer Corporation.) a) Infrared absorption spectrum of heroin hydrochloride. b) (i) Infrared absorption spectrum of lysergic acid diethylamide (LSD) (80 mg). (ii) Infrared absorption spectrum of sucrose.

8.4 Tell-Tale Vestiges of Drug Usage in Hair

In a news report describing a new technique for studying the hair of people suspected of using drugs, it is said that the test detects drugs such as cocaine, marijuana, and heroin, all of which leave the bloodstream quickly but deposit tell-tale by-products in a person's hair, which remain indefinitely. It is written that:

> *Hair testing can reveal even modest use such as a weekly diet of one line of cocaine or two marijuana cigarettes, as claims Charles Black, Vice President of a Company called Psychemedics.*[15]

Human hair grows about 1.25 centimeters a month. This means that shoulder-length strands might betray a history of drug consumption over the past two years. To be able to understand several related techniques for studying human hair for past drug use, it is necessary to understand a process, known as **immunoassay**. To understand this process we start with a definition of the **immune system** of the body. The term immune was originally a legal term used to describe Roman citizens who did not have to perform a given service for the state. It comes from two root words: *im* meaning not and *munus* meaning service. In medical science, it came to be used to describe someone who could not be affected by a disease and later for the immune system responsible for combatting disease in humans. Thus, specialists talk about the immune system of a person being either strong or weak. In a procedure known as immunoassay, the immunologist seeks "to identify a substance (such as a protein) through its capacity to act as an antigen." The term **antigen** is used in medical circles to describe any substance in the body which stimulates the immune system. In non-technical language, an antigen can be regarded as a foreign or invading object finding its way into the body. When it is aware of the antigen in the body, the immune system produces what is known as an **antibody** (a defender molecule) which attacks the foreign antigen. In their discussion of immunoassay techniques in the forensic science of drug identification, Loveland and Williams have made the following comments:

> *To raise (create) antibodies to relatively small molecules such as cocaine, heroin, amphetamine, cannabis, it is necessary to couple the drug molecule or its metabolite chemically, usually to a protein or polysaccharide carrier with a molecular weight greater than a thousand. This combination (the antigen) is then injected into an animal such as a rabbit.*

In the bloodstream of the animal, the antigens stimulate the production of antibodies. After a sufficient time has elapsed, blood is taken from the animal and the antibodies are separated from the blood and stored ready for use in an immunoassay. The next step in the process is to create what is known as a solution of labelled antigens. A known amount of the antigen is reacted with a radioactive isotope such as iodine 125. This step is taken so that the concentration of the antigen in a solution can be measured at very low concentrations using a radioactive measurement instrumentation. When using such labelled antigens, the technique is known as **Radio-Immune Assay** (**RIA** for short).

In the assessment of the presence of a drug in a sample of urine, the first step is to mix a known amount of the antibodies with a known excess of antigen labelled molecules. The antibodies bind onto the antigen to form complexes, but there are labelled antigen molecules left over. The conditions of the assay are usually arranged so that after this mixing process, 50 % of the antigen molecules are left free of antibodies. Broad describes the next step in the assay as follows:

> *Into this mix is now added a solution (urine) containing an unknown amount of antigen whose molecules are not labelled. The aim is to find out how much unlabeled antigen there is in the urine. The antigen could be a target drug such as cannabis. When the unlabeled molecules enter the solution, there is competition for the antigen binding sites on the surface of the antibodies. The newly arrived unlabeled molecules displace the labelled ones. As would be expected from an orderly nature, this displacement is in direct proportion to the concentration of the unlabeled antigen (the drug being evaluated) added. After an incubation period to establish an equilibrium, the mix is centrifuged or subjected to a separation process. The result is two fractions, one containing all the antibodies that are complex to labelled or unlabeled antigens and the other containing all other molecules. The amount of labelled antigen in both fractions can be estimated. The smaller the labelling response from the antibody complex fraction, the more unlabeled antigen added in the urine.*[4]

The process has to be calibrated with known substances. Loveland and Williams state the following:

> *Dr. Smith of the Metropolitan Police Laboratory has developed assays which can detect most illicit drugs and their metabolites at levels as low as 10^{-9} grams per milliliter of fresh or treated blood. The amount of blood required for RIA is barely more than 0.05 milliliters and procedures have been developed using reagents to carry out the test costing only a few cents per test.*[13]

Because so many of the methods for detecting drugs in fluids have to be calibrated using known standard material, the standards organizations of many countries provide their forensic scientists with reference standards. Thus Rasberry, Chief Officer of Standard Reference Materials for the National Institute of Standards in Technology in Washington D. C., reported on the availability of reference standards for measuring cocaine in urine in a recent review article.[16] He stated the following:

The National Institute for Standards in Technology has an ongoing program in co-operation with the College of American Pathologists to provide the drug abuse testing community with urine-based reference materials. The first such material was SRM1507, a freeze dried urine pool certified for the principal urine metabolite of marijuana (20 plus or minus 2 ng/ml). Scientists in the organic analytical research division have recently completed work on SRM 1508, a freeze dried human urine pool with certified concentrations of cocaine and its principal urinary metabolite, benzoylecgonine. Each unit of this standard reference material consists of four bottles of freeze dried urine; one bottle for each of the three levels of cocaine, and the metabolite benzoylecgonine, plus one bottle of blank urine. The SRM was certified using two independent methods that agreed within a statistical tolerance. One of the methods was based on a gas chromatography mass-spectrometry procedure and was similar to those used in drug-testing laboratories to compare positive results from preliminary screening analysis. Information values are provided for the methyl ether of ecgonine, another important metabolite of cocaine in each of the three levels.

Now that we have explored the physical principles of immunoassay techniques, we can review some of its uses. The hair to be studied is chopped into segments that correspond to periods of time. As stated earlier, hair grows at about 1.25 centimeters a month so that for very long hair one can actually look at drug consumption over the last two years. The president of the company Psychemedics points out that Insurance companies are interested because hair analysis would give hospitals a record of drug use to provide more effective treatment. Employers who want to test workers in sensitive jobs can also review any possible drug usage over a period of several months. (It has already been reported that one worker was able to regain his job after a hair analysis proved that the previous urine test had given a false positive result.)

This technique has also been used to study hair from the poet John Keats, who was writing at a time when opium was freely available. It was known that he took opium to combat pain. Baungartner obtained a lock of Keat's hair (it was common practice in the time of Keats to give a lock of hair as a momento to a friend or relative). Baungartner was able to show that the hair contained a record of the opium used by Keats.[17]

8.5 Were the Witches of Salem High on LSD?

The formula for the hallucinogenic drug LSD is shown in Figure 8.12(a). In chemistry book discussions of this drug, you will be able to read that:

> *LSD has been the subject of extensive study. Numerous technical papers have been published about it. Its chemical structure and properties are well-known, yet the mechanism by which it produces its effect is not known. The amount of material required to cause hallucinogenic effect is very small – a tiny fraction of milligram being an effective dose of LSD in man.*

A particularly pernicious effect of LSD is that a recurrence of the hallucination derangement of the brain can occur for some time after the last dose has been taken. The hallucination can be triggered by the simple act of drinking a glass of water. When I first started teaching at Laurentian University in the late 1960s, it was no secret that certain students were dropping acid (the slang term used for taking LSD). At least one of those students suffered permanent brain damage and dropped out of University. The witches of Salem, referred to in the title of this section, are famous figures in American history. The alleged witchcraft took place in Salem Village, near Boston, Massachusetts in 1692. In the words of a reviewer of the book "The Devil in Massachusetts:"

> *In the grim winter of 1692, 20 citizens, including the beloved matriarch Rebecca Nurse, a controversial tavern keeper Bridget Bishop and a pipe-smoking female tramp were sent to the gallows accused of consorting with the devil by a group of hysterical young women.*[18]

In an article on the Salem affair, Watson describes the behavior of the girls who accused the older women of witchcraft in the following way:

> *By February (of 1692), something was gripping the young girls of Salem village, something neither doctors nor ministers could define. Betty Paris, the local minister's own daughter, was in a trance, hands frozen in place uttering the most hideous gargles and growls. Prayers did no good. 'Our father which art'set her screaming. Soon her cousin Abigale began crawling around the house, under chairs barking like a dog stomping her feet. 'Their arms, necks and backs were turned this way and that,' another minister wrote, 'so it was impossible for them to do of themselves and beyond the power of any epileptic fits or natural disease to affect.' These hysterical girls went on to accuse these older women of witchcraft resulting in the hangings.*[18]

Various explanations such as psychological factors or devil worship have been put forward to explain the behavior of the young girls. But recently, Mary Kay Matossian has made a good case for the fact that the hysterical girls may well have been suffering from ergot poisoning.[19] **Ergot** is the raw

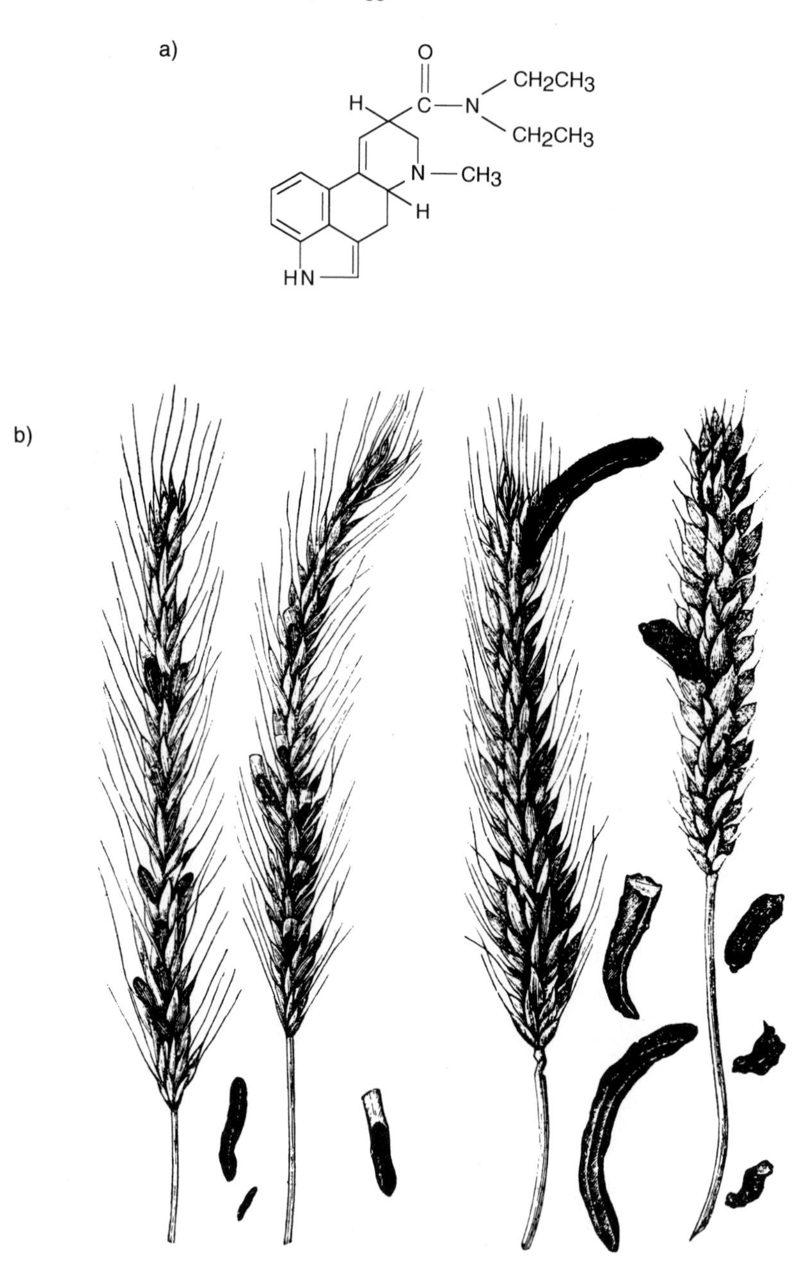

Figure 8.12. The drug LSD is extracted from ergot, a seed-like body produced by a fungus which can grow on cereal crops, especially on rye in a wet summer.[21] a) Chemical formula of LSD. b) Ergot bodies look like grains of the cereal crop on which they grow. (Reproduced from the *American Scientist*.[19])

material from which LSD is derived and several outbreaks of mass hysteria in society reported in medieval times may well have been examples of **ergot poisoning**. In the life cycle of the fungus, clavix purpurea, which usually grows on the grain crop rye, a small seed-like body, called the scerotia (a word which means hard), is developed. This hard little body is called an ergot. Because it is seed like it is very difficult to separate from the actual heads of grain of the plant, as can be seen from the information summarized in Figure 8.12 (b).

The study of fungi is known as **mycology**. This word is derived from the Greek word for a mushroom. The dried sclerotial bodies of an ergot fungus grown on rye contain several alkaloids. Therefore, the term ergot can actually refer to several alkaloids produced by this fungus. In large doses the ergot alkaloids become poisonous and are described technologically as mycotoxins. Medically, the ergot alkaloids are used to treat the uterus after childbirth. In therapeutic doses it induces contractions of the uterine muscle. If taken in overdoes, it can effect the blood circulation system so that gangrene sets in. There have been several historic episodes of ergot poisoning from eating contaminated rye. After reviewing the historic background to the Salem witch episode, Matossian makes a strong case for the fact that although the witches themselves were not high on LSD, there is the possibility that the accusers of the witches were suffering hallucinations after eating rye flour contaminated with ergot alkaloids.[19–21]

8.6 Drugs and Athletic Performance

In recent years there have been several scandals involving the use of drugs by athletes, which can affect their athletic performance. At first sight this does not appear to be a criminal activity. However, individuals who win major sporting events in today's society derive high incomes from appearances on television advertisements to endorse products, such as running shoes and other sporting goods. This type of revenue can amount to several million dollars. Therefore, to use drugs to win an athletic prize is really an act of theft. One of the more widely publicized use of drugs to improve performance, and subsequently ones chance of winning a major sporting competition, is the case of the Canadian Ben Johnson. Johnson was accused of using anabolic steroids to improve performance.[22] The dictionary definition of steroid is:

Any of numerous compounds containing a 17 carbon four ring system and including the steroids of various hormones and glyocides.

Anabolic steroids are described as:

Any of a group of usually synthetic hormones that increase constructive metabolism and are sometimes taken by athletes in training to increase temporarily the size of their muscles.

In an article on "The Perils of Doping In Athletics," Chris Woods makes the following statement:

The reason for the popularity of steroids is simple, they work. The synthetic hormones allow athletes of both sexes to train harder, recover more quickly from injury and build up larger, stronger muscles. However, there is growing evidence that the long-term use of steroids in athletics may have long-term effects such as kidney failure and the stimulation of fatal cancers.[22]

Most of the drug testing of blood samples or urine samples taken from athletes is carried out using either straight forward gas chromatography or a combination of gas chromatography and mass spectroscopy.[23,24]

Drug testing carried out at various athletic events has captured widespread public attention in the last several years. One of the methods used for such testing is gas chromatography. Figure 8.13 shows typical results of a chromatographic analysis of a urine sample. This particular chromatogram shows

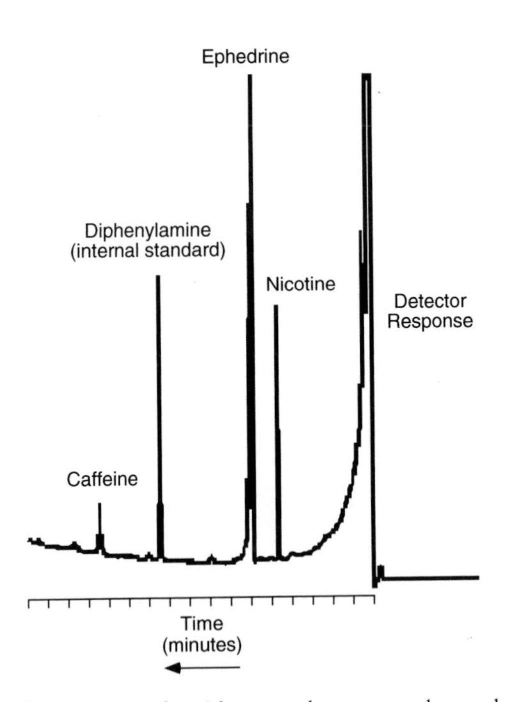

Figure 8.13. Analysis of a urine sample with a gas chromatograph reveals the presence of ephedrine, which dilates respiratory passages to increase oxygen uptake. Note also the presence of nicotine, which may indicate that the person is a smoker, and caffeine, indicating that the person may have consumed coffee recently. (Used with the permission of the Perkin-Elmer Corporation.[6])

the presence of ephedrine, which dilates respiratory passages to increase oxygen uptake. The presence of nicotine and caffeine in the chromatogram also suggests that the person tested had a cigarette and coffee recently. Modern equipment is so sensitive for detecting trace amounts of chemicals that confusion can sometimes arise from the innocent drinking of a fluid that contains natural stimulants. Thus, at the games in Seoul, Korea, Linford Christie, a British athlete, was accused of taking **pseudoephedrine** to improve his performance on the running track. However, this chemical is often found in cough and cold cures as a decongestant. Further investigation into the case of Christie demonstrated that the minute traces of the stimulant that the drug testers had found in the urine came from certain batches of ginseng, a herbal root he had taken with the full approval of the British Olympic association. Christie was cleared for further participation in the Olympic games. Drug testers say that modern GC/MS equipment is so sensitive that it cannot be fooled by what is known as **masking agents**. These are chemicals that can theoretically cover up the use of drugs by altering the data. Liz Johnstone of the Drug Control center at King's College, London, England states:

We can spot these masking compounds anyway.

She explained that one type of masking agent is **probenecid**, a drug taken for gout. "Probenecid can sometimes alter the way people excrete steroids into their urine and so mask their presence. But we can detect probenecid easily enough," said Liz Johnstone.

It has recently been shown that athletes can actually benefit from two cups of coffee. A research worker at Christ Church College in Canterbury discovered that if he gave 18 male athletes three hundred and fifty milliliters of strong coffee an hour before a 1,500-meter run on a tread mill, these small amounts of caffeine helped the athletes to shave an average of 4.2 seconds off their time. Further experiments showed that athletes given strong coffee ran 0.6 kilometers an hour faster over the last minute of the race, equivalent to putting ten meters between them and their rivals in the closing stage of a race.[25]

The International Olympic Committee bans athletes with **caffeine** levels above 15 micrograms per milliliter of urine – the equivalent to 5 or 6 cups of coffee within one or two hours of an event. The finding that two cups of coffee may significantly affect performance may lead to a revision of the International Olympic Committees standards. The high rewards associated with competitive sports will require that drug testing authorities in sports maintain constant vigilance and exploit the latest equipment being developed for measuring drugs in both urine and the bloodstream.

8.7 Drugs and Distracted Horses

Human athletes are not the only competitive creatures whose performances can be enhanced by the use of drugs. Race horses can be given drugs to make them run faster. Therefore, horses are screened for the possible use of performance enhancing drugs. In Figure 8.14, some of the activities of the horse racing forensic laboratory are illustrated by the job description of several openings in the horse racing laboratory. As can be seen from the job description of the analyst, the same type of equipment is used to study the urine and blood samples of horses that is used when studying human urine and blood samples. The analyst uses gas chromatography, high performance liquid chromatography, and immunoassay techniques. (Incidentally, the novel symbol used for this organization is a good example of the creativity of the symbolic artist). The very high sensitivity of modern tests for drugs and their metabolites has recently lead to an acrimonious conflict between Britain's Jockey Club (the organization that supervises horse racing in Great

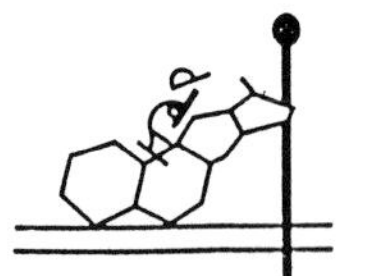

The Horseracing Forensic Laboratory

One of the world's leading anti-doping laboratories for equine sports

PO BOX 15
SNAILWELL ROAD
NEWMARKET
SUFFOLK CB8 7DT
TEL: 0638 663867
FAX: 0638 665232

THE HORSERACING FORENSIC LABORATORY REQUIRES THREE SCIENTISTS TO FILL THE FOLLOWING POSTS

ANALYST

(Scientific Administration)

Ref: SSA1

Will form part of the Support Services Division with responsibility for compliance with statutory government legislation and for the collation of information relative to the Laboratory's operational and research functions.

The successful candidate will probably have:

- An honours degree (or equivalent) in Chemistry or Biochemistry.
- Computer literacy.
- Awareness of HSAW and COSHH regulations.
- Imagination and innovation.
- An interest in information technology.

ORGANIC CHEMIST

(For Synthetic Work)

Ref: RDRC1

Will be based in the Research Division and be responsible for the provision of a service to the Laboratory primarily in the areas of synthesis of drug metabolites, stable isotope labelled substances and drug protein conjugates.

The successful candidate will probably have:

- An honours degree in Chemistry or Equivalent.
- Three to five years' experience in synthetic work.
- Experience in the use of modern analytical techniques for structure elucidation.
- Self motivation.
- Capaiblity of collaboration with other scientists.

ANALYST

(Operations Division)

Ref: ODA1

Will form part of the team involved in the analysis of horse body fluids for drugs in connection with doping control.
The work involves the application of a wide range of modern analytical chemistry.
The successful candidate will probably have:

- An honours degree (or equivalent) in Chemistry or Biochemistry.
- The ability to work in a team.
- The capability of working under pressure and to deadlines.
- Some familiarity with computers.
- Some experience with analytical techniques such as GC, HPLC and immunoassays (although an advantage, this is not essential).

The salaries for these posts will be commensurate with experience. The company operates a contributory pension scheme and offers other attractive benefits. Please phone or write for an application form and further details indicating which post interests you. Closing date for applications: Monday, 3 September 1990.

Figure 8.14. Illegal use of drugs to enhance performance is not only a problem in human athletes; horse racing experts must be constantly on the lookout for drugged horses! (Reproduced with the kind permission of the Horse racing Forensic Laboratory Limited.)

Britain) and the Aga Kahn, a rich race horse owner. The conflict arose in November 1989, when one of Aga Kahn's horses, Alysa, was tested for drugs after winning a major horse race. The forensic laboratory detected a metabolite of camphor in the urine of the horse. The use of camphor with a race horse is forbidden, because it may help them to breathe more deeply. A dictionary definition of **camphor** is as follows:

> *It is a tough, gummy, volatile, fragrant, crystalline compound $C_{10}H_{16}O$ obtained from the wood and bark of the camphor tree and used especially externally as a liniment and mild analgesic, as a plasticizer and as a insect repellent.*[7]

The director of the forensic laboratory, Neville Dunnett, claims that the test showed that the horse had been exposed to camphor, whereas a team of experts hired by the Aga Khan, led by Robert Masse, the chief dope tester for the Montreal Olympics, believed that the tests carried out by the horse racing forensic laboratory were flawed. The tests on the horse's urine were carried out with GC/MS equipment that could not detect less than a 125 nanograms of the camphor metabolite per milliliter of urine. Masse, using more efficient methods, said that he found small amounts of 3-hydroxycamphor in the urine of 6 horses fed **borneol**, a chemical relative of camphor sometimes found in feed, straw or wood shavings. The conflict essentially revolves around the fact that some experts maintain that the very low level of camphor metabolite found within the urine of the horse could have come from the straw and material in the horses stable. Others deny this possibility, claiming that to find the concentrations discovered in the urine, the horse would have had to eat two tons of carrots or similarly large amounts of hay or wood shavings to produce the amount found in the urine! As a result of this conflict, the Aga Kahn has withdrawn his 90 horses from Great Britain and from British racing, costing that industry several millions of dollars.[26]

In a chance conversation at a scientific conference early in 1991, I was given an alternative explanation for the presence of camphor metabolites in the horse's system. Over a cup of coffee, I had a discussion with an eminent chemist from Great Britain, and he offered the following explanation of the conflict over the Aga Kahn's horses. Apparently, in a line-up of horses for a race, if a male horse detects, through odor, the presence of a lady horse in the same line-up, his mind may wander from the task of winning the horse race. An old trick amongst horse trainers for such situations is to rub a well-known proprietary brand of camphor based decongestant ointment under the nostrils of the horse. (This is an ointment that helps people to breathe when they have a cold.) Therefore as a direct result of inhaling this medication, the camphor metabolite would show up in the drug test. Probably, however, no one would own up to such a practice when an inquiry was held. Whatever the true story, it is obvious that the forensic scientists will be actively involved in the prevention of fraud at the race track when horses compete for large sums of money, just as they will always be involved in testing athletes to prevent drug-based fraud.

References

1. H. J. Walls, *Forensic Science. An Introduction to the Science of Crime Detection*, Frederick A. Praeger, New York, 1968.
2. L. C. Nickols, *The Scientific Investigation of Crime*, Butterworth, Toronto, 1956.
3. *Laboratory Aids for the Investigator*, Published by the Center of Forensic Sciences, Public Safety Division, Ministry of the Solicitor General, 25 Grosvneror Street, Toronto, Ontario, M7A 2G8, 1990.
4. J. Broad, *Science and Criminal Detection*, McMillian, London, 1988.
5. The data of Figure 8.3 are taken from the commercial literature of Gaw-Mac Instrument Company, 100 Kings Road, Madison, NJ, 07940.
6. J. S. Swinehart, R. C. Gore, *Perkin-Elmer Infrared Application Study 6: Forensics*, commercial literature prepared by the instrument division of Perkin-Elmer Corp. Norwalk, CT 06852.
7. R. W. Pease Jr. (Ed.), *Webster's Medical Dictionary*, Mirriam Webster, Springfield, MA, 1986.
8. R. A. Boese, "Heroin. If it's There, We'll Find it," *Industrial Research Development*, December 1978.
9. I. S. Lurie, "Forensic Drug Analysis by HPLC," *American Laboratory*, October 1980, 35.
10. B. K. Krotoszynski, J. M. Mullaly and A. Dravnieks, "Olfactronic Detection of Narcotics and Other Controlled Drugs," *Police*, January/February 1969.
11. "Forensic Analysis by Liquid Chromatography," Commercial literature by the Waters Associates Inc. Maple Street, Milford, MA 01757.
12. "Analysis of Illicit Diamorphine Preparations by High Pressure Liquid Chromatography," *J. of Chromatography 104* (1975), 205.
13. M. Loveland, R. Williams, "Science Against Drugs," *Science Spectrum 213* (1988), 7.
14. J. D. Michaud and D. W. Jones, "Thin Layer Chromatography for Broad Spectrum Drug Detection," *American Laboratory*, November 1980, 104.
15. "Test Analyses Hair to Detect Drug Use," *High Technology*, February 1988.
16. S. Rasberry "Reference Material," *American Laboratory*, June 1991, 74.
17. L. Crawford, "Roots of Evil," *Science 231* (1986), 67.
18. Trade literature from the Quality Paperback Club describing the book *The Devil in Massachusetts* by M. L. Starkey, 1993.
19. M. K. Matossian, "Ergot and the Salem Witchcraft Affair," *American Scientist 70* (1982), 355.
20. B. Watson, "Salem's Dark Hour – Did the Devil Make them Do It?," *The Smithsonian, 23* April 1992, 117.
21. M. K. Matossian by R. Porter, *New Scientist*, 9 December 1989, 55.
22. C. Wood, "The Perils of Doping," *MacLeans*, 27 July 1992, 48.
23. "Doping and Anti Doping in Sports," (new story), *Sudbury Star*, 1 August 1992, A 10.
24. S. Connor, "Urine Analysts at Seoul Makes Athletes Sweat," *New Scientist*, 8 October 1988, 2.
25. T. Moore, "Take Time for a Winning Cup of Coffee," *New Scientist*, 18 July 1992, 8.
26. W. Brown, "Horses Hobbled by Flawed Dope Test," *New Scientist*, 22 December 1990, 8.

Chapter 9

Poisoned Arrows and Dangerous Bulgarian Brollies

Chapter 9

Poisoned Arrows and Dangerous Bulgarian Brollies

9.1 Poisoned Points

The very word **toxicology**, the scientific term for the study of poisons, hints at the fact that chemical warfare is not a modern invention. The word **toxin** for a poisonous substance comes from the Greek work *toxicon*, which is a poison into which arrow heads were dipped. The word is related to the Greek word *toxicos*, which means bow. In fact, modern archery enthusiasts are called **toxophiles**.[1,2]

Although the use of poison arrows in warfare is a very old practice, one of the most recent innovative methods of political assassination involved a point of a similar kind. In British colloquial English the term "brollie" is used to describe an umbrella, standard equipment for a traveller in Great Britain because of the amount of rain. In 1978 a Bulgarian journalist, Georgi Markov, was assassinated by a poison-tipped brollie forced into his thigh. Markov was a dissident member of the Communist Party who, having become disillusioned with communist ideology, moved to Bulgaria in 1968 and then to England in 1971. He was active in broadcasting anti-communist party programmes to Bulgaria via Radio Free Europe. It is suspected that his death was ordered by the Bulgarian Secret Police. The details of the attack are reviewed in an article in *New Scientist* by Griffiths and co-workers.[3] In their words:

> *while waiting at a bus stop near Waterloo Bridge on Thursday, December 7, 1978, Markov felt a sharp stab in the back of his right thigh. On turning, he saw a man picking up an umbrella. The man apologized and hailed a taxi. When Markov arrived at work, he showed the wound to a friend because he was in pain. The wound looked like an inflamed spot. He gradually began to feel ill and was admitted to hospital late the next day. Markov's condition deteriorated, and clinically his illness looked like septicemia or blood poisoning with a fever and raised white-blood cell count. Bacterial cultures of samples of his blood failed to demonstrate the presence of any disease-causing organism. Markov finally died of heart failure four days after the incident. The postmortem revealed something interesting. The*

pathologist found a metal pellet of 1.53 millimeters diameter. It was taken from the site of the wound. The pellet contained two channels of approximately 0.3 millimeters diameter at right angles to each other. Although the doctors found no toxin in Markov's body tissue at the time, the plant poison ***ricin*** *seemed the likely culprit, since it produces similar clinical symptoms.*[3]

Griffiths and co-workers call poisons such as **ricin** "super poison," because they can kill when administered in quantities measured in micrograms.

Where would Markov's attacker attain such a dangerous poison? Would he have to deal secretly with a pharmacist to gain access to a limited supply of the substance? The surprising answer is no. The poison comes from the seeds of the **castor bean plant**. A single seed of this plant is capable of killing a person. The plant is grown widely in North America as a decorative plant. When my wife and I lived in Chicago, we used to visit friends in a suburb of Chicago who always had a beautiful castor bean bush growing in the middle of their flower bed. I used to grow the plant myself until I found out how lethal a poison was contained in its seeds. Many plant compounds are deadly poisons. Later in this chapter, we will look at some of the dangerous house plants that can cause serious problems if they are inadvertently ingested.

Before discussing specific poisons in detail, it is necessary to realize that at low levels of ingestion, substances that are potentially lethal can actually help to cure severe illness. Thus, a medical dictionary defines **curare** as "a dried aqueous extract from the vine *Strychnos toxifera*, and other plants, used in arrow poisons by South American Indians, and in medicine to produce muscular relaxation." The varying effects of any drug in medicine at different dosage levels is described and quantified by medical experts and toxicologists using the concept of a **therapeutic dosage window**. Thus, for the widely used drug aspirin (acetyl salicylic acid), there is a dosage below which there is really no effective therapeutic value. Then, there is a dosage range in which the medication is useful for treating illness. Above this level, ingestion can cause death by poisoning. The range of dosage that can have useful effects is called the therapeutic dosage window. Again, one should be aware that this range is not the same for all individuals and that, because the chemistry of the body is different in different individuals, what constitutes a therapeutic dose in some individuals can prove dangerous to others. For example, it is estimated that one in 20 people will experience internal bleeding and other problems when taking aspirin. When we come to look at plant poisons in detail, we will find that drugs such as digitalis, given for therapeutic reasons, can become dangerous if too much of the drug is taken.

The existence of variations in the behavior of drugs at various dosage levels is discussed in the interesting article, mentioned above, on the drug used in the Bulgarian brollie murder. The title of the article, written by Griffiths and co-workers is "Proteins that Play Jekyll and Hyde." Ricin, the poison suspected of being the lethal agent in the Bulgarian Brollie attack, is a protein which kills by rupturing the wall around the body of a cell. One molecule of the poison is sufficient to damage the cell wall and cause the cell to die. The term **"Jekyll and Hyde"** refers to the character in the novel "The Strange Case of Dr. Jekyll and Mr. Hyde" by Robert Louis Stevenson. This novel was first published in 1886. In the story, Dr. Jekyll, who represents the good in man, is transformed periodically into an evil creature called Mr. Hyde, after drinking a chemical cocktail. When the effects wear off, Mr. Hyde is transformed back into Dr. Jekyll. Eventually, the change from one personality to the other becomes unpredictable and in time becomes a nightmare for everyone involved. Thus, in the title of their article, Griffiths and co-workers draw attention to the fact that many compounds from plants can behave like Dr. Jekyll or Mr. Hyde at different dosage levels. It is rather interesting to note that some commentators on the Robert Louis Stevenson novel suggest that it was written after the author started to take opium to kill pain, and that the bizarre behavior of Mr. Hyde represents the influence of the drug distorting the behavior of the person.

Police journals from as long ago as 1910 have reported criminal use of extracts from the castor bean and another related plant, the **jequirity plant.**[3] In the words of Griffiths and co-workers, such plant poisons have probably been used since much earlier times. For instance, a poison closely related to ricin can be extracted from the berries of **mistletoe**.

In Figure 9.1 the variability in effect of potentially poisonous substances as reported by Timbrell, is shown.[1] It should be noted that the bottom part of this graph has a logarithmic scale. In Figure 9.2 the way in which the toxicologist records the effects of drugs to delineate the therapeutic window is shown. This curve was obtained from animal tests on a particular compound. Here, the animals were surrogates for humans. However, as Timbrell tells us, as early as 200 years B. C., Niccander of Colophon, physician to Attalus, king of a country in the Middle East, was allowed to experiment with poisons, using condemned criminals as subjects. As a result of his studies, he wrote a treatise on antidotes to reptile poisons and to other substances. Niccander mentioned 22 specific poisons, including the two compounds white lead and litharge (another lead compound), which we will discuss later. He also mentioned aconite, hemlock and opium. The graphs of Figure 9.2 summarize the type of technical data used to determine the therapeutic window of a substance. Important aspects of the data are the 50 % dosage levels for effective, toxic, and lethal doses. Again, it should be noted that the bottom scale of this graph is what is known as a **logarithmic scale.** The vertical axis is a scale widely used in science, known as a **probability scale.**[4] When comparing the toxicity of two different compounds A and B, one constructs a graph such as that shown in Figure 9.3. Again, in terms of the language used by the toxicologists, although the two compounds A and B have the same 50 % value, A is considered less potent than compound B.

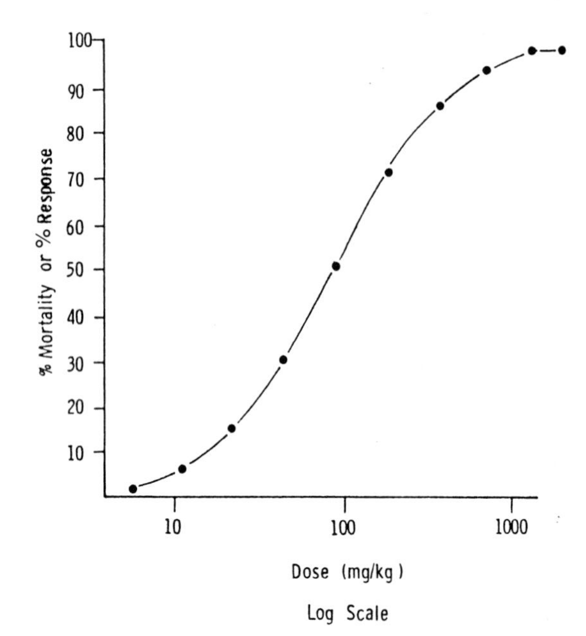

Figure 9.1. A typical dosage–response curve for a potentially poisonous substance. The percentage response of the organism or system or the percentage mortality of organisms in a group exposed to the compound is plotted against the log of the dosage. (Reproduced from *Introduction to Toxicology* by J. A. Timbrell, © Taylor and Francis, 1989.)

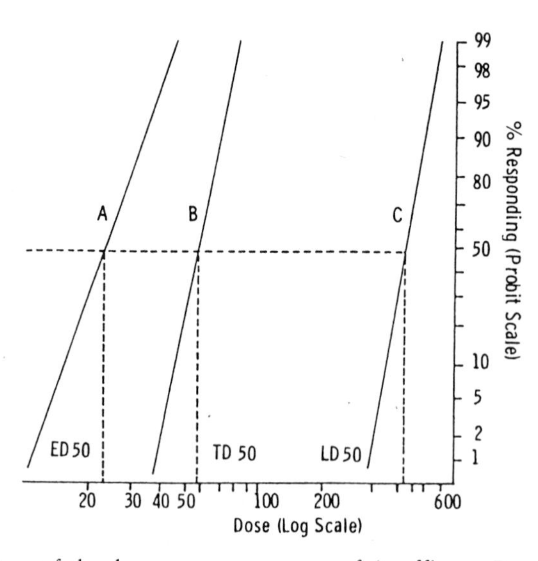

Figure 9.2. Comparison of the dosage–response curves of A: efficacy: B: toxicity: C: lethality. The effective, toxic, or lethal dose for 50 % of the animals in the group (ED50, TD50, LD50, respectively) can be derived from the curve. The relationship between ED50 and TD50 is a measure of the margin of safety of the compound. (Reproduced from *Introduction to Toxicology* by J. A. Timbrell, © Taylor and Francis, 1989.)

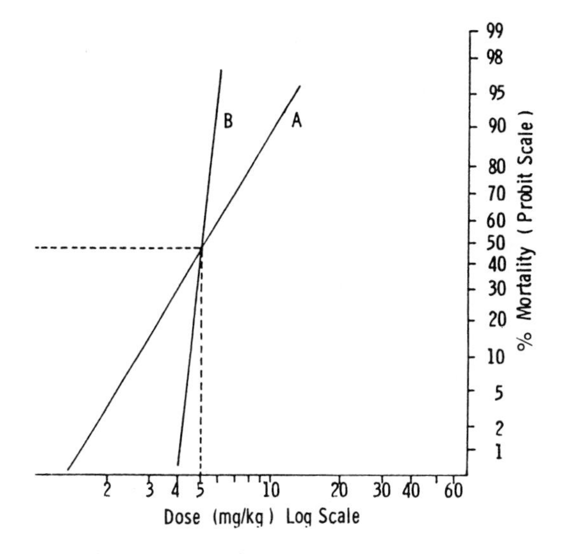

Figure 9.3. Comparison of the toxicity of two compounds A and B. Although they both have the same LD50 compound A is less potent than compound B. (Reproduced from *Introduction to Toxicology* by J. A. Timbrell, © Taylor and Francis, 1989.)

Griffiths tells us that the forensic toxicologists are engaged in a tireless search for means of tracking various poisons in the body of suspected victims, and that his own group is looking for methods to detect poisons, such as ricin at autopsies. They conclude their article on ricin-type poisons with the following statement.

> *The Markov case has been extensively documented from tabloid press to textbooks, but secret intelligence services are not the only ones to use these toxins for sinister purposes. The ability to detect these poisons in postmortem investigations is then extremely valuable. Who knows how many unfortunates have fallen foul of these super poisons?*[3]

If the scientists were not able to detect ricin in the body of Markov, how did they arrive at their conclusions? The conclusion rests upon that fact that, although there are variations in the amount of poison needed to cause death, when using a given poison the various types available have very different levels of toxicity. This is shown by the data summarized in Figure 9.4, taken from an article by Emsley and Pallister.[5]

In discussing which poisons could have been used to kill Markov, Emsley and Pallister state the pellet removed from Markov offers several clues. A cavity of this size could hold only about 0.4 milligrams of material, and for the deadliest inorganic poisons, such as cyanide and arsenic, a minimum of 100 milligrams are required to kill a human. Around 50 milligrams would be needed of the deadliest organic poisons, such as fluoroacetates and dimethyl mercury. Even of the nerve gases, such as Sarin, Tabun, and the deadly V agents, one would need at least one milligram. Suggestions that a germ warfare agent was used can virtually be discounted, because of the absence of recognizable lesions (wounds) attributable to such agents. This, therefore, leaves a choice between biological toxins and radioactive emitters. **Biological toxin** is the term used to cover a range of lethal agents extracted from, for example, snake and spider venoms or from bacteria. Many of these poisons are commercially available. Again Emsley and Pallister note that Markov's symptoms before he died – high fever, and recurring comas – are consistent with the effects of a bio-toxin. The two holes in the pellet are a key factor pointing to bio-toxins. One hole would have been sufficient to carry a lethal dose of the deadliest radioactive agents, such a Plutonium 239, and it would have been relatively easy to detect. Two holes would be required to carry a lethal dose of a bio-toxin together with its supporting medium. Over all, the scientific community seems to have reached the conclusion that the actual poison used was ricin.

Now that we have explored some of the basic concepts of toxicology and the relative deadliness of various poisons, we will look at the way in which several poisons have been used to commit crimes both recent and ancient.

Although the main focus of this book is on the use of science to fight criminal activities, a second major task of forensic scientists is protecting the public from inadvertent poisoning by illegal mercantile or industrial activity.

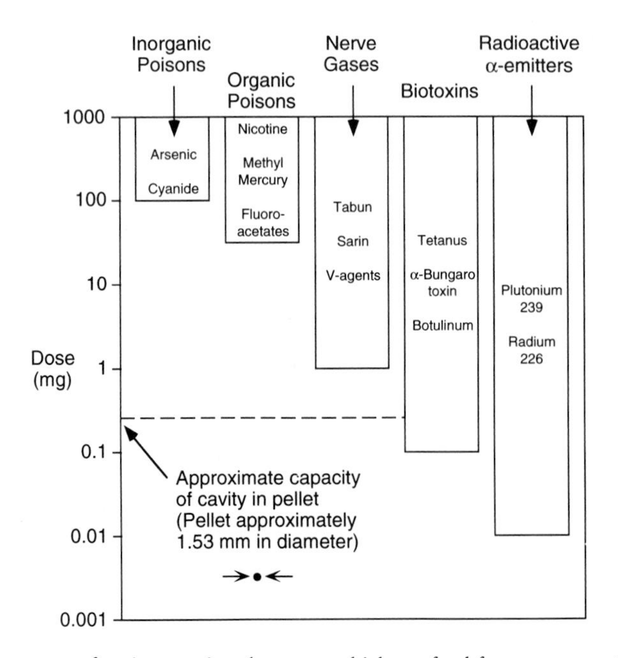

Figure 9.4. Amounts of various toxic substances which are fatal for an average (70 kg) person.

For example, in May 1981 there was an unusual outbreak of a pulmonary disease reported around Madrid, Spain. In total, there were more than 20 thousand cases of an illness, which involved serious symptoms and resulted in 351 fatalities. A toxic substance was suspected, and forensic scientists finally established a connection between the disease and the use of cheap cooking oil. To protect the olive oil industry in Spain, the government had ordered that canolla seed oil (known as rape seed oil in Europe) used for industrial purposes, must be treated with aniline to make it unsuitable for cooking. (After the treatment of a compound such as cooking oil to make it no longer suitable for its original use, the product is said to have been **denatured**). Although the case was never properly solved, it is believed that some people had been trying to treat the doctored Canolla oil to remove the aniline and had sold the resultant product as cheap cooking oil for human consumption. Incomplete processing of the denatured oil left it with poisonous contaminants. Before the case could be completely solved, the contaminated oil disappeared from the market.[1]

The public must be protected from the possible side effects of new drugs, and in this chapter we will look briefly at some of the problems, which have arisen with new drugs. Sometimes the poisoning of the public from industrial activity is due to ignorance, sometimes due to deliberate cynical use of material, which is inherently dangerous. The general subject of industrial toxicology is much bigger than can possibly be covered in this first text on forensic science, but we will touch on one or two salient points and review case histories that indicate the kind of problems that we face in society from mass poisoning – either intended or inadvertent – by industrial activity.

9.2 Omelettes, Flypapers and Weed Killers – A Look at how Arsenic has Been Used in the Commission of Crime

In a review article on the use of arsenic in poisoning episodes Emsley states:

> *Popular myth greatly exaggerates arsenic's power to kill. For centuries it was the perfect poison, although it is much less toxic than many organic chemicals. It was undetectable and the poisoner needed relatively little of it – less than 1/4 of a gram in order to dispose of their victim. The symptoms of arsenic poisoning are easy to mistake for other ailments and the poison is fairly easy to administer. So-called **white arsenic**, (**arsenic trioxide**), was the form commonly used. This is not very soluble in water but enough will dissolve in a glass of wine to do the job.*[5]

Emsley states that the golden age of arsenic dawned in 1830 and was continued another 120 years. The last conviction for murder by arsenic in Great Britain was of a U. S. sergeant, Marcus Marymont, in 1958. He poisoned his wife Mary. Apparently, he had obtained the arsenic from the U. S. Airforce base, where he worked. He poisoned his wife after she had discovered that he was having an affair with another woman. The court sentenced him to life in prison. Neutron activation analysis of Mrs. Marymont's hair showed regular bands of arsenic at weekly intervals, indicating that her husband had tried several times to poison her when he was at home on leave.

We sometimes forget that our food safety laws have only evolved in recent years as authorities become aware of the dangers of uncontrolled food production. In 1900 seventy out of a total of 6,000 people who had been made ill by drinking beer that contained 15 ppm of arsenic died. The words used in the title of this section, omelettes, flypaper, and weed killer, refer to sources of arsenic poisoning in famous detective stories or in the actual committing of a murder. Dorothy L. Sayers, the famous detective story author, built a plot of murder around an arsenic laced omelette. In the story, called "Strong Poison," the victim shares an omelette with the murderer. The victim died but the murderer survived. The plot of the story revolves around the idea that one can build up an immunity to arsenic by eating small doses on a regular basis. In the novel the murderer places himself on an arsenic diet for a while and then the omelette containing enough arsenic to kill a person under normal circumstances did not affect him.[5,6]

To our generation the idea that one would be able to eat some arsenic to build up ones resistance to the substance is a strange idea, but this was quite commonplace in Victorian England. In Victorian times, there was a spectacular murder case in which a woman was accused of poisoning her husband using arsenic trioxide. Although she was convicted of the crime her death sentence was commuted (changed) to life imprisonment because of the doubt that the amount of arsenic in her husband may have killed him because he was a regular taker of arsenic for "medicinal" reasons.[7] Emsley has reviewed the medicinal use of arsenic compounds. He tells us that in small quantities, arsenic compounds can promote growth. In his article he states that arsenic is known to be very good at promoting growth in animals. Farmers use a compound containing it to help fatten their pigs and poultry. This compound is called **roxarsone**. Scientists discovered this chemical's ability to stimulate growth after it was tested as a treatment for the disease coccidios in chickens. In 1985 Emsley stated that roxarsone was used widely, particularly in the United States. Pigs and hens fed roxarsone put on about 3 % more weight than if fed without the additive. The cash value of this extra meat more than offsets the cost of the special feed. The animal excretes the arsenic very rapidly, so that when the farmer discontinues the doped feed a few days before slaughter, the residual amount of arsenic in the animals tissue falls rapidly to an acceptable level of one part per million or less. Emsley also tells about a medication known as "Dr. Fowler's tasteless ague and fever drops." In the late 1780s Dr. Fowler and his assistant found out that if they put small amounts of arsenic trioxide and potassium carbonate into a mixture, it reduced fever. This medication remained in use until the early 1920s.

In late Victorian times people believed that it was an aphrodisiac and a tonic for tired businessmen. Aphrodite was the Greek goddess of sexual love, fertility and beauty. An **aphrodisiac** substance is one that increases desire and enjoyment of sex. Humans have sought after aphrodisiacs in every generation and it is amazing how they have persuaded themselves that many strange compounds have an aphrodisiac effect. Thus, in many parts of the world there is a belief that powdered rhinoceros horn is an aphrodisiac; a factor which contributes to the modern illegal hunting of the rhinoceros. One of the late Victorians who believed that arsenic was an aphrodisiac was a person known as James Maybrick.[7,8] Maybrick was a prosperous British cotton broker who died in 1889. In the words of Max Haines, the crime writer:

James Maybrick used arsenic, regularly covorted with several mistresses, was a frequenter of horse races and worked hard at his business. All the while he complained about not feeling well. He married a woman 23 years younger than himself. In April 1889 his wife Florence purchased a quantity of fly paper from a druggist. She told the druggist that her kitchen was inundated with flies, whereas later at her trial the kitchen staff gave evidence that there were no flies to be seen in the household.

In the 1880s two different kinds of fly paper were available in British homes. One was covered with sticky material so that when a fly landed on it became stuck and died on the paper. The other type of paper was treated with arsenic trioxide. Two members of the Maybrick staff observed Florence soaking the fly paper in water. In her own defence Florence said that she was extracting the arsenic compound from the fly paper for cosmetic purposes. Strange as it appears to our generation, arsenic compounds were often used as a treatment for the complexion in Victorian times. At about the time that Florence was working with fly paper, her husband died exhibiting all the symptoms of arsenic poisoning. In the words of Emsley:

Mrs Florence Maybrick's husband clung on for three days by which time the level of arsenic in his body was quite low.

The forensic scientists had no trouble in finding arsenic in his medicines and on a jug that Florence had used for preparing the invalid's food. Doubt over whether the possibility of arsenic trioxide from the fly paper would have killed him when he was an arsenic eater led to her sentence being reduced to a prison term. She lived to be 79 and died in poverty and obscurity in October 23, 1941.[8] The Maybrick case made headline news in 1993, when it was claimed that a diary had been found, which proved that James Maybrick was actually "Jack the Ripper." Jack the Ripper is the name given to an unknown person who carried out five spectacular murders in the autumn of 1888 in London. The murders were also known as the White Chapel murders because they occurred in a district frequented by prostitutes known by their clients as White Chapel. Over the years, 72 different individuals have been suspected of being the notorious killer. The suspects range from the Duke of Clarence (the deranged grandson of Queen Victoria) to a kosher butcher.[9]

In September 1993, an American publisher refused to have the supposed diary of Jack the Ripper released in North America on the grounds that an American expert had shown the claimed diary to be a forgery. The forgery expert stated that the handwriting was not Victorian style, and that tests on the paper and the ink showed that the document was written in 1920, not 1888.

In an earlier chapter we reviewed the possibility that Napoleon, the First Emperor of France, had died from arsenic poisoning. Another historic personality who may have been a victim of arsenic poisoning is the famous scientist and philosopher René Descartes (1596–1650). Descartes was also known by his Latin name Renatus Cartesius. Descartes was the inventor of a form of geometry in which the location of a point in space is specified by its distance from a given point using x, y and z as the address of the point. This form of geometry is known as **Cartesian geometry** and the address of a point in three dimensional space is known as the Cartesian coordinates of the point. Descartes is also known for a very famous statement that he made, "cogito, ergo sum" this is a Latin phrase which means "I think; therefore I am."

In 1649, Descartes, who was 53 at the time, was invited to go to the court of Queen Christiana of Sweden. He was to act as a tutor to the Queen. Four months after he arrived in Sweden, Descartes caught a chill that turned into pneumonia and ten days later he was dead. Recently, evidence has emerged that Descartes may have been poisoned by members of the Court of Queen Christiana, because they were afraid that he would influence Queen Christiana to become a Roman Catholic. Since Sweden was a Protestant country at the time, such a religious conversion could not be tolerated.[10]

Recently, W. Glenn has reviewed the possibility of arsenic poisoning being a problem for industrial workers. In an industrial environment arsenic compounds usually enter the body through inhalation or ingestion of fine particles or fumes and is quickly and widely distributed by the blood to all tissues. He points out that arsenic can cause the degeneration or inflammation of the nerves in the arms, legs, hands, and feet. This is characterized by pain and burning tenderness in the afflicted limbs, and the affected person has difficulty walking. The condition develops gradually, beginning with tingling, numbness, and weakness. Glenn refers to the fact that empidemiological surveys (**epidemiology** is a study of patterns of disease amongst the general population or among a specific group of workers) going back to 1948 have shown that arsenic poisoning can be a problem with workers in metal refineries or amongst workers who use pesticides and herbicides (weed killers), which contain arsenic. Thus, a study of 920 copper smelter workers in Rouyan, Quebec, uncovered lung cancer rates five to ten times higher than would be expected amongst an equivalent number of people taken from the general public. Modern industrial hygiene practices have tended to reduce problems from exposure to arsenic compounds in industry, but massive liability lawsuits can hinge on proving that a worker was exposed to arsenic compounds, which caused health problems.[11]

The term bronze is used to describe various alloys (mixed compounds of metals) of copper with other metals. The alloying of copper with metals such as arsenic and/or tin can result in a substance, which is much harder than copper and can be used to make devices such as knives, axes, and swords. The term bronze age is used to refer to a period of history between the stone age and iron age. The manufacture of bronze weapons began around 3,500 B. C. In a recent article on such bronze alloys, Martin Harper of the London School of Hygiene and Tropical Medicine has pointed out that copper ores naturally contain arsenic, and smelting would have reduced the arsenic to a very low level, because of the volatility of arsenic oxide. Thus, in modern metallurgical operations, the amount of arsenic oxide, which is usually around 9 % in the copper ore, is reduced to 0.2 % in the final copper. However, studies of the most ancient bronze weapons show that they contain as much as 7 % arsenic, which indicates that the metal workers must have added arsenic to their bronze. Indeed, ancient Greek writers such as Aristotle talk of "earths" being added to copper to improve its hardness. The addition of up to about 4 % of arsenic increases the hardness of alloys; however, alloys with more than 2.5 % arsenic tend to be brittle. This would mean that, although the sword made from bronze would be hard, it would also snap if it contained more than 2.5 % of arsenic. Archaeological studies point to the mystery that while the arsenic alloys was very useful in making weapons, within 400 years of the development of bronze weapons, the use of arsenic alloys were phased out in favour of tin alloys. In view of the fact that tin ores had to be imported, whereas "arsenic earths" were readily available, the switch from arsenic-based bronze to tin based-bronze is an archaeological mystery. Harper points out that the most likely reason for the change was the health hazard posed by arsenic. Arsenic-induced neurosis may lead to the weakness of the legs and feet (see earlier comments in this section by Glenn). Harper comments that it is probably more than a coincidence that the metal-worker gods of the Indo-European people [Hephaisto in Greece, Vulcan in Rome (Volcanoes were the imagined smithies of the blacksmith gods) and Wieland in Germanic countries] were all lame. Harper suggests that their myths record the fact that people were aware of the effect of arsenic poisoning, which can still be detected amongst smelter workers today.[12]

9.3 Bitter Almonds and Sudden Death

Perhaps the most notorious poison is cyanide, which is featured in many detective novels. Potassium cyanide kills quickly and leaves a faint odor of **bitter almonds** on the body of the victim. Potassium cyanide looks like table salt, and can be mixed in with other white materials without being distinguishable. In 1982 there was wide-spread panic after seven Chicago area people died from cyanide-spiked extra-strength Tylenol capsules. This case was never solved. As a consequence of the crime, product tampering in the United States is now a federal offense. In a review of a famous case in which Sue Snow, a 40 year old bank manager in Washington, was killed, D. D. Jackson comments:

> *When ingested, cyanide prevents cells from using oxygen. It looks like table salt and a small dose can kill rapidly; it is the perfect poison for murderers.*[13]

Because of the way in which Sue Snow died on June 11, 1986 – suddenly and without recovering consciousness, an autopsy was carried out. During the examination, one of the assistants to the pathologist detected a faint odor of bitter almonds emanating from the body. Apparently just before she died, Sue Snow had taken two extra strength Excedrin capsules for a throbbing headache. When the Police checked the capsules left in her bathroom, they found that the capsules contained cyanide. On June 17, 1986, a 42 year old woman named Stella Nicholl called the Police and reported that 12 days earlier her husband Bruce, 52, had died suddenly after taking extra strength Excedrin capsules. Because her husband had volunteered to be an organ donor, a sample of his blood serum had been preserved and tests of the blood samples showed the presence of cyanide. In this investigation the Police tracked down 5 bottles of Excedrin, which had been contaminated with cyanide. On examining these spiked capsules, a young chemist noted that mixed with the white chemicals were a few specks of green. The tiny green crystals were identified as an algae killer used in home fish tanks. Later this was to turn out to be an important clue. It appeared that whoever had spiked the Excedrin capsules had used a bowl, which earlier had been used for crushing algaecide pellets. Police looking for clues in the home of Bruce Nicholl found that his wife used this product in the fish tank she had in her home. The police became even more suspicious when they found out that Stella had taken out an insurance policy of $31,000 on the life of her husband, and that if his death was accidental she could collect an extra $105,000. Furthermore, she had purchased two additional $20,000 policies on his life in the year before he died. In total, she stood to receive $176,000 if her husband died accidently. The police were curious to find that Stella had called the doctor several times to question his findings that her husband had died a natural death from emphysema. Eventually it was proved that

Stella had laced several bottles of Excedrin medication and placed them back in the store. Sue Snow was an innocent, random victim of a scheme to collect insurance money.

As part of the investigations the FBI crime lab found that Stella Nicholl had taken out a library book on poisons and that 84 fingerprints were found in the book, "Deadly Harvest." The biggest concentration of her fingerprints were on the pages discussing cyanide.[13] After this spectacular case, and that of the Tylenol poisoning, the pharmaceutical companies have abandoned the use of capsules in non prescription drugs.

Although cyanide is rarely used now in criminal poisoning, the use of cyanide continues to be popular in detective stories. Thus in an episode of "Murder She Wrote" broadcasted in the U. S. in the fall of 1993, the victim died from a cyanide-laced sauce. The source was in fact a type of sugar glaze on the victim's dish from which the cyanide dissolved without any indication of its presence.

Peach pits contain cyanide – another fact that has been exploited in detective stories. One should never crack open peach pits and eat the stones, nor should they be allowed to stay in the fruit when making jam. Apple pips (seeds) also contain small amounts of cyanide compounds, and for this reason there are no orchards in the gardens of prisons or mental institutions. Spies carried cyanide hidden on their body or in hollow teeth so that they could commit suicide if captured, to avoid having information tortured out of them.

9.4 Curiosity (and a Poisonous House Plant) Killed the cat!

At the beginning of this chapter we discussed briefly the fact that some very potent poisons come from plants to be found around the garden. Some house plants are also very poisonous.[14,15] For example, Johnson has recently reviewed the sad case of a 28 year old woman, who died after she had eaten the tubers of the glory lily plant, *Gloriosa superba*. These lily tubers look like sweet potatoes.[14] Another case reported in a book about poisonous plants is that of a 16 year old girl, who died after eating 12 flowers from a plant known as the autumn crocus (*Colchicum atumnale*). Both of these plants contain the poison known as **colchicine** (the full chemical name is acetyltrimethylcolchicinic acid). This poison damages blood vessels and nerves, and stops cell division. A fatal dose of colchicine lies somewhere between seven and 60 milligrams. Johnson points out that this not much when you consider that a dried seed of the autumn crocus can contain up to 3.5 milligrams, while as little as 20 grams of the tubers of the Lily plant provide 60 milligrams.

In their book on toxic house plants, Spoerke and Smolinkski review 109 species of poisonous house plant and assess their toxicity, the symptoms of poisoning from specific plants, and possible treatments. They also review specific poisoning cases. They give the advice in their book that:

cats partial to a little greenery to supplement their diet should avoid plants of the philodendron species, which are highly toxic.[15]

9.5 Was Van Gogh a Victim of Plant Poisoning?

Vincent **Van Gogh** was a Dutch painter who lived from 1853–1890. In his early days he led a varied career. In turn, he was a salesman for an art gallery, a tutor of French, a theology student, and an evangelist amongst coal miners in a Belgian town. He became very active as a painter. Like many other artists, he never made much money when he was alive, but in 1990 one of his paintings sold for 82 million dollars. Towards the end of his short life he suffered from mental problems. As a consequence, he spent a year in a hospital at Arles in the South of France. In one of his periodic fits of depression he cut off his ear and sent it to a prostitute whom he frequented. A possible clue as to what could have created Van Gogh's problems is the fact that his later paintings were often dominated by yellow objects. In one of his letters he wrote "how beautiful is yellow." His mental problems finally overwhelmed him and he committed suicide.[16] There have been many studies of the behavior of Van Gogh and many explanations have been put forward. One of the simplest suggestions is that he was suffering from syphilis, a **venereal disease**. (A venereal disease is one spread by sexual contact.) A medical dictionary definition of **syphilis** is:

a chronic contagious venereal and often congenital disease that is caused by a spirochete of the genus treponema. If left untreated it is characterized by a clinical course in three stages over many years. The final stage can often be severe mental illness.

Historic personalities whose strange behavior has been attributed by some scholars to advanced stages of syphilis are Hitler and Henry VIII. It was known that Van Gogh had led a wild youth and he may well have contracted syphilis. Cutting off his ear and hallucinating may be an indication of the physical disease of the brain.

In early 1990, a report from the Swedish medical center in Englewood, Colorado, suggested an alternate explanation of the Van Gogh's behavior.[17] After an examination of medical records and his letters, the workers at the Swedish medical center suggest that the Dutch painter had an advanced case of **Ménière's disease**. This disease is named after the French doctor Prosper Ménière, who described it in 1853. The disease is characterized by a painful ringing or buzzing in the ear and attacks of disabling dizziness. In their report, research workers from Englewood said the voluntary admission of Van Gogh to the asylum at St. Remi showed that Van Gogh hoped to find help for his attacks of vertigo (dizziness). At that time everyone else thought his problem was a form of epilepsy. However K. Arenberg, a senior author of the Swedish Medical Center, reports that Van Gogh's rational behavior at the asylum before as well as after attacks should forever banish the notion that he was an epileptic or mad. In a letter written in 1888, Van Gogh complained of being assailed by auditory hallucinations. Arenberg suggests that the episode in which he cut off his ear is typical of patients with advanced Ménière's disease, who will severely damage their ear in an attempt to stop the persistent ringing sounds. There are recorded cases of patients who severed their ears or poked holes in their eardrums to end the noise.

J. Aronson, a specialist in clinical pharmacology at the Radcliffe Infirmary, in Oxford, England, has yet another explanation for the deranged behavior of Van Gogh.[18] He finds it significant that a painting of Dr. Gachet, who treated Van Gogh for his mental problems, shows the Doctor holding a foxglove flower in his hand (this was the painting that sold for the record price in 1990). Apparently, Gachet himself thought that Van Gogh was suffering from a combination of turpentine poisoning, from the fumes he inhaled when painting, and sun stroke from painting out in the open in Southern France. Aronson does not think that this diagnosis is very probable. He suggests that the foxglove shown in the painting of Dr. Gachet indicates that Van Gogh may have been suffering from overdoses of **digitalis**, a drug extracted from the foxglove. Extensive intoxication with digitalis can result in pain in the eyes when exposed to light.

Foxgloves have been used in traditional medicine for centuries. William Withering published a book in 1785 in which he wrote medical accounts of 163 cases of **dropsies** (accumulation of fluid in the body tissues because of a malfunctioning heart), which he treated with formulations of the leaves of the purple foxglove. For hundreds of years extracts of digitalis were used to treat tuberculosis of the skin and as a medication for epilepsy. We now know that the active ingredients in the foxglove plant are compounds known as cardiac glycosides, which influence the heart. Part of their chemical structure are glucose-type molecules. Digitalis not only has an effect on the heart, but it can have side effects on the brain. Thus, according to Witheringson it was used by women of the poorer class in Derbyshire, who drink large draughts of foxglove tea as a cheap means of obtaining the pleasures of forgetfulness. Witheringson also states the foxglove, when given in very large and quickly repeated doses, occasions vomiting, purging, giddiness, confused vision. For instance, vision may be blurred or blind spots may appear in the visual field. Objects may appear green or yellow and they may be accompanied by halos of light. Aronson tells us that he has seen patients suffering from digitalis intoxication, who create paintings showing a field of regular blue with splashes of color superimposed.

After reviewing all of the evidence, however, Aronson is not sure that Van Gogh did suffer from digitalis poisoning. He suggests another possibility. Apparently, in the 19th century a favorite drink in Paris was made from the plant wormwood. It was drunk by many poets and artists. Actors drank it for its mildly "Euphoric, aphrodisiac and hallucinogenic effects." The scientific term for the wormwood plant is *artemisia absinthium*. In France the drink is known as **absinthe**. The artist Degas created a painting called "L'absinthe" showing a person with a glazed look in their eyes. Charlesworth tells us that absinthe was made by steeping wormwood, anise and fennel (plus nutmeg, juniper, hyssop etc.) in 85 % alcohol, adding water and distilling the brew. Eventually, one ended up with a liqueur whose alcohol content was 75 %. Charlesworth goes on to comment:

> *excessive absinthe drinking produced terrifying hallucinations and an enfeebled condition known as absinthism. Absinthe-generated crime was enormous, and the liqueur was banned in France in 1915.*[19]

Wormwood contains substances called Thugones, which can cause auditory and visual hallucinations, paranoia, mania and convulsions. Van Gogh was known to have drunk absinthe, sometimes in large quantities and, in the words of Aronson:

> *it is not too farfetched to believe that pharmacological effects (of absinthe) may have caused his illness and influenced his painting style.*[18]

In the conclusion of his review of the possible causes of Van Gogh's illness, Aronson asks:

> *wherein lay the source of Van Gogh's striking creations? The inner workings of an unusual artistic mind? Certainly, but perhaps not only that. Turpentine? Unlikely. Digitalis? I doubt it. Absinthe? Perhaps. We shall never know whether any of these really influenced him, nevertheless it is interesting to speculate.*[18]

Before moving onto a discussion of food poisoning we should note that nutmeg – defined in the dictionary as the aromatic kernel of an East Indian tree – was apparently added to absinthe because of its supposed aphrodisiac properties. Charlesworth comments:

> *nutmeg contains no known psychoactive compounds, its intoxicating effects are so far unexplained. Possibly the volatile oil (that gives nutmeg its odor) is structurally related to amphetamines.*

My mother used to grate nutmeg onto my rice pudding when I was young. I don't recall it having an aphrodisiac effect; perhaps I was too young to know.

Charlesworth also tells us that nutmeg was also thought to induce abortions. She concludes her article on legal drugs with a cartoon character saying:

Remember dear, one man's aphrodisiac is another woman's prophylactic!

9.6 Lethal Hamburgers and a New Wrinkle on Botulism

Early in 1993 there were several deaths in the United States caused by hamburgers sold by a chain of vendors, which had not been sufficiently cooked. The death-causing poison in this case was a secretion from a bacteria known as **Verotoxigenic coli**. This bacteria is often referred to by the initials **VTEC**. It produces a toxin that can break down the lining of the intestine and damage the kidneys. People who develop this syndrome frequently report that they ate ground beef prior to their illness. For this reason, the illness is sometimes called **hamburger disease** or **barbecue syndrome**. However, people have become ill after eating other kinds of under-cooked meat and poultry, and after drinking unpasteurized milk or unchlorinated water. At the time that this book was written, the relatives of the people that died from the poisonous hamburgers were suing the hamburger vendors, who in turn were suing the meat company that provided the ground meat. To understand the basis of the claims and counterclaims in such law suits, it is necessary to understand why these particular bacteria can become a real source of trouble in ground beef. The term *e. coli* is short for the word escherichia coli. These bacteria are often present in the human intestine and some varieties of the bacteria can be present in soil and in water. The presence of escherichia coli in rivers and in lakes is usually an indication that the water is contaminated with human sewage. The term escherichia comes from the name of the doctor Theodore Escherich 1857–1911, who was very active in the study of illness in children. In 1886, he published a book in which he discussed the relationship of intestinal bacteria to the physiology of digestion in infants. The VTEC bacteria live in the intestines of cattle and other meat animals such as pigs and sheep. When the animals are slaughtered, the bacteria can get onto the outer surfaces of the meat and when the meat is ground, VTEC can be spread over the newly exposed surface. The level of contamination with VTEC depends upon the way in which the animals are slaughtered, the procedures used to butcher the meat and the process used to grind the meat when making such items as hamburger patties. Hence, the reason that the hamburger vendors are suing the hamburger makers. Meat patties of ground beef which are contaminated with VTEC can be rendered safe by cooking them thoroughly until there is no pink meat in the middle of the pattie. The people suing the hamburger sellers claim that they had undercooked the patties. It is reported that after the deaths occurred, hamburger vendors increased the cooking time and temperature in order to render the meat safe for consumption.

In recent years the problems of barbecue syndrome have certainly become more widespread. But the dramatic increase in the number of cases (in Canada, 25 in 1982 compared to 2,500 in 1989) is also the result of an increased awareness of the disease and the development of better laboratory methods to diagnose the syndrome.[20] Some people who are infected by VTEC do not get sick at all, some feel that they have a bad case of the flu, and yet others experience severe or even life threatening symptoms. Probably in the past many people have died from this type of food poisoning and the victims' relatives have been unaware of the source of the illness. As people become more aware of the problem, there will certainly be more lawsuits and more forensic activity to track down the source of mysterious illness caused by the eating of food.

In Canada, the health and welfare authorities advise:

> *Hamburgers and other ground meat must be cooked completely through so that the meat is brown and the juices are clear rather than pink. You can eat most roasts and steaks a little rare as long as they are well cooked on the outside. However rolled roast must be cooked like ground meat so that no pink remains. Poultry must also be cooked until there is no pink left near the bone.*

There may well have been murders committed in the past by deliberately feeding contaminated meat to victims, but in the past, criminal activity of this kind has been very difficult to prove.

Another devastating form of food poisoning is called **botulism**. This type of food poisoning occurs when people eat food in which the bacterium *clostiridium botulinum*, is present. Originally, the name comes from the fact that this type of poisoning used to be very common in badly prepared sausages (in Latin *botulus* means sausage). The botulism bacteria flourish in the absence of oxygen and are, therefore described as **anaerobic bacteria** (**anaerobic** means without oxygen). Modern outbreaks of botulism are usually traced to the improper canning of fruit and other foods in the domestic environment. Commercial canneries take precautions to prevent botulism by heating the material when preserving it. Boiling food for a few minutes destroys the toxin produced. Death from botulism can occur very quickly. In recent years botulism toxin has been extracted and injected into muscles near the eye. By this method, skin wrinkles can be made to disappear by paralyzing the appropriate muscles! (at a cost of $450 per injection in 1993). It should be noted that children under the age of one should not be given honey or maple syrup, because these products contain spores of the bacteria causing botulism. In young children, for some reason that is not entirely understood, the spores can produce viable bacteria, which then produce the toxin. The death from such poisoning can be similar to sudden infant death syndrome. The scientists draw a distinction between the two types of poison attacks involving botulism. Thus, in adults where the spores of the bacteria cannot become viable, the death is due to a poisoning from toxin already present in the food. In the case of a child under one year old, the attack is an infection in which the toxin is produced by the active bacteria in the child's body.

9.7 Seafood Zombies

Another very potent food poison found in some foods is called **tetrodotoxin**. This is a powerful neurotoxin about 500 times stronger than cyanide. Tetrodotoxin induces a profound paralysis. In the word of N. Saunders:

> *Tetrodotoxin leads to a complete immobilization where the victim literally hovers between life and death. The condition of stupor, reduced heartbeat and barely detectable respiration is so deathlike that many victims of tetrodotoxin poisoning have been certified as clinically dead only to revive miraculously on the autopsy slab, in the morticians parlour or, to the consternation of the pallbearers, on the way to the grave!*[21]

Individuals can be poisoned by tetrodotoxin by eating a puffer fish. This fish is a delicacy in some parts of the world. Tetrodotoxin poisoning may be a factor underlying the existence of zombies. Before looking at the possible role of tetrodotoxin in the creation of zombies, it is necessary to explain the difference between the popular image of a zombie, as shown in countless films, and a real-life zombie. In the words of Saunders:

> *countless films and books have portrayed the living dead (zombies) as gory animated corpses breaking through the soil to cause havoc and terror amongst simple country folk. This gross distortion has become the pervading public image of what in truth is a complex but fascinating system of spiritual belief, sacred knowledge, political expediency, and effective social control.*[21]

The word zombie is of African origin and was a name used by West African slaves taken to the Caribbean Islands. In the voodoo cults of these slaves, the zombie was the snake god. In many religions, the snake is a symbol for new life, because of the fact that periodically snakes shed their skins and apparently create a new body. In the voodoo cults, a zombie was a spirit that could enter a corpse and reanimate it. The word ***voodoo*** is another word of African origin which has different meanings in different languages. In a West African language known as Fon, it means "spirit," whereas in another language, Ewe, it refers to a subsidiary god or a demon.[22] In popular imagination, the term voodoo has been corrupted, and today it conjures up the idea of an evil witch doctor sticking pins in a doll made to look like someone that the doctor wishes to hurt. Miles away, because of the voodoo curse, the unwitting victim writhes in pain as each pin is pushed into the doll. In fact, voodoo is a religion derived from ancestor worship. It is practiced throughout the Caribbean, parts of South America as well as in the Southern United States. Devotees believe in one supreme god, as well as in lesser gods that demand ritual service while acting as protectors, guides, and helpers for individuals and families. Voodoo priests have the power to turn people into zombies as a form of social discipline. Zombification was achieved by giving the victim what was known as a "**coup poudre**" (a phrase from the French language meaning "powder that hits"). Davis was able to establish that a major ingredient in this powder was taken from the puffer fish. The person given this powder appeared to die. In some cases the zombie victim was buried alive only to recover at a later date to wander the island, listless and alone – forever outcast. This aspect of zombiism was given a great deal of publicity when news stories involving a zombification victim, Clairvius Narcisse, were printed in the Western press. He was buried alive in 1962 after being given a "coup poudre." It is claimed that he was later dug up and kept in a drugged state as a slave laborer for the people who had given him the drug. Narcisse was a citizen of Haiti at the time of his zombification. In countries where law administration systems are weak, there may well have been many deaths from these zombification powders.

Public-health authorities must constantly be aware of products that may be harmful or lethal. Sometimes the experts differ as to the safety of a consumer product. For example, the widely used artificial sweetener **saccharin** was banned in 1977 in Canada and the United States because of some experiments, which showed that the artificial sweetener could cause cancer in laboratory animals. However, other scientists claimed that the animal experiments were not appropriate tests for judging the effect in humans, because the dosage used in the animal tests was too high. The chemical is now cleared again for use as an artificial sweetener. (Not all individuals like saccharin as an artificial sweeteners; approximately one in ten people experience a bitter aftertaste when using the sweetener.) The saccharin controversy has been reviewed by Timbrell.[1]

The drug industry is very active in testing possible side effects of newly developed drugs. One of the specialists involved in this type of research is called a **teratologist**. The Greek word for a monster is *tero*. A teratogenic chemical is one which causes development malformations and monstrosities in organs, particularly in the developing fetus. The first major example of a drug induced toxicity that hit the general population was the use of thalidomide. The drug is now a well established human teratogenic.

Thalidomide is a sedative that was used in the 1950s and 60s in the treatment of morning sickness during pregnancy. Early experiments seemed to indicate that it was an effective and relatively non-toxic drug. However, as its use by pregnant women became more widespread, doctors found that the drug was associated with characteristic limb deformities in newborn children. In these infants, the arms and legs were foreshortened. Further work indicated that these deformities were associated with the use of thalidomide from weeks nine to 24 of pregnancy.

In the early 1990s, the possible use of thalidomide as an effective sedative in men was explored, and it appeared to offer some use in the treatment of AIDS. However, in view of its past history, experiments with the drug are preceding with great caution. The question of who was responsible for the problems caused by taking the thalidomide has been the subject of many lawsuits. Heavy financial settlements have been paid by the drug companies to help support the lifestyle of the victims of the drug.[1,23]

9.8 Deadly Mushrooms and Killers from the Crypt

The use of poisonous mushrooms to dispatch victims is a frequently used plot in crime novels. Thus, in a Clint Eastwood movie made in the 1970s, the leading character was a Yankee soldier hiding from the southern army at a girls' school. After he had dallied with the affections of several of the female teachers and several of the girls, they decided they had had enough of his duplicity. They planned and cooked him a special meal with poisonous mushrooms. After a few bites of the mushrooms, his activities in the girls' school ended suddenly.

In everyday speech, the word mushroom is used to describe many different types of fungi. Fungi are plants, which must live off other plants, because they do not make their own **chlorophyll** (a green chemical that enables plants to utilize the sun's radiant energy to make sugars and carbohydrates). In everyday speech it is common to refer to edible fungi as mushrooms, and the term **toadstool** tends to be used for those which are considered either poisonous or inedible. This classification of the two types of fungi has no scientific basis. The most common edible mushroom, has the scientific name **Agaricus campestris**. In nature this field mushroom is found in open pastures, grassy places, and well-manured fields during most of the summer. It is cultivated commercially in caves, dark cellars, and specially constructed mushroom houses.

Perhaps the most famous edible fungus is known as the **truffle**. These plants grow essentially underground and give off a distinctive odor. Their underground location may be determined by animals trained for this purpose. Pigs and dogs are both used in the hunting of truffles. Perhaps pigs could also be trained to smell out drugs and other contraband!

In an encyclopedia article on poisonous mushrooms it is pointed out that the number of poisonous fungus species is probably less than one hundred.[24] Many mushrooms formerly considered doubtful or poisonous have been found to be edible. The original misconception regarding these mushrooms probably resulted from observations of sickness following the consumption of mushrooms that were no longer fresh, and which contained poison similar to those generated in putrefied meats and vegetables. In the same article it is noted that members of the fungus family **amanita** are extremely poisonous. They contain organic toxins, which destroy cells in the central nervous systems, blood vessels, kidneys, liver, and in muscle tissue. The ways in which different toxic products of poisonous mushroom kill when ingested are also discussed in reference 24.

Some mushrooms are a source of hallucinogenic chemicals. Kate Charlesworth comments that the mushroom **fly agaric** is the presumed source of a hallucinogenic drug called soma. She tells us that some of the visual imagery of "Alice in Wonderland" is a suspiciously accurate description of the type of hallucination induced by this drug. Perhaps the mushroom on which the caterpillar sits as it talks to Alice is not just a picturesque image, but hints at the source of the visual inspiration for the stories in this book and suggests that Lewis Carol may have nibbled at "magic mushrooms" himself. Fly agaric, or fly mushroom, is common in the open woods, wood margins and at next to roads. The popular name for this mushroom comes from the fact that it kills flies.

The second part of the title of this chapter probably conjures up images of toothsome vampires emerging from their tombs at midnight to stalk innocent victims for their blood. However, the type of killer that I had in mind when preparing this section, is an invisible cloud of dangerous fine particles. Specialists in inhalation hazards distinguish two types of dangerous dusts. The first type is simple dust, such as freshly shattered quartz dust, which causes silicosis in miners. The other is what is known as an aerosol. The term aerosol is the technical name for a dust cloud. The expression "viable aerosol" indicates that this type of dust contains living organisms, such as bacteria or viruses. Some of the medical textbooks say that viruses cannot exist as free-floating fine particles; however, they can pose an inhalation hazard if they are encapsulated in the dried mucus from coughs and sneezes. In a hospital, invisible clouds of viable aerosols created by a sneeze can travel through the ventilation system to cause new illnesses in patients. Infections acquired in hospital are described as **nosocomial diseases**. We can expect to see a great deal of litigation against the hospital authorities by relatives of patients who die from nosocomial diseases. As the general population and lawyers become more aware that many cases of nosocomial disease can be prevented by better hygiene in the hospital. A detailed discussion of the way an aerosol spreads nosocomial diseases is beyond the scope of this text.

The unseen threat of a viable aerosol that I had in mind when creating the title of this section, was the dust cloud created when workers enter the crypt of a church to move old tombs. Smallpox has been eradicated as a disease, but health officials advise workers who move ancient bodies in crypts that, if the people died from smallpox, there is a remote possibility that dried-out sores on the body could contain dormant bacteria, which could turn into viable aerosol. Recently, authorities in the Baltic states were worried after the body of a person who had died of the plague was stolen from a cemetery. The grave robbers were in danger of inhaling plague-bacteria dust, rubbed from the dehydrated body.

There have been several suggestions that dangerous aerosols may be a possible explanation for the supposed curse of the ancient Egyptian pharaoh, Tutankhamen. This pharaoh was one of the last kings of the 18th Dynasty of Egypt. His tomb was discovered by Lord Carnarvon and Howard Carter in 1922. Soon after the discovery of the tomb, Lord Carnarvon died of pneumonia. Since that time, several people associated with the discovery have died violent deaths. These deaths have generated the idea that Tutankhamen uttered a curse on anyone who disturbs his tomb. The fact that Lord Carnarvon died of pneumonia could indicate that he had inhaled an aerosol, which either contained viable bacteria or lethal fungal spores. The tomb had been sealed for centuries and it may be that infectious agents of dangerous fungal spores were stirred up when the tomb was disturbed. Inhalation of this killer dust could have caused the mysterious deaths of the people involved in the discovery and those who later worked with artifacts from the tomb. One of the problems of accepting this explanation of Tutankhamen's curse is that Howard Carter, the co-discoverer of the tomb, lived to a ripe old age. This could, however, be explained by the fact that Howard Carter had spent many years working in the various tombs and archaeological sites of Egypt, so he may have developed an immunity to "the disease of the curse" from frequent low doses of fungal spores and other dangerous tomb dust. Whatever the reason for the supposed curse of Tutankhamen, anybody working with dusty objects should be aware of the health hazards of dust. We will discuss several of these dust-caused diseases in later sections of this chapter.

Even if there are no killer dusts in the air, poisonous unseen gases which can kill people can be generated by combustion processes. Thus, in January 1986, a young woman suffered nausea and headaches for weeks. She blamed her problems on a pregnancy. However, a blocked chimney had created the health risk. Soot had partially clogged the furnace system and had allowed carbon monoxide to leak back into the house, creating health problems for its inhabitants.[25] **Carbon monoxide** is formed when coal, gasoline, or diesel oil are burned in confined spaces without adequate ventilation. Similarly, it was reported that 15 people became sick during an ice-hockey game held in an enclosed rink. All the people who were sick were either hockey players or coaches. The primary symptoms were nausea, weakness, severe headaches, and even unconsciousness. At first it was thought that this might have been an outbreak of mass hysteria, but blood tests showed acute carbon monoxide poisoning. The problem was traced to the use of a gasoline-powered ice-resurfacing machine. Tests showed that this machine pumped large quantities of carbon monoxide into the enclosed skating rink. Enough carbon monoxide was in the indoor air to make the fast breathing hockey players suffer carbon monoxide poisoning.

Every now and again during a cold spell, there is a news item reporting that people have been killed by carbon-monoxide poisoning from using charcoal burning barbecue equipment inside the house to keep themselves warm.

Another source of carbon-monoxide poisoning is the muffler system of old cars. Sometimes an old exhaust system will leak carbon monoxide below the car body. The gas then moves through holes in the body work into the car to poison the passengers. Carbon monoxide has been used in murderous crimes. A 1980s article in the newsletter of the Ontario Science Center entitled "The Crime of the Decade" told the story of a murder carried out using carbon monoxide. The story describes how police were called to the home of a well-to-do corporate executive, who claimed he awoke to discover that his house was full of exhaust fumes, and to find his wife lying unconscious inside their car, which had been left with its engine running in the adjoining garage. The woman was found too late to be revived, and her death was apparently suicide. However, when the police entered the house they found a dying blaze in the fireplace. Subsequent investigation turned up soot stains on the wife's bed and underneath the bedroom window. A partially burnt garden hose was found in a garbage pile behind the house. Specialists from the Toronto Forensic Science Center were asked to stage an experiment at the executive's house. The garden hose was connected from the tail pipe of the car to the master bedroom. The room was soon filled with exhaust gases and a chemical analysis showed that the air of the room had an eight per cent concentration of carbon monoxide. This is enough to kill a person. When the hose was removed from the room, it sprayed soot in the same spot under the bedroom window where soot stains were found on the fateful night. Further chemical analysis of the hose in the garbage pile revealed it was made of material that can be burned in a fireplace. The executive was convicted of poisoning his wife in the master bedroom, and then moving the body out into the garage. He was sentenced to life in prison.[26]

9.9 Crazy Cats and Mad Hatters

Perhaps one of the worst episodes of food poisoning occurred in **Minamata**, Japan.[29] The Chisso Corporation, a chemical company, which manufactures industrial chemicals, dumped mercury into the bay. The first sign that this dumping was causing a problem were poisoned cats, which went crazy and jumped into the sea. It was later discovered that this crazy behavior was due to damage of the nervous system of the cats, caused by eating mercury polluted fish.

It is unfortunate that in popular discussions of poison episodes, oversimplified scientific terminology, such as "**mercury pollution**" is used; often the chemical nature of the problem is even disguised by calling the sickness by names such as "Minamata." The metal mercury, the only metal that is a liquid at room temperature, is not as dangerous as many mercury-containing compounds. Thus, in an overview of mercury poisoning, Glenn makes the following statements:

> *Inorganic compounds of mercury are highly toxic by ingestion, inhalation or absorption. Following exposure to mercury salts, the kidneys are the organ of most concern. Skin problems and allergic reactions are common.*[27]

The distinction between **inorganic and organic compounds** is made by the chemists mainly to describe whether the material in question is combined with other inorganic materials or with carbon and hydrogen. Thus, the compound of mercury with sulfur is an inorganic salt, while a compound known as methyl mercury is an organic compound, which is extremely toxic. Because compounds of mercury vary in their toxicity, some compounds are relatively safe whereas others are extremely dangerous. An important aspect of the poisoning of people with mercury compounds is the fact that, up to 20 years ago, mercury itself was considered to be relatively harmless if it was immobilized in river sediments.

However, partly because of the stimulus of the Minamata episode, scientists discovered that chemicals in the mud of a river or lake can convert inorganic mercury into methyl mercury. This organic mercury compound can then be incorporated into plants and eventually the fish.[28,] Although inorganic mercury in the river bed was considered relatively safe, scientists have always been aware that mercury vapor in a room, over a long period of time, can have disastrous results. We will discuss this fact in greater detail later in this section. The final lawsuits in the Minamata case were so punitive financially that the Japanese government had to step in to save the company from bankruptcy. They did so because the collapse of the company would have created great economic hardship in the surrounding countryside. The question of liability in such situations is complex and hinges on whether the company knew that it was causing the problem. Often this type of situation is made more difficult by the reluctance of the company to admit the reality of the problems.[29] We will discuss a similar case in more detail when we look at mass deaths allegedly caused by asbestos.

In all fairness it should be pointed out that the diagnosis of mercury poisoning is not always easy. Thus Glenn comments:

> *the medical effects are explicit though insidious; they are often confused with other emotional or health problems including stress.*[27]

As enumerated in an industrial reference text in mercury poisoning:

> *psychic and emotional disturbances are characteristic; the victim becomes excitable and irascible, especially when criticized. He loses the ability to concentrate mentally and becomes fearful, indecisive or depressed, and may complain of headache, fatigue, weakness, loss of memory and either drowsiness or insomnia. Objectively he exhibits a fine tremor and is unsteady in attempts to perform fine motions. The tremor may effect hands, head, lips, tongue or jaw. His writing is affected with letters omitted or even becomes illegible.*[30]

Commenting on these symptoms, Glenn states that even though he did not suffer from mercury poisoning, he experienced many of the symptoms as "an editor who frequently misses his deadlines. I found the litany of behavioral symptoms disturbingly familiar." Glenn points out that the behavioral symptoms listed in this description of a mercury-induced industrial disease fits the character of the mad hatter in "Alice in Wonderland." At the time that Lewis Carol was writing his book it was well known that many workers in the hat industry showed this type of behavior. Again, in the words of Glenn:

> *This was because they were undoubtedly poisoned by the mercury used to treat the furs in hat factories. Hundreds of men and women who cut felt or shaped hats in factories at the turn of the century were seized by real tremors, psychic problems and aggressive mood swings.*[27]

Glenn also points out that there are many historic examples of mercury poisoning recorded in the history books. Thus, the Chinese **alchemist** (the name used for an early scientist who was part wizard and part scientist) Ko Hun wrote, 4,500 years B. C. that

> *holding mercury in one's hand would ward off evil spirits.*

Ko Hun practiced what he preached and suffered from severe mercury poisoning! Glenn also tells us that the famous eighth century Arabian chemist Abu Jabiribnhayyan, whose name was westernized to simply Gerber, carried out many experiments with mercury. His writings were so confused that his name has given us the word jibberisch in the English language.

Glenn comments that the confusion in Gerber's writings arose because he was an early victim of chronic mercury poisoning.

Lenhin tells us that Charles II of England may have met his death from experimenting with mercury. Two of England's greatest scientists may also have suffered from mercury poisoning, because they experimented with mercury in rooms without ventilation. Sir Isaac Newton carried out many of his experiments on the properties of light using large pools of mercury. In one of his books he describes how he repeatedly distilled kilograms of mercury in an enclosed room. At about the age of 45 Newton's productivity began to decline abruptly. His letters of the period show a characteristic shakiness in the handwriting and he accused others of persecuting him. He was noted for his outbursts of rage against even his closest friends. Sites, of Harvard University, has compared the behavioral patterns and symptoms of Newton with those anticipated in a person suffering chronic mercury poisoning and concludes that Newton did suffer from mercury poisoning.

Michael Faraday used a great deal of mercury in his electrical experiments. In 1862, when he was 70, he wrote that he felt he had to retire and he noted that he suffered from loss of memory and physical endurance of the brain, hesitation, uncertainty of convictions, inability to draw from the mind the treasure of knowledge previously received. In the middle of his farewell lecture, he was unable to continue talking. There seems to be little doubt that he suffered from mercury poisoning.

When I was a student studying physics in high school in the late 1940s, we used cheap electrical switches of copper wire, made by our school teacher. The ends of the wires were dipped into mercury filled wells in wax blocks. Like children the world over, we enjoyed chasing drops of mercury across the surface of the bench to see how many could be made to collide with each other or shatter into a thousand drops. I shudder to think how much **mercury vapor** we inhaled. A very serious problem with such old school laboratories is mercury that fell to the floor and disappeared through cracks. Many of these old buildings are still in use and can pose a health hazard to people working there. As students we were probably safe, because we only spent two or three hours a week in such laboratories. Perhaps our teachers were more at risk than we were. The very real problem with older laboratories in some countries has recently been illustrated by a study of the air in a laboratory building in Cambridge, England.[31] Many of the very famous experiments in physics, which led to the splitting of the atom, were carried out in the building known as the Cavendish Laboratory in Cambridge, England. In 1970 the physicists moved out of this building and it was converted for use as a social-science building. In 1990, 43 members of the staff were tested for mercury poisoning. 14 were deemed unexposed while 23 were rated as exposed and six had levels characteristic of people working with mercury in the electronics or chemical industry. When the air quality in some of the offices was measured it was found to contain mercury vapor above the safe level set by the health and safety executive. As a consequence of these studies, the people working in the building now have their urine checked every three months to monitor mercury-exposure levels.

It is incredible how lax people can be who handle mercury. For example, in September 1993, the city of Hamilton, in Ontario, Canada, declared a state of emergency after school children were found playing with mercury. They were throwing it at each other and even selling it in small plastic bags for two dollars. Authorities suspected that as many as a hundred students at eleven schools may have been exposed to the mercury, which was stolen from a laboratory in a closed industrial plant. The surprising aspect of this story is that the mercury was left in the derelict factory, because mercury is valuable. When I worked in a laboratory in Chicago someone was stealing mercury there. The director of research arranged for a hidden camera to be set up to photograph the thief. We never solved the problem because the thief stole the camera! When people are contaminated in an old building containing residual mercury under the floorboards, it is very difficult to establish who is liable for the problem. Monitoring the workers seems an inadequate approach to the problem. At the very least it seems that the owners of the building are responsible for taking up the floor boards and searching for other sources of mercury, for instance, in porous walls, until the air in the building no longer contains dangerous levels of mercury vapor.

In August 1989 a four-year-old boy in Michigan developed severe rashes, itching, sweating, and a personality change, ten days after the interior of his home had been painted. At first the doctors were mystified by this sudden illness, then an astute observer recognized the symptoms of mercury poisoning. Where could the mercury be coming from? Tests of the latex paint, which had been used to paint the room showed that it contained a mercury compound intended to prevent mildew and also to suppress slime in the paint when it was stored before being used (such compounds are called slimicides).[32] Tests showed that the latex paint used to decorate the room contained 2.5 times the amount of mercury compound recommended for use as a slimicide. Health officials were able to track down 19 other families in Oakland County, Michigan, whose homes had been painted with the same brand of paint within the previous six months. Air inside the newly painted homes contained levels of mercury up to 20 times higher than the acceptable residential concentration. The highest concentrations were found in the most recently painted rooms. In this case the simple solution was to ventilate the rooms. Not in all countries is it mandatory to label paints. As far as I am aware, there have been no legal suits against the manufacturers of paints, which contain higher than permitted levels of mercury compounds. Part of the problem is that, as already pointed out, the symptoms of mercury poisoning are very difficult to distinguish from problems caused by other factors. However, as the public becomes more sophisticated one can expect to have lawsuits in which manufacturers of paint containing illegal quantities of chemicals, such as mercury based slimicides, are subject to legal action.

Dianna Gibson worked as a dental assistant. In the late 1970s, she began to have health problems. For no apparent reason she continuously walked into the sides of doors and picked up wrong instruments. She began to stutter and a tingling sensation invaded her arms and legs. The diagnosis of her illness seemed to be as varied as the doctors she consulted. Amongst other things, she was diagnosed as having multiple sclerosis, epilepsy, and schizophrenia. By July 1978, after 11 years, Gibson was forced to quit her job. She moved back to her native Winnipeg from Calgary. One day she read an article on the growing evidence that working with mercury-based compounds in dentist offices was hazardous to dental personnel. Gibson realized that she had all the symptoms of chronic mercury poisoning. This was confirmed by medical tests.[33] A study of Manitoba dental offices in 1983 showed that 14 % of dental personnel surveyed had unacceptably high levels of mercury in their urine. The government report pointed out that a serious hazard is generated by the accidental spilling of mercury onto dental office carpeting. Now there is pressure to prevent mass poisoning of dental assistants. As yet, it is not clear how effective the monitoring has been. Other studies of health hazards in dental practice suggest that mercury contamination may cause spontaneous abortions and stillbirth amongst female dental workers. Dr. Schipiro of the University of Pennsylvania studied a group of three hundred middle-age dentists. He found that 30 % of the dentists had nerve damage that could be attributed to low level mercury poisoning. Mercury fillings in teeth are known as amalgams. An amalgam is a mixture of two or more metals. Mercury forms very strong amalgams with different compounds. It has been suggested that mercury amalgams in teeth can cause mercury poisoning in individuals, but there is no conclusive evidence. It should be noted that in the early 1990s there was considerable concern expressed in densely populated countries, such as Great Britain, that the mercury vapor released from a crematorium could be causing health problems in the general population. The debate continues!

In the early 1990s officials in Brazil and other South American countries became concerned about mercury poisoning in villages along the Amazon river. Thus, Forendo Banches, a doctor based in the city of Santarem, analyzed the amount of mercury in hairs from 20 members of a fishing community. The world health organization safety limit for mercury is two micrograms per gram of hair. All 20 people were over the safe limit, including a child of 16 months. Some of them were three times over the limit. The mercury was entering the river system from gold mining in the various tributaries of the Amazon. It has been estimated that 1,200 tons of mercury have been dumped in the Amazon in recent times. The contamination of the water is spreading into Venezuela, Paraguay, Argentina and Bolivia.[34,35] Officials are predicting a major disaster amongst people living along the river banks and amongst wildlife in native reserves, such as the Pantanial Wetlands of Brazil. Already some fish caught in that area have been found to contain 24 times the level of mercury that the world health organization considers safe. Although the problem of the river-system contamination is a sufficient tragedy in itself, the gold miners are also suffering very high levels of mercury poisoning. It is very difficult to control gold mining activities in sparsely settled areas of Brazil, and the mercury poisoning of miners and fisherman is likely to be an ongoing story in coming years.[36]

9.10 Poisonous Plates and Sweet Wine

In 1978 when Don and Fran Wallace of Seattle were stationed by the United States Air Force in Italy, the couple bought a charming set of dishes in a small village. Soon after the purchase Don became irritable and aggressive. He lost more than 30 pounds and suffered from insomnia, and pain in his wrists and forearms. In 1981 the Wallaces moved back to Seattle, where the wife Fran became gravely ill with body aches, anemia and dehydration. After searching through medical books, Don insisted that they be tested for lead poisoning. The two were found to be severely poisoned. Finally, Wallace traced the source to the Italian dishes. Two coffee mugs were especially dangerous. They released over three hundred times more lead than the federal standards permitted. The problem with pottery is the finish applied to the dishes. If the pottery is covered with a glaze and heat treated at a high temperature, the poisonous material in the color finish of the pottery is safely sealed from the fluids placed in the cups. However, dishes bought in some parts of the world are not fired at the right temperatures and can be lethal for those who drink from them.[37] The couple's experience with contaminated pottery is not unique. In the United States in 1985, doctors of several critically ill patients with symptoms of lead poisoning traced the problem to the traditional pots bought in Mexico.

Another source of lead poisoning is modern scouring pads and dishwashers, which apparently break up the lead oxide glaze, leaving large amounts of lead to be dissolved by any slightly acidic liquid. If you look inside cups washed regularly in a dishwasher you can see the pattern of scour marks caused by the spinning of the cups on the support pins of the dish-washing machine. Investigations found that some coffee placed in damaged cups contained more than a thousand parts per million of lead after a few minutes. The persons who carried out the investigation, Dr. D. M. Wallace and his colleagues, say that the findings demonstrate a potentially wide-spread health hazard, both in the countries of origin of the pottery and the countries of importation.[37,38] In Canada, Michael Cline and his colleagues at McGill University investigated some cases in the early 1970s.[39] One poisoning, in which a child died, was traced to an earthenware jug in which the child's mother kept a continuous supply of apple juice. The slightly acidic juice was leaching lead out of the glaze of the jug.

Government regulations forbid the importation of improperly fired pottery, but it is hard to stop the entry of souvenirs purchased in various parts of the world. It is interesting to note that James Lind who, in 1753, recommended lemon or lime juice as a preventative for scurvy amongst sailors warned that the juices should not be stored in earthenware jugs. A recent case in Great Britain involved someone who wanted to go back to nature. He had resurrected an apple cider press and became poisoned with lead, because the old press was lined with lead sheets. Lead poisoning, like mercury poisoning, is very difficult to diagnose unless tests are carried out on the blood. Unfortunately, lead poisoning also looks very similar to alcoholism and many people are misdiagnosed. Even physicians are not always aware of the problem. One medical doctor poisoned himself by drinking a cola-type beverage every evening for two years from a mug his son had made for him.[39] There are no known cases of deliberate murder, which can be traced to fruit juice in inappropriate vessels, but if they have been committed it would be very difficult to detect, unless one knew where to look for the lead poisoning!

The sweet wine part of our title to this section comes from the fact that the Romans deliberately added lead compounds to their wine to sweeten it. As we shall discover later in this section, one of the problems with lead compounds as a source of poisoning is that many of them taste sweet. Another long-term effect of lead poisoning is sterility. Some scholars even believe that the fall of the Roman Empire could possibly be traced to wine poisoned by lead compounds, with resultant wide-spread sterility in the ruling families of Rome.[40,41]

Not only Roman society probably had many problems that can be attributed to the use of lead cooking vessels and sour wine; even British history may have been drastically altered by the possibility that Queen Anne suffered from lead poisoning. The early deaths of all of Queen Anne's 17 children may have been caused by lead absorbed from wine and from the cosmetics she used. A widely used pigment in cosmetics was lead carbonate, and it is believed that Queen Elizabeth I ruined her complexion by using makeup containing lead carbonate. With true royal authority Queen Elizabeth solved the problem of the deterioration of her complexion by banning all mirrors! Dr. Roger Rowes, in a communication published in a British medical journals, said that the 17 babies of Queen Anne could have been harmed by lead while they were still in the womb. He cites case records by a physician of the time, which show a high frequency of miscarriages and deaths soon after birth. Thus, Dr. Rowes notes:

Between 1645 and 1715 the quality of German and French wines suffered from the ravishes of atrocious weather and the 30 Year War encouraging wide spread adulteration of wine with litharge (oxide of lead) to improve their flavor. Secondly Queen Anne covered up her blotchy face with cosmetics, which were quite likely to have been compounded from lead salts.[41]

Queen Anne gave birth to only 5 live children, none of whom lived to be adults. Since she had no heir to the throne, her sister Mary became Queen and brought her husband William of Orange to England to become her consort and joint ruler. Many scholars think that the reign of Mary and William was an important change in British History.

The fact that lead compounds can cause problems to the fetus of pregnant women has figured in a very important lawsuit.[42] For many years companies excluded women of child bearing age from jobs where occupational hazards involved the possibility of being exposed to lead. This is because it was well-known that the fetus could be damaged by lead. In the United States a major lawsuit over this exclusion of women from certain jobs was fought out in a court case known as the "Automobile Workers vs Johnson Controls Incorporated." Over a period of seven years the workers fought a legal battle with the company, which established that the company must protect workers and not remove them from the work place. The company, Johnson Controls, had thirteen factories where electrical storage batteries for automobiles and other purposes were produced. The production required large quantities of lead. Johnson Controls strictly excluded women capable of bearing children from any job where lead readings reached specified levels. The company officials had taken this step because of medical evidence, indicating that contamination of a mother could cause serious damage to the nervous system of a fetus she carried. Finally, the Supreme Court of the United States ruled that the bias in the Johnson Controls worker policy was unfair. Fertile men but not fertile women were given a choice as to whether they wished to risk their reproductive health for a particular job. As a result of this ruling companies must now protect workers rather than remove them from the job. As stated in a *Time Magazine* article on the Supreme Court decision, the victory was bittersweet for some of the plaintiffs. For instance, in 1984, Joyce Quals, 41, was involuntarily transferred from a high risk area at Johnson Controls, where she welded posts onto batteries to a safer work place, where she cleaned and installed vents in motorcycle batteries. The move halved her salary. To get back to the higher-paying post, Quals underwent a tubal ligation. She subsequently married and now regrets that she cannot bear children. Quals commenting on the supreme court decision said, "nothing really would make up for it, but now other women won't have to be pushed into the corner to make a similar decision between sterility and pay. This ruling will help others."[42]

In the late 1980s Mr. and Mrs. Spells and dozens of other families launched a lawsuit against the New Orleans Housing Authority. They claimed that the houses in which they lived had been painted with paints containing lead pigments. This paint was flaking and children ingested the flakes of lead paint suffered lead poisoning. The Spells originally discovered their problem when they took their two year old son to a clinic in New Orleans. Ray had lead poisoning and has since been hospitalized 19 times for therapy that helps his body excrete lead. The housing project in question is a few square blocks of three story brick buildings built in the 1930s. Lead-based paint was used inside the apartments and on the outside; on screen doors, porch railings and on the delicate grill work that gives New Orleans its classic French look. In a report on the problem of lead poisoning in children, Pollack gives the figures that eight to eleven percent of poor black children in the cities have over 25 micrograms of lead per deciliter of blood, which is the level that the Centers for Disease Control defines as lead poisoning.[43]

However, the specialists are still arguing as to what is a dangerous level in the blood, and some experts would like to see the level defining lead poisoning fall to 10 to 15 micrograms per deciliter. In a 1988 report prepared for the Congress of the United States, Mushak estimates that of the 2.4 million pre-school children in urban and suburban areas of the United States, 17% have blood levels over 50 micrograms per deciliter. Pollack states that an estimated three million tons of lead remains in paint accessible to children scattered amongst 25 to 40 million housing units. Experts are pushing for lead removal programs to make this type of housing safe.

Another source of lead poisoning in urban and rural areas is contaminated drinking water. In urban areas the problem is old plumbing, which is not only made from lead, but is often joined up with a lead based solder. Thus, Nell Hals discovered the problem of poisoned water when the water main on her street in St. John, New Brunswick, burst in 1990. A city worker who came to repair the main suggested she replace the lead pipe from the city water system that runs into the house. Mrs. Hals called the water department and they basically said patronizingly "Well dear it is not a big problem, there are lots of lead pipes everywhere and nobody has died of it and there is no need to worry." Mrs. Hals was not amused and had the water coming into her house tested for lead. After a double check from the department of the city, it was established that the lead levels in the water was 28 times the acceptable amount! Blood tests showed that her husband and their five children then aged seven to 15 also had elevated levels of lead. The family had to spend $600 of their own money to replace the lead pipe. They did not have sufficient funds to replace all the bad plumbing in the house, but took precautions to run the water for 5 minutes before drinking it. Problems of lead ingestion from paint flakes and plumbing are not confined to North America, the problem is worldwide.[44,45] In my home town of Hull, England, the problem is aggravated by people installing water softeners on their household water supply. The water of that city is so hard (containing large amounts of dissolved calcium carbonate) that, over a period of time, the lead plumbing becomes lined with a calcium deposit that protects the inhabitants from lead poisoning. Installation of a water softener results in the dissolution of this protective coating with subsequent dissolution of lead compounds, which attack the nervous system of the inhabitants.

Dr. Hershovitz, Deputy Chief of the Lead Poisoning Prevention Branch of the Center for Disease Control in Atlanta, Georgia, says the following about lead poisoning:

> *The most common and devastating environmental disease of young children in the United States today, affecting between three and four million youngsters – 17% of the population of U.S. youth. The more we learn about leads effect on children and fetuses, the lower the blood level at which adverse affects can be documented.*[44]

Dr. Hershovitz states that Studies show that children exposed to moderate levels of lead in pre-school years were seven times more likely to fail to graduate from high school and six times more likely to have a significant reading disability than children who had lower lead exposures.[44] Water fountains can sometimes have a high lead content; the Environmental Protection Agency was alerted to the problem of drinking fountains when water from two U.S. Navy fountains was found to have extremely high concentrations of lead. Concentrations as high as 830 parts per million, more than 16 times the safe level recommended by the Environmental Protection Agency were measured. Apparently, the problem was caused by lead-based solder used in some of the connections.[46]

Lead can be a problem in rural areas when drinking water is collected as run-off from roofs, which are lead lined or which have been painted with paint containing a lead pigment. Some states in the United States are introducing laws mandating that property owners remove lead paint from apartments and houses in which pre-school children live. Massachusetts enacted such a law in 1971, but by 1987 fewer than 20 thousand of the states 1.2 million lead-painted units had been cleaned. A commission investigating the case stated that problems in renovating the houses included the lack of awareness and money.[43] Perhaps as citizens become aware of the poison threat they face in their homes the pressure of legal suits will accelerate the clean up.

Paint chips are not the only lead-based hazards. In Northern Ontario a few years ago, several milking cows died mysteriously. Finally, the problem was traced to the edge of a meadow where the cows munched their lunch. Someone had dumped a car battery there. Cows like salt and the abandoned car battery had become a salt lick even though the sulfuric acid probably did not do wonders for the cows' digestion. The cows had died of lead poisoning after staggering around the field with the typical symptoms. Before the problem was solved there must have been several deliveries of lead loaded milk to the local dairy, which presumably was drunk by children as well as adults. The people who abandoned the car battery would probably be surprised if they were charged with killing the cows and endangering public health – but that is the charge that they deserved.

Another case of lead poisoning of cattle occurred in Britain and in the Netherlands in the late 1980s. Poisoning of the cattle was eventually traced to a 188-ton consignment of rice bran imported from Burma, intended to become an ingredient in cattle food. The rice bran was imported into Antwerp, Belgium. The bran was contaminated accidentally, when a ship hatch broke lose and quantities of lead and zinc concentrate, which were stored in an adjacent hold of the ship, spilled into the bran. The importer of the metal concentrate in Antwerp notified the person handling the rice bran – a Mr. DeBruin by telephone (this is denied by DeBruin who maintains that he was only advised of the presence of zinc in the contamination.) The contaminated feed was mixed with high protein fodder, and sold to animal feed companies. The feed was distributed to more than 100,800 farms in Britain and the Netherlands; farmers that had fed their livestock were unable to sell their milk and beef. The people involved in this particular case are being prosecuted in Holland for their part in poisoning the cows.[47,48]

For centuries explorers setting out from Western countries tried to find a sea route to China by sailing around the North of Canada. The search for such a sea route is known as the search for the North-West Passage. In Figure 9.5 a map of northern Canada showing the final route discovered by the Norwegian explorer Roald Amundsen is shown. It took Amundsen nearly three years to complete the journey, and the North-West Passage has not proved to be a viable route for cargo ships. One of the explorers who tried to find the North-West Passage but failed was the British Admiral, Sir John Franklin. In 1845 he set out for Northern Canada with two ships and an expedition of a 129 sailors. After three years no one had heard anything from the expedition and a search for members of the expedition was started. However, they had vanished without trace. In 1855 a physician of the Hudson Bay Company, Dr. John Rae met some Eskimos on Bouthia Peninsula, circa 300 miles south of Beechee Island, who had in their possession some of the artifacts of the Franklin personnel. They had obtained these by trading food to the survivors as they had made their way south. Eventually, the graves of several of the members of the expedition were found, and Rae learned that everyone had died trying reach a Hudson Bay post on the mainland of Canada. Near the first skeletons of the crew found in the 1850s Rae found loaded guns and provisions – food that was still well preserved and presumably edible. Why, if the people had food and ammunition, did they perish? The death of all of the sailors remained a mystery for over a century. Then, in August 1984, an expedition was able to dig up three of the sailors who had died and been buried on Beechee Island. The scientists chipped through nearly 5 feet of gravel and permafrost for four days to reach the dead sailors. Owen Beattie, an anthropologist, studied the exhumed bodies and found that they contained unusually high amounts of lead. This lead

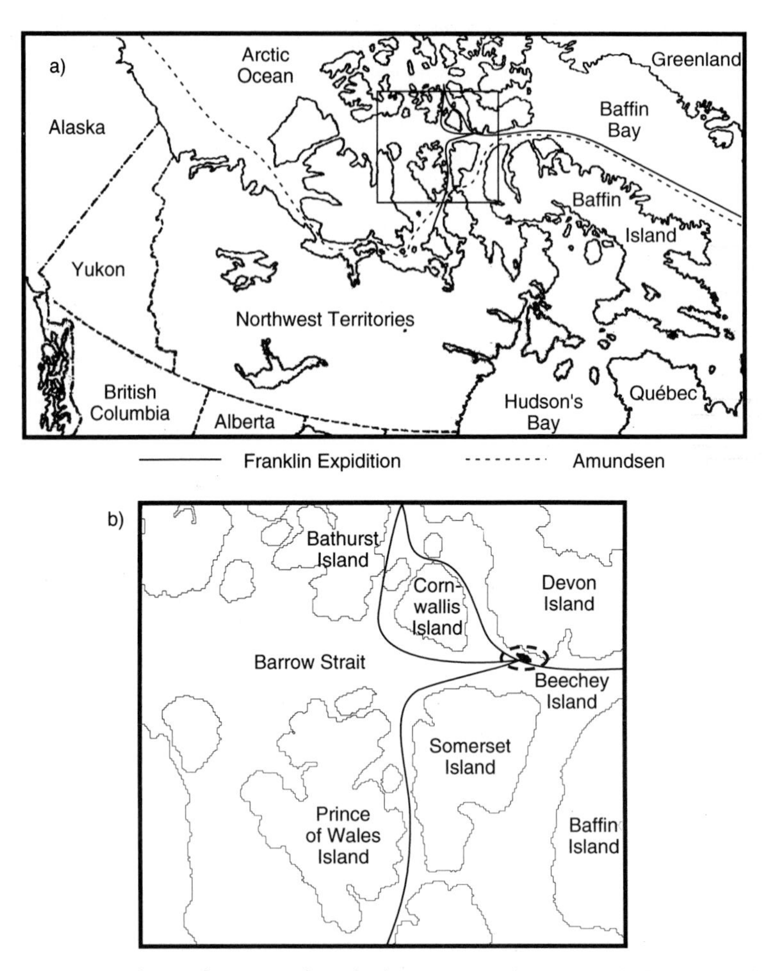

Figure 9.5. For centuries explorers tried to find a way to China around the north of Canada. Amundsen finally completed a journey ending near what we now call Alaska but many sailors died in failed attempts to find this so-called "Northwest Passage." a) The main map shows the Northwest Passage route finally discovered by Amundsen and the path followed by the Franklin Expedition as they tried to reach the safety of the Hudson Bay station on mainland Canada. b) The inset shows the location of the graves of members of the Franklin Expedition found on Beechy Island in 1984 and studied by Owen Beattie.[49]

apparently came from eating food that had been stored in imperfectly soldered tin goods and was taken on the journey as provisions. Scientists now believe that the failure of the Franklin expedition and the death of all its members was caused by acute lead poisoning, which not only caused the crew to behave in unpredictable ways, but probably also made them very susceptible to common illness such as pneumonia.[49]

9.11 The Synergistic Killers

Asbestos is a fibrous-type mineral that can be woven into cloth or made into products such as pipes and floor tiles. The name is rather curious, it means "something that will burn forever." The Greek root words in asbestos are *a*, meaning not, and *bestos* meaning to quench or to put out. The name comes from the fact that the Greeks and Romans burned oil in a shallow bowl to give out light. In these lamps a ball of fibrous material that floats on the oil surface to created a wick. Moss can be used for this purpose, but it tends to burn away as the oil is used up. The Greeks discovered that if a piece of asbestos was used as a wick, the lamp burned as long as there was oil in the lamp, leaving the wick unharmed.

The fibrous nature of asbestos is shown in Figure 9.6. **Asbestos** is a general name for a group of minerals and much confusion has arisen in a discussion of the health hazards of asbestos, because of a failure to differentiate between the major types of asbestos. The two main types are known, respectively, as blue and white asbestos. **Blue asbestos** has the technical name crocidolite. It is mined in South Africa and Australia. Canadian asbestos, also known as **white asbestos**, is a different mineral, known by the technical name of chrysotile. The physical and chemical structure of these two main compounds varies, as illustrated by the information summarized in Figure 9.7. As can be seen from this diagram, the amphiboles type of asbes-

Figure 9.6. Asbestos is the general name for a group of minerals which have fireproof properties. Blue asbestos, shown on the left, is mined mainly in South Africa and Australia. (Note the straight sharp fibres.) On the right, a sample of Canadian chrysotile asbestos, commonly known as white asbestos, has curly, wool-like fibres.

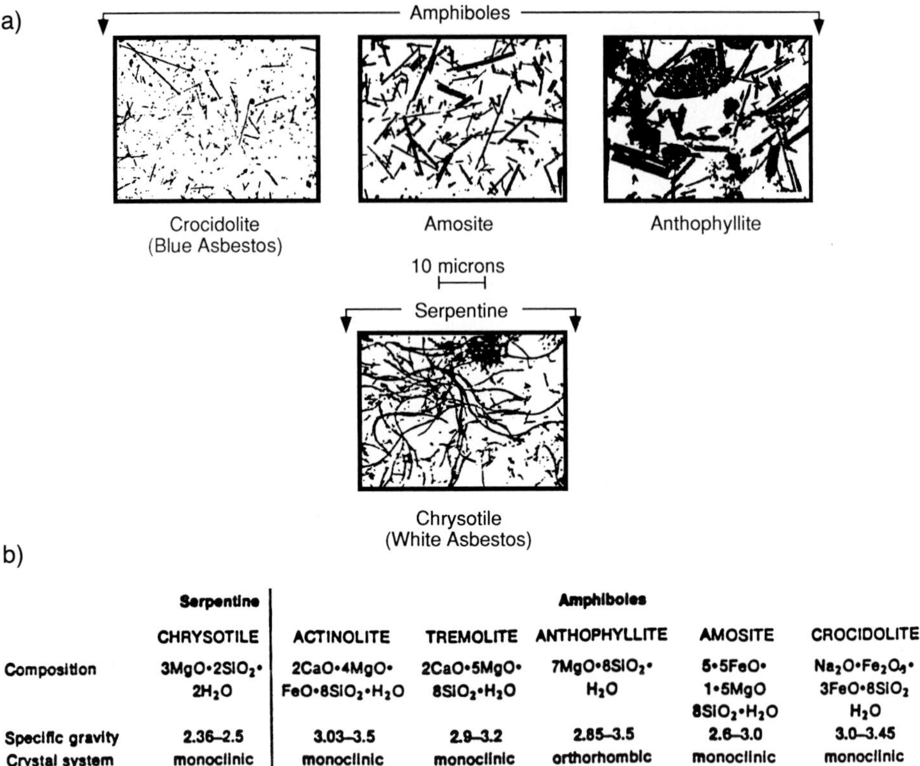

	Serpentine	Amphiboles				
	CHRYSOTILE	ACTINOLITE	TREMOLITE	ANTHOPHYLLITE	AMOSITE	CROCIDOLITE
Composition	$3MgO \cdot 2SiO_2 \cdot 2H_2O$	$2CaO \cdot 4MgO \cdot FeO \cdot 8SiO_2 \cdot H_2O$	$2CaO \cdot 5MgO \cdot 8SiO_2 \cdot H_2O$	$7MgO \cdot 8SiO_2 \cdot H_2O$	$5 \cdot 5FeO \cdot 1 \cdot 5MgO\ 8SiO_2 \cdot H_2O$	$Na_2O \cdot Fe_2O_3 \cdot 3FeO \cdot 8SiO_2\ H_2O$
Specific gravity	2.36–2.5	3.03–3.5	2.9–3.2	2.85–3.5	2.6–3.0	3.0–3.45
Crystal system	monoclinic	monoclinic	monoclinic	orthorhombic	monoclinic	monoclinic
Refractive indices	1.49–1.57	1.62–1.68	1.60–1.65	1.60–1.66	1.66–1.70	1.69–1.71

Figure 9.7. Asbestos minerals can be divided into two main families, which are known as Amphiboles and Serpentines. a) The physical appearance of several types of asbestos fibres. b) Chemical formulae and physical properties of the two main types of asbestos minerals.

tos is mainly needle-like, whereas the serpentine group of minerals, to which chrysotile belongs, is made up of tangled sets of fibers. It is now known that the health hazard of the fibers of the different types of asbestos depends on how fine the fibers are, and whether they are curly or straight. As the asbestos is handled, it becomes finer, because the fiber is made up of many fibrils joined together like a bundle of reeds, which breaks apart as the asbestos is used. Thus, the health hazard for asbestos miners is less than the health hazard for those who install fireproof partitions made of asbestos, for instance, in ships. Those workers most at risk are the ship wreckers who take the asbestos out of the ships. (For the same reason the workers who have taken asbestos out of buildings were more at risk than the people who installed the material). Amphibole asbestos was preferred for making fireproof pipes and building material because, due to its long straight fibers, a slurry of this material drains better than a chrysotile slurry. Chrysotile, on the other hand, was preferred for weaving fireproof blankets and clothing for industrial workers. Chrysotile is not as dangerous as amphibole asbestos, basically because its curliness helps prevent penetration into the lung when inhaled. Also, the material is more soluble in body fluids than blue asbestos. Many lawsuits have been launched by people who worked with asbestos or who lived close to asbestos mines. The number is so large that the pending payments will probably threaten the stability of many companies that formerly manufactured asbestos products. Perhaps even the major insurance institution Lloyd's of London will be endangered.

The reader may be wondering what the health hazards of asbestos has to do with the title of this section. The explanation of the title lies in the data summarized in Figure 9.8. This data was collected by two epidemiologists, Elmes and Simpson, who studied the death rates of workers who installed asbestos-based insulation in a Belfast (Northern Ireland) shipyard.[50] To be able to discuss the significance of the data of this graph it is necessary to examine in detail how epidemiologists search for the cause of the disease. It is also important to realize that there are conflicting scientific views over the interpretation of some epidemiological data. Some scientists are uncomfortable with epidemiloogical studies, which often link an observed pattern of disease with a possible cause, without offering any apparent clearly delineated relationship between cause and effect.

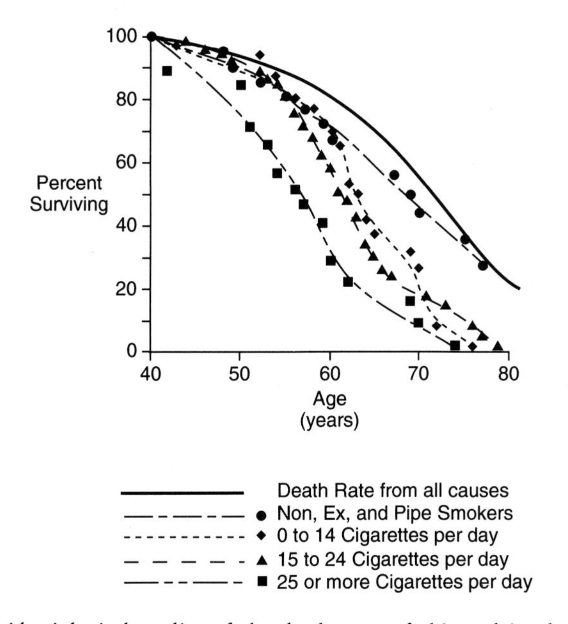

Figure 9.8. Epidemiological studies of the death rates of shipyard insulation workers in Belfast, Northern Ireland, indicate that heavy smokers give up at least 10 years of life expectancy to indulge in their nicotine addiction. (Reproduced with permission from the *British Journal of Medicine 34*, 174.)

Modern science has reached its pinnacle of achievement by linking cause and effect. When no obvious direct linkage between a suggested cause and a pattern of events can be established, quickly the disbelievers jump on the scene, demanding more proof! One of the earliest events in the history of epidemiology illustrates the problems of proving a suspected linkage between observed patterns of events and possible causes. In 1849, a severe epidemic of cholera struck the city of London, England. John Snow, a physician, plotted the location of the cases of cholera on a map and noticed that most of the people who died had drunk water from the Broad-Street pump. This water coming from the Thames river had the taste and odor of sewage. Snow went to a meeting of the Parish council and advised them that they could stop the epidemic by taking away the handle of the pump, so that no one could use it. At the time this seemed to be a very strange piece of advice, because scientists had yet to discover the existence of bacteria and their role in spreading disease. We now know that cholera bacteria thrive in sewage-contaminated water and hence drinking the water spread the disease. In this case the advice was followed, but many years passed before the spreading of cholera was understood by medical experts. The work of the epidemiologist in his study of the patterns of events leading to possible causes is very similar to the way in which detectives work with circumstantial evidence to link cause and effect in the committing of a series of crimes. For this reason, epidemiologists are sometimes called "the disease detectives."[51]

In recent years there have been many efforts to track down the causes of cancer. A major weapon in this search is the construction of occurrence maps of a given type of cancer. Thus, in Figure 9.9 occurrence maps for two different types of cancer in Great Britain are shown. When looking at the occurrence of nasal cancer the specialist recognizes that locations of high cancer of the nose occurs in cities where furniture manufacture is a main industry. Workers in these industries are now pushing for recognition that some types of sawdust can cause nasal cancer. This suggestion was rejected initially by the makers of furniture on the grounds that sawdust is widely recognized as a nuisance dust and does not pose a health hazard to the workers. However, technology changes. The manually operated saws and drills used on natural woods may create harmless sawdust. However, high-speed drills, power sanders, and other equipment generate a great deal of heat at their working point. In materials made from wood chips and phenolic glues, carcinogenic chemicals may be created in the thermal breakdown of the material. These chemicals can then be adsorbed onto the sawdust to turn it into a cancer causing inhalable dust. As a result, sawdust of the 1990s may be much more hazardous than sawdust of the past.

As one attempts to find the causes of industrial cancers there are some surprises. Thus, in Great Britain truck drivers showed an excess of bladder cancer, which may be due to exposure to diesel fumes. Most people in the early days of occupational health and hygiene studies expected diesel fumes

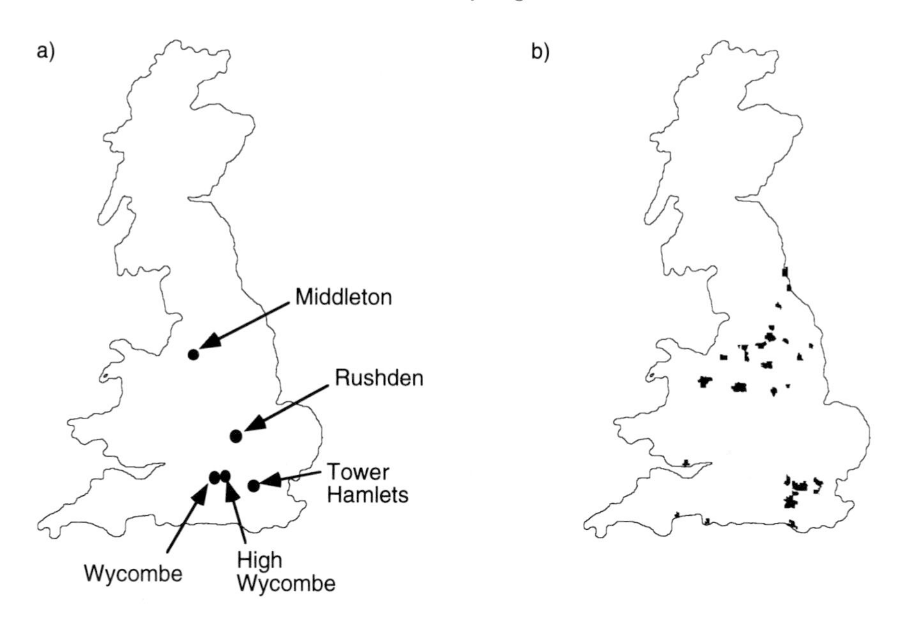

Figure 9.9. Epidemiologists create maps showing patterns of disease and then suggest possible causes for the structure of the observed pattern. a) Occurrence of nasal cancer in men in England. b) Occurrence of bladder cancer in men in England. (Adapted from *New Scientist*.[52])

to create lung cancer, not bladder cancer. The occurrence of bladder cancer may be explained by comparing soot fine particles in diesel exhaust with dandelion seeds. They float around in the air even though they are relatively large, so that instead of moving into the lung during breathing, they are trapped in the upper parts of the respiratory tract. They are then removed by natural cleaning mechanisms. The chemicals are carried through the respiratory system to the bladder, causing bladder cancer.[52–54]

As already mentioned, many traditionally trained scientists are uncomfortable with epidemiological evidence because of the absence of a direct cause and effect link. Another problem for traditionally trained scientists is that sometimes a problem such as cancer is caused by several interacting factors; thus some workers may be prone to industrial illness because they are concurrently suffering from malnutrition or they may have a lifestyle that contributes to the illness. A particular point of contention is the possibility that some people are predisposed to cancer due to their genetic make up. Thus, Gilbert Omenn of the School of Public Health at the University of Washington, in the United States, discussed the case of workers in a large plant run by ICI (Imperial Chemical Industries) in Huddersfield, England. The workers had used dyes in their industrial activities. A study found that of 23 dye workers in the factory who developed bladder cancer, 22 had a genetic problem that caused the absence of a chemical in the body, which could deactivate a carcinogenic chemical found in the dyes. Grant Brewen argued that industry today should monitor possible genetic predisposition to industrial hazards. This viewpoint is quite controversial in many industries, because it raises the possibility that persons can be denied access to a work place because they are at risk from a genetic predisposition. The unions argue that the work place must be made safe for every worker, regardless of his genetic makeup. In view of the way in which the courts have dealt with the lead-exposure problem, it seems likely that this is the position which will be taken by the courts in the Western world as trial cases on various industrial hazards are put before them.

In the early 1960s considerable controversy was generated by an epidemiologist who put forward the claim that cigarette smoking was causing an epidemic of lung cancer. The first one to draw attention to this pattern of disease was Richard Doll. A famous graph used to illustrate the explosion of the modern epidemic of lung cancer is shown in Figure 9.10. This graph shows that over the period 1910 to 1960 the overall death rate fell steadily, with the exception of deaths from lung cancer. These rose in a dramatic manner. In the 1980s death from lung cancer was so common that it is hard for us to realize that, prior to the beginning of the Second World War in 1939, lung cancer was a rare disease. To demonstrate the way in which the incidence of lung cancer has changed within a generation, it is worth quoting at some length the experience of Alton Ochsner as recorded in the scientific journal, *American Scientist*.[55] Alton Ochsner trained as a doctor in the period around 1919. In his discussion of the incidence of lung cancer he tells us:

> *In 1919 during my junior year at Washington University, a patient with cancer of the lung was admitted to the Barnes hospital. As was usual the patient died. Dr. George Dock a professor of medicine who was not only a clinician and scientist but also an excellent pathologist, insisted upon the two senior classes witnessing the autopsy. Dr Dock stressed that the condition was so rare that we might never see another case as long as we lived. Being young and impressionable I was very much impressed by the rarity of this condition. I did not see another case until 1936 when at Charity hospital in New Orleans I saw nine cases in six months. Having been impressed with its rarity by Dr. Dock 17 years previous I wondered what was responsible for this apparent epidemic. Because all the patients were heavy smoking men who had begun smoking during World War I, and because I learned that cigarettes were not consumed frequently until World War I, I had the temerity at that time to postulate that the enormously increased incident was due to cigarette smoking. This observation and many subsequent ones were based upon retrospective studies which were criticized by statisticians as being of relatively little value.*[55]

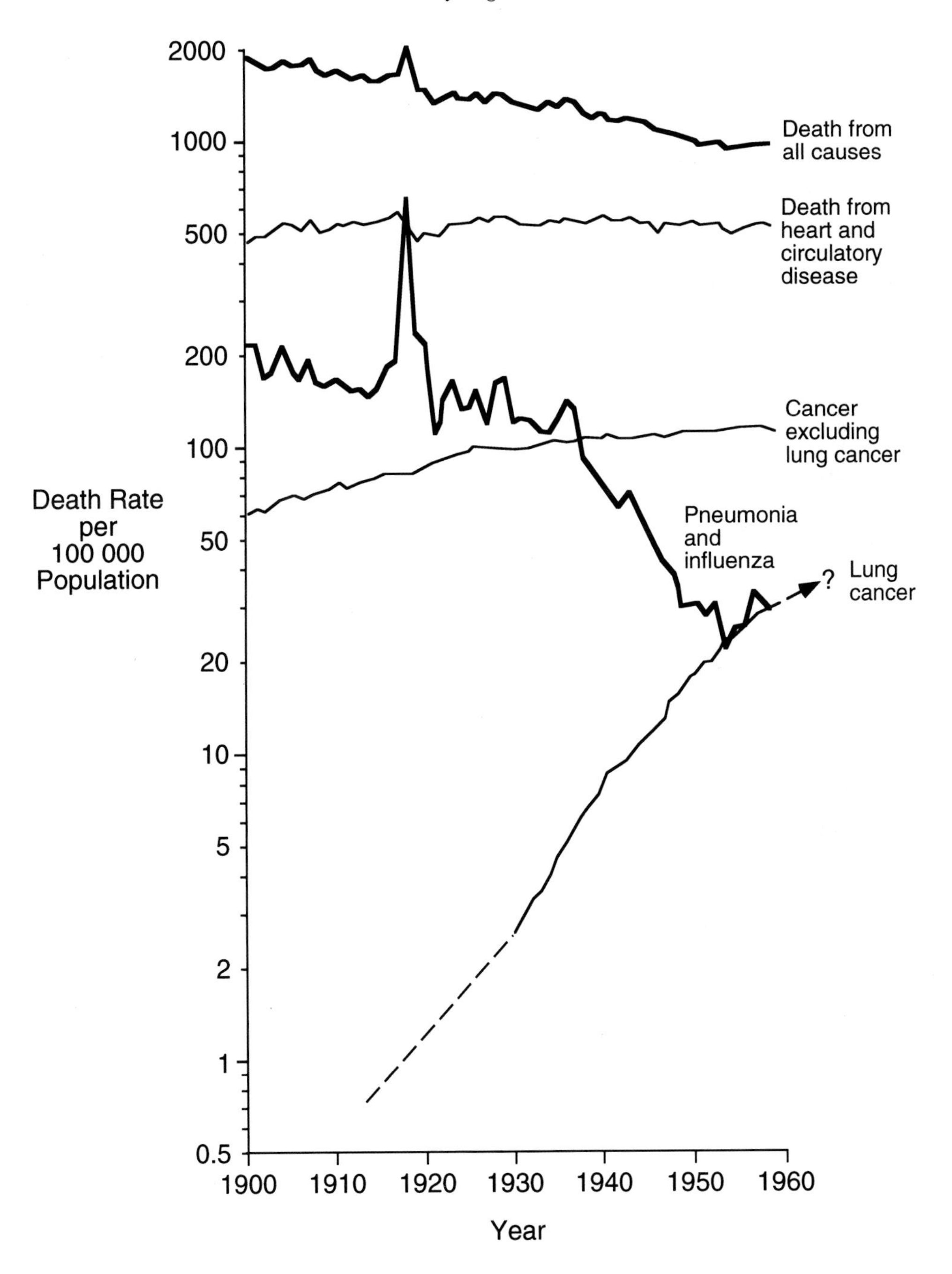

Figure 9.10. The evidence that lung cancer death rates are linked to smoking tobacco products is epidemiological in nature. That is, it is a pattern of occurrence in search of a cause. (Adapted from *New Scientist*.[52])

It is interesting to note that these initial observations by a doctor working in the field were criticized because of their lack of statistical rigor and that, while the statisticians waited for evidence, the epidemic spread.

A recent article on the value of scientific evidence in the law court reports that it "took a decade or more to establish links between cancer and cigarettes." Although that statement was written in 1990, there are still people who reject the link between cigarette smoking and lung cancer because there is no direct causative evidence linking a chemical in the smoke to a cancer. In the law courts there have been attempts by relatives of heavy smokers who have died of lung cancer to claim compensation from the tobacco companies, but as of 1994 none of these law suits were successful.[56,57]

Epidemiologists have continued to study the link between smoking and lung cancer. Thus, in 1993, Richard Doll celebrated his 80th birthday at a conference in London. At that conference Doll announced the latest results of a 40-year study of the smoking habits of more than 3,400 doctors. Half of the heavy smokers aged 35 at the start of the study died before their 70th birthday. Only one third of light smokers, and one fifth of the non-smokers had died. Doll pointed out that persons who smoke 60 cigarettes a day and live to be 95 do exist but they are rare. They probably survive because they have an excellent immune system and health inherited from their ancestors. Apparently, those who persist in smoking cigarettes should make sure they have the right heredity to fight the potential illness! In the analysis of his data Doll points out that three out of 200 heavy smokers might expect to reach age 90 compared with nine light smokers, and 30 non-smokers.

In the initial data on lung-cancer studies women seemed to have a better survival record than men, but this apparent advantage of gender has disappeared as the data continues to accumulate. The initial lower death rates recorded for women appear to be due to the fact that women did not take up cigarette smoking in large numbers until the Second World War whereas men started the habit during the First World War. There is apparently an approximately 20-year delay before the smoke creates a lethal harvest of death.

Once there was evidence that lung cancer and cigarette smoking could be linked, epidemiologists started to look for an interaction of cigarette smoke with other industrial hazards. This is illustrated by the data of Figure 9.8. Before we can review the comments of the significance of the data of Figure 9.8, as put forward by the original investigators, it is necessary to review the concept of synergism. A medical dictionary describes **synergism** as:

> *the interaction of discrete agents such as drugs so that the total effect is greater than the sum of the individual effects.*

Thus, barbituates and alcohol are said to interact synergistically to produce a more severe effect than would be anticipated from a simple addition of the effect of the two substances. When discussing the effect of asbestos on the insulation workers (as manifest in the graph of Figure 9.8) Elmes and

Simpson make the following statement with respect to lung cancer:

> *It is related to smoking in a synergistic manner so that the combined risk is the product of multiplying rather than adding the risk of the two factors alone. In the normal working population, including smokers, half the men are dead by the time they reach the age of 72. Taking the insulation workers as a group, half are dead before they reach the age of 61, but the non-smokers amongst the insulation workers live almost as long as the general population, indicating that the risk from smoking in the general population is approximately the same as the risk from asbestos dust in non-smokers. However, the combined risk in heavy smokers is a 50 % mortality at 53 years, a life expectancy loss of nearly 20 years.*[50]

In fact, since the work of Simpson and Elmes in 1977 there has been a greater appreciation of the fact that passive smoking can be a significant factor in industrial disease. Probably most of the people listed as non-smokers in their studies were exposed to considerable passive smoking in the lunchrooms and working environments shared with their colleagues who smoke. Proving what causes the synergism of cigarette smoke and asbestos in assessing industrial hazards is a difficult problem but the following appear to be significant contributing factors to the interaction of the two hazards. First of all, the nicotine in the cigarette smoke paralyzes the hairs lining the airways of the lung (called the cilia) so that they can no longer effectively remove deposited dust. In a healthy person these hairs in the mucus of the airways pass dust particles up to the mouth in the same way that a crowd of people will pass an injured person over their heads to a first aid station. As the damage to the cilia becomes progressively worse the lining of the airway is eroded. The site of the erosion is often where the lung cancer starts. Asbestos fibers absorb carcinogenic chemicals in the cigarette smoke and activate these molecules so that they react more vigorously with surfaces they come into contact with. Cancer results from an uncontrolled growth of normal cells, which is triggered by physical and/or chemical damage to the programming parts that direct new growth. Thus, the very long, thin fibers of blue asbestos appear to cause direct damage to the cells of the lung, physically penetrating into the cell and disrupting the growth instructions stored in the genes inside the cell. Furthermore, the asbestos fibres introduce carcinogenic chemicals directly into the cells.

In a curious twist of fate some people who suffer industrial lung damage are told that the main problem is their smoking habit, even though the official courts have yet to prove a direct link between cigarette smoking and cancer.[58] In court cases which are concerned with the evaluation of industrial hazards, it is obvious that the key question to be answered is, "Who is an expert and what is the value of evidence presented in a court case?" In the next section we will look at the problems of deciding who is an expert witness within the domain of the law courts.

9.12 Frye the Witness

Dr. Douglas Ubelaker is a **forensic anthropologist**. This a relatively new area, which embraces the application of anthropology to the study of bones and human remains in the court of law. In his book on forensic anthropology, Dr. Ubelaker tells us how he was called as an expert witness in a case in which a victim had been hit on the head with a tire iron. The force of the blow had left a ring-like hole in the skull into which the tire iron fitted perfectly. At the trial Ubelaker states that:

> *the defence attorney wasn't very happy to see me. He put me through a Frye procedure which essentially placed the science of forensic anthropology on trial away from the judge in an attempt to disqualify my testimony against his client. The attempt failed. Even though the killer's attorney seemed to lack a serious appreciation for my discipline it was soon readily apparent why he had tried to stop me. When I finally did get to testify before the jury I stood before them with the skull in one hand and the iron bar in the other and demonstrated how the lug wrench fitted directly inside the punch hole. It was damning evidence.*

What did Dr. Ubelaker mean by being put through a **Frye procedure**? At the time of the court case mentioned by Ubelaker, it was standard practice in the United States to base the witnesses expertise on the acceptance of his work by his peers in the field of science, according to the precedent set in the Frye trial.[59] In 1923 in a Federal appeals court, a judge ruled that the lawyers for a convicted murderer named Frye couldn't present the results of a systolic blood pressure test (an early version of the lie-detector test), because it had not won general acceptance in the scientific community. Since then judges have tended to equate general acceptance of scientific evidence with acceptance by the expert's peers. Testimony satisfies the Frye rule if the work behind it has been published in a scientific journal, such as *Nature* or the *Journal of the American Medical Association*. These scientific journals publish papers only after they have been peer reviewed, that is, approved by other scientists in the field (the term **peer** is defined in the dictionary as an equal, someone of the same rank). My personal experience with the application of the Frye rule, when I was asked to be an expert witness in a lawsuit in the United States are recounted in Chapter 1.

From 1923 to 1975 the Frye rule was the basis of court procedures in the United States. A major challenge to the Frye rule came in 1975 when Congress established a rule requiring judges to ensure that expert testimony was credible. Commenting on this situation in an article on the validity of scientific evidence, Freedman points out the rule set in 1975 was so vague that it did not specify that the judges couldn't use peer review as a guide, and many federal judges simply continued to rely on the Frye rule, while others groped for new guidelines.[56] Finally, a precedent was set by the U. S. Supreme Court. The case involved a lawsuit filed by the parents of Jason Dawbert and Eric Schuller, children born with stunted limbs after their mothers took the anti-nausea drug Bendectin during pregnancy. This drug was manufactured by Merrill Dow Pharmaceuticals. It was taken by 33 million pregnant women between 1956 and 1983. When the controversy of the drug's possible harmful effects began, Merril Dow took the product off the market. In defense during the lawsuit, the company pointed to 30 published studies showing no statistical link between the drug and birth defects in humans. The parents who had launched the lawsuit countered with testimony from several scientists, including a California health department official specializing in pregnancy risks, whose research indicated that such a link might exist. However, because the experts for the parents had not published their results in scientific journals two lower courts relied on the Frye rule to bar their testimony. The parents appealed until the case went to the supreme court. In June of 1993, the United States supreme court ruled by a seven to two majority that a judge has to rule on scientific testimony based not on peer review, but on the judges own evaluation of the science in question. The majority opinion on the Bendectin case was written by Justice Harry Brackman, who stated that the judges should have responsibility for examining an expert's credentials, ensuring that the reasoning or methodology underlying the testimony is scientifically valid, and determining that the conclusion presented in the testimony is relevant and reliable. In effect, the court ruled that the inadmissibility of scientific evidence is too important an issue to be left to scientists![56] The scientific community is divided on the impact of this change in the assessment of the expertise of a scientific witness. Gerald Halton, a well known Harvard physicist and Stephen J. Gould, along with a number of other prominent scientists who were involved in the Bendectin case, filed a brief before the supreme court urging them to overturn Frye. They pointed out that many revolutionary scientific discoveries, were originally rejected for many years by peers, who did not understand the new work. On the other hand, the National Academy of Sciences, the American Association for The Advancement of Science, and The American Medical Association all filed "friend of the court" briefs supporting Merril Dow on the grounds that only the scientific community is qualified to distinguish pseudo-science from the real thing. John Dwyer, a professor of environmental law at the University of California at Berkley, who also holds a Ph. D. in chemistry states:

> *The ruling of the Supreme Court with respect to Bendectin and the value of scientific evidence is legally correct but it is bad policy.*

In his opinion, for every case where the new rules will admit a theory that would have been excluded under Frye but that is later proven right, there will be ten cases where it will admit theories that are later proven wrong. Critics of the new ruling charge that it will further tip the balance in product liability cases toward plaintiffs. Juries naturally tend to take pity on victims and give them money. George Levi, a chemist and law professor at Syracuse University states:

> *No matter how many respected scientists the defense brings in, the plaintiffs need only one fringe scientist to sow enough doubt in the jury's mind to allow them to find for the plaintiff.*

In the words of David Freedman:

> *It may be years perhaps even decades before enough cases come through the courts to build up a coherent body of new precedent regarding scientific testimony.*[56]

We began this section with a quote from the book by Ubelaker on Frye; perhaps we should give the last word on "experts" to the same scientist. In his book Ubelaker states:

> *When expert witnesses are called it is not uncommon to find equally respected, equally qualified authorities testifying to exactly opposite conclusions from the same information. If the issue is mental competence it is almost guaranteed the jury will be treated to diametrically opposing portraits by psychiatrists with equally impressive credentials. Neurologists commonly contradict each other in tough cases and so do orthopedists. The role of the expert witness in many fields is frequently little more than the art of advocacy dressed up in a white coat.*[59]

Some may regard this as a relatively harsh judgement on the role of witnesses in a court case, but it illustrates graphically how establishing the expert stature of a witness is likely to become an increasingly acrimonious debate in the law courts. After I had given a lecture on a television show on forgery, broadcast in Ontario, I was asked to give an opinion as to the verity of a signature on a will. I declined involvement in the case pointing out that since I had never published anything on the art of checking the verity of signature it would be most unlikely that my opinions would be accepted in court. Perhaps now that I have written this book the situation may have changed – especially since in the next chapter we will explore the question of checking forged signatures!

References

1. J. A. Timbrell, *Introduction to Toxicology,* Taylor and Francis, London, 1989.
2. J. A. Timbrell, *Principles of Biochemical Toxicology,* Taylor and Francis, London, 1982.
3. G. Griffiths, A. Lith, M. Green., "Proteins That Play Jekyl and Hyde," *New Scientist,* 16 July 1987, 59.
4. For a non mathematical explanation of the special scales used on probability graph paper see B. H. Kaye, *Chaos and Complexity. Discovering the Surprising Patterns of Science and Technology,* VCH Verlagsgesellschaft, Weinheim, 1993.
5. J. Emsley, D. Pallister, "Bulgarian Brollie Baffles Germ Warfare Boffins," *New Scientist,* 12 October 1978, 92.
6. R. Philmore, "An Inquest on Detective Stories," *Discovery* April 1938, 28.
7. J. Emsley, "Whatever Happened To Arsenic?," *New Scientist,* 19 December 1985, 10.
8. M. Haines, (news story) "She Killed Jack the Ripper," *Sudbury Star,* 9 July 1993, 16.
9. The Case of Jack the Ripper is Reviewed in Gardener Associates, (Eds.), *Great Mysteries of the Past,* Readers Digest Association, Montreal, 1991, 356.
10. The career and possible death from arsenic poisoning of René Descartes is reviewed in Gardener Associates, (Eds.), *Great Mysteries of the Past,* Readers Digest Association, Montreal, 1991, 118.
11. W. Glenn, "Arsenic: A New Threat From an Old Poison," *Occupational Health and Safety 3* (1987), 12.
12. M. Harper, *British Journal of Industrial Medicine, 44* (1987), 652. See also "Arsenic and Old Ore," *New Scientist,* 22 October 1987, 35.
13. D. D. Jackson, "Who Killed Sue Snow?" *Readers Digest,* March 1991, 77.
14. J. Johnson, "Croaking on a Crocus," *New Scientist,* 2 February 1991, 69. (This is a review of the book quoted in reference 15).
15. G. Spoerke Jr., S. C. Smolinkski, *Toxicity of House Plants,* CRC Press, Boca Raton, 1991, p. 244.
16. For a review of the life of Van Gogh see H. R. S. Phillips, (Ed.), *Funk and Wagnall's New Encyclopedia Vol. 24,* Funk and Wagnall, New York, 1979, 225.
17. "Van Gogh Wasn't Mad," (news story) *Sudbury Star,* 25 July, 1990, C8.
18. J. Aronson, "Colored Vision? On the Pharmacological History of Van Gogh," *New Scientist,* 30 June 1990, 80.
19. K. Charlesworth, "Legal Drugs," *New Scientist,* 16 September 1993, A7.
20. "Backyard Chefs Should Beware the Dangers of Barbecue Syndrome," *Canadian Occupational Safety,* May/June 1993, 5.
21. N. Saunders, "Law and Order in the Land of the Living Dead," *New Scientist,* 7 August 1986, 47.
22. See the entries on voodoo and zombie in the *Mirriam Webster New Book of Word Histories,* Mirriam Webster, Springfield, MA, 1991.
23. M. Hoffman, "A Second Chance for Thalidomide," *American Scientist 81* (1993), 425.
24. P. S. Phillips (Ed.), *Funk and Wagnall's New Encyclopedia,* Funk and Wagnall, New York, 1979, p. 57.
25. (news story) "Carbon Monoxide Fumes Endanger Unborn Child," *Sudbury Star,* 6 January 1986, 11.
26. "The Crime of the Decade," *New Science 7* (1983).
27. W. Glenn, "Mercury – to Those Who Ignore History," *Occupational Health and Safety 4* (1988), 18.
28. L. J. Goldwater, "Mercury in the Environment," *Scientific American 224* (1971), 15.
29. J. Putman, "Quicksilver and Slow Death," *National Geographic,* October 1972, 507.
30. A. Tucker, *The Toxic Metals,* Ballentine Books, New York, 1972.
31. "The Mercury Physicists left Behind," *New Scientist,* 1 December 1990, 22.
32. J. E. Bishop, "Data Site Perils of Mercury in Latex. Paint," *The Wall Street Journal,* 18 October 1990, B 4.

33. H. Nikiforuc, "Hazards in Dental Offices," *MacLeans*, 7 May 1984.
34. "Fools Gold Refining With Mercury in Brazil," *Discover*, December 1986, 14.
35. B. Homewood, "Mercury Poisoning Confirmed Along Amazon Villages," *New Scientist*, 9 November 1991, 18.
36. "Gold Diggers Poison Brazil's Wild Paradise," *New Scientist*, 12 September 1992, 8. The chemical mechanics underlying mercury poisoning by gold mining activities are reviewed in W. Sligliani, W. Salomons, "Our Fathers Toxic Sins," *New Scientist*, 11 December 1993, 38.
37. M. Weisskoef, "Lead Astray: The Poisoning of America," *Discover*, December 1987, 68.
38. "Lead in Your Coffee," *New Scientist*, 26 September 1985, 21.
39. J. J. Chisom Jr., "Lead Poisoning," *Scientific American 224* (1971), 15.
40. S. C. Gilfillan, "Lead Poisoning and the Fall of Rome," *Journal of Occupational Medicine 7* (1969), 53.
41. (news story) "Lead Poisoning may Have Changed Events," *Sudbury Star*, 15 June, 1992, A 5.
42. "Weighing Some Heavy Metal," *Time*, 1 April 1991, 42.
43. S. Pollack, "Solving the Lead Dilemma," *Technology Review*, October 1989, 22.
44. (news story) "A Household Poison that will not go Away," *Globe and Mail*, 8 February 1992, 8.
45. "Asian Children Risk Lead Poisoning," *New Scientist*, 26 September 1985, 19.
46. "Fountains of Lead," *Discover*, May 1988, 20.
47. P. Spinks, "Dutch Buyer Failed to Test Cattle Feed for Lead," *New Scientist*, 2 December 1989, 26.
48. F. Pearce, "Lower Limits for Lead in the Pipeline," *New Scientist*, 19 September 1992, 4.
49. The Mysterious Case of the Disappearing Franklin Expedition is reviewed in Gardener, ed. *Great Mysteries of the Past*, Reader's Digest Association, Montrèal 1991, 62.
50. Elmes, Simpson, *British Journal of Medicine 34*, 174.
51. N. Hirschhorn, W. B. Greenough III, "Cholera," *Scientific American 225* (1971), 15.
52. "Cancer Fears for Pastry Cooks," *New Scientist*, 19 June 1986, 28.
53. J. Christopher, K. Michael, F. Pearce, "Is your card Marked for Cancer?," *New Scientist*, 5 June 1988, 24
54. The problems of the health hazards of inhaled diesel soot are reviewed in B. H. Kaye, "The effect of Fractal Structure on the Health Hazards of Respirable Dusts," J. Otter, M. Fayed (Eds.), in preparation.
55. A. Ochsner, "The Health Menace of Tobacco," *American Scientist 59* (1971), 246.
56. D. H. Freedman, "Who's to judge?," *Discover*, January 1994, 78.
57. J. Castro "Caveat Fumator-Tobacco Firms win two Cases," *Time*, 7 September 1987, 46.
58. C. Angus, "The Ontario Mining Industry's Dirty Little Secret," (news story) *Toronto Star*, 20 November 1993, B1, B5
59. D. Ubelaker, H. Scammell, *Bones. A Forensic Detectives Casebook*, Harper Collins, New York, 1992.

Chapter 10

Forgery and Fraud

Experienced Documents Examiners

The Forensic Science Service gives scientific support to police forces in England and Wales in solving serious crime. As a Documents Examiner you will be involved in a wide range of cases requiring expertise in all aspects of work in this field from handwriting comparison to inks and paper analysis. In due course you will be required to report on these cases and give evidence in court.

We are looking for Documents Examiners with a science degree and at least two years experience.

Home Office, Birmingham

Scientists

...forensic Studies of Documents

Scientists are required in the Document Section of the Forensic Science Laboratory, which carries out investigations for the Police relating to questioned documents and handwriting.
The successful candidates will be concerned with photographing documents by infra-red, ultra-violet and luminescence methods and the examination of paper, inks, handwriting and typescript for evidence. Initially, whilst training, they will assist senior colleagues, but will ultimately prepare reports and attend court hearings as expert witnesses.
Candidates, normally aged under 27, should have a good honours degree or equivalent in a scientific subject.

Science Graduates

Focus on a career in forensic science

Forensic work is an exacting science that forms a crucial part of the criminal investigations carried out by the Metropolitan Police. It is one of the few areas of crime detection that is carried out by civilians within the Service. The work is specialised and challenging offering science graduates who enjoy stretching their minds daily, an opportunity to develop exciting and rewarding careers.
You will work in the forensic science laboratory in London, one of the best equipped of its kind, enjoying a world-wide reputation for developing new and improved analytical methods.

Documents Division

To carry out all examinations relating to forged and altered documents, the dating of documents or parts of them, and the comparison of handwriting and typescript. The Officer may be required to give evidence in Courts of Law. They may also be involved in research into various aspects of document examination.
Candidates must possess a good honours degree in a science discipline. Appointment will be at Scientific Officer level. (1 vacancy).

Head of Document Examination

We require a document examiner with proven managerial skills to head our Documents service. The successful candidate will have a science degree and several years' experience in this field, including case-reporting and giving evidence in court in a wide range of cases from handwriting comparison to inks and paper analysis.

Home Office
Document Section, Birmingham

... to be concerned with photographing documents by Infra-red, ultra-violet and luminescence methods and the examination of paper, inks, handwriting and typescript for evidence. Initially, whilst training, the successful candidate wil assist senior colleagues, but will ultimately prepare reports an attend court hearings as an expert witness.

Candidates should have a good honours degree or equivale in a scientific subject.

Chapter 10

Forgery and Fraud

10.1 Are You Sincere?

One of the earliest written stories in the Bible involves deception and fraud. In chapter 27 in the book of Genesis, Jacob impersonates his brother to steal the blessing of his blind father. Isaac advises his older son Esau: *You see that I am old and may die soon. Take your bow and arrows, go out in to the country and kill an animal for me. Cook me some of that tasty food that I like and bring it to me, after I have eaten it I will give you my final blessing before I die.* This conversation was overheard by Rebekah who favored her younger son, Jacob, over Esau. She told Jacob: *I have just heard your father say to Esau "bring me an animal and cook it for me."* As the story unfolds, Rebekah tells Jacob: *Listen to me and do what I say. Go to the flock and pick out two fat young goats so that I can cook them and make some of that food that your father likes. You can take it to him to eat and he will give you his blessing before he dies. But Jacob said to his mother: You know that Esau is a very hairy man and I have smooth skin, perhaps my father will touch me and find out that I am deceiving him. In this way I will bring a curse on myself instead of blessing.* At this point in the story Rebekah told Jacob to leave it to her and Jacob goes for the goats. She then took Esau's best clothes, which she kept in the house and put them on Jacob. She put the skins of the goats on his arms and on the hairless part of his neck. She handed him the tasty food and sent him to his father. When Jacob came into the presence of his father Isaac asked: *Which of my sons are you?* Jacob answers: *I am your elder son Esau, I have done as you have told me. Please sit up and eat some of the meat that I have brought you so that you can give me your blessing.* Isaac says to Jacob: *please come closer that I can touch you to know if you really are Esau.* Jacob moved closer to his father who felt him and said: *Your voice sounds like Jacob's voice but your arms feel like Esau's arms.* Isaac was about to give Jacob his blessing but asked again are you really Esau? *I am*, answered Jacob. Then his father told him: *come closer and kiss me, my son.* As Jacob came up to kiss him, Isaac smelled his clothes so he gave him his blessing. *The pleasant smell of my son is like the smell of the field which the Lord has blessed.* What is interesting about this story is that, as told in the Bible, the story-teller obviously accepts the fact that, although the blessing of his father was obtained by deception, even God is bound by the transaction of the giving of the blessing. Today, in Western courts of law such deception would result in a legacy being declared null and void. Note that the father had his doubts, particularly with regard to the voice of the deceptive son, but that the odor signature and the tactile structure of the arms helped to complete the fraudulent encounter between Jacob and Isaac.

This story serves to point out that it is a very old streak of man's nature to lie and steal to obtain financial advantage. In this chapter we will explore some of the major frauds that have occurred in society, as well as some historic frauds, such as the Shroud of Turin. We will also look at how forgery played an important part in a famous French lawsuit, in the conviction of Dreyfus. Another illustration that I intended to use as an opening phrase in this chapter had to be given second place, because what I thought to be a true origin of a word turned out to be not true. (I hate to use the word fraudulent, because in this situation I doubt there was any intent to deceive.) As outlined earlier, one of my hobbies is that of writing dictionaries. A study of the origins of words is described by the technical word **etymology**. This term comes from the Greek word meaning the truth or the real origin of something. Many years ago I was told that word sincere came from Latin root words *sine* meaning without and *cera* meaning wax. It was claimed that the word **sincere** originally meant "without wax." I was told that the use of the phrase originated from a practice pursued by some thieving artists who tried to sell inferior statues containing chips and cracks in the marble. These were covered up with white wax molded into the flaws of the statue. Thus, when a housewife demanded to know if a statue was without wax she was asking whether this was a genuine article without fraudulent touch-up. When sitting down to write this chapter I thought I had better check up on this word origin in an etymological dictionary.[1] I was disappointed to find that my dictionary defined "sincere" as pure and unadulterated, unmixed from a Latin word *sincerus* meaning clean. Whatever the real origin of the word sincere, it is obvious that the manufacturing and sale of falsified objects is a very old practice. In this chapter we will explore how modern forgers create fraudulent statues and other art objects. How scientists have been involved in the fight against forgery is illustrated by a story that physics students learn in the beginning of their studies. Archemedes was a Greek scientist who lived from 287 B. C. to 212 B. C. The King of Syracuse (a Greek colony in Sicily) asked Archemedes to study a crown, the manufacture of which he had recently commissioned, and which he suspected was not made of pure gold. Archemedes knew how much a given volume of gold should weigh and he realized that, if he could measure the volume of the completed crown and weigh it, he would be able to find out if it was pure gold. The difficult part of the problem was measuring the volume of the crown. It is said that one day when Archemedes got into the bathtub he noticed that the water line in the tub rose up by an amount equal to the volume of his body submerged below the water. He realized that all he had to do now was to put the crown in water and measure the rise of the water in the vessel. He could work out the volume of the crown from the rise in the water, and determine whether or not the crown was made of pure gold. We are told that Archemedes was so delighted to have discovered this new principle of solving density problems that he raced through the streets of Syracuse stark naked, yelling *Eureka*. This is a Greek word meaning "I have discovered it." Using his new found scientific ideas, Archemedes was able to show that the crown was not made of solid gold and that it probably had some lead in the middle or had been alloyed (mixed) with another metal to reduce the amount of gold in the crown. We are told that the metallurgist who made the crown had his career terminated quickly and painlessly![2]

In November 1967 the *London Sunday Times* was offered the opportunity of publishing what was claimed to be the personal diary of Benito Mussolini, the former dictator of Italy. Some experts had already endorsed the genuineness of the handwriting and Mussolini's surviving son Victorio had vouched for the accuracy of the document. However, to carry out a further check on the authenticity of the diary, the *London Times* called in the world's expert on forgery, a man called Julius Grant, who operated from an office in London, England. Grant was a chemist who for over 30 years had specialized in the exposure of forgeries.[3] When Grant examined the diary something about the feel of the paper troubled him. Therefore, he snipped a tiny corner from a page and subjected it to chemical analysis. As it turned out, the paper contained straw pulp. This kind of pulp paper was not manufactured in Italy before 1936, and Grant denounced the document as a forgery. Commenting on his experience over the years Grant states:

> *There are hundreds of forgers now serving prison terms because they used the wrong kind of paper!*

In view of this comment we will begin our exploration of forgery by looking at the way in which the forger can make mistakes by failing to understand the technology and history of paper manufacture.

10.2 Paper, the Forger's Nemesis?

In Greek mythology Nemesis is the goddess of retribution. The word paper is defined in a dictionary as:

> *a material in the form of thin sheets manufactured by compression of any of various types of vegetable fibers.*

Today this definition must be extended to include synthetic fibers. The basic process of making paper has not changed in more than nearly 2000 years. The process involves two stages: the breaking up of the raw material in water to form a suspension of individual fibers and the formation of felted sheets by spreading this suspension on a suitable porous surface from which water can drain. One of the major sources of cellulose fibers used in the making of paper is wood. Wood consists of two main ingredients, cellulose and a substance known as lignin. Lignin gives wood its stiffness and cements the cellulose fibers together. In the simplest process for making paper, pieces of wood are held against a rapidly rotating grindstone. The fibers are ripped away from each other by the grindstone, but much of the lignin remains with the fibers. In the chemical process for making paper, the cellulose fibers are separated from the lignin by digesting the wood chips in a chemical solution. One of the major processes involves the chemical reaction of the wood chips with sulfate to produce fibers. Historians tell us that paper was first made in China, in the year 105 A.D. by Tsi'Lun. It was probably made from the bark of the mulberry tree, while some of the finest paper is made from linen rags. From a scientific point of view, linen is made up of the cellulose fibers found in the flax plant. Paper made from flax is superior to that made from other plants, because of the length of the cellulose fibers, which occur naturally in that type of plant. For this reason, even today, the best high quality paper, such as that used for bank notes is made from linen rags. Cotton fibres are also used in good quality paper.

To turn a suspension of the cellulose fibers into a finished piece of paper, the suspension is poured over a wire mesh screen. The water drains through the holes of the screen and the fibers form a felt on the surface of the wire screen. The thickness of the paper can be controlled by how much pulp suspension is poured through the wire mesh. Very early in the development of paper making, manufacturers weaved a pattern into the wire mesh used to drain the fiber slurry. The resulting pattern on the newly made paper sheet could easily be seen in the finished product by holding the paper up to a light. This pattern is known as the watermark. Paper companies keep a list of all the water marks used to identify quality paper made all over the world. In an episode of the detective series Matlock, broadcast in the fall of 1993, the fact that a will had been forged was established by examining the watermark of the paper on which the will had been prepared. This paper was purchased two months after the date on which the will had supposedly been prepared! In Figure 10.1 the appearance under the microscope of various types of paper are shown.[4] Newsprint has an open structure and absorbs the ink before it dries. This is why one's hands can become quite dirty from reading a newspaper fresh off the press. To obtain high-quality colored pictures on paper, its surface must become non-absorbent. for this reason, non-absorbent paper like that used in the printing of *Time Magazine* and other glossies contain up to 20 % of a special type of clay. The appearance of the top and underside of clay treated paper is shown in Figures 7.4(a) and (b). The ink used on paper can vary enormously, as we shall discover in a later section of this chapter. In Figure 7.4(c) a typical picture of the ink from a printing process, which has left many very fine particles of ink in between the fibers of the paper, is shown. In the machine manufacturing of paper, after the felt of fibers has been dried, it is put through a process which is known as calendering. Once again we have a word that has one meaning in every-day English and a different meaning in technical English. The process for finishing paper takes its name from the Greek word for a roll. The confusion arises from the fact that the earliest time calendars were written down on rolls of manuscript. The way in which the surface of a piece of paper changes during the stages of production, coating and finally calendering is shown in Figure 10.2. The contours of the maps of Figure 10.2 are rather like the contour maps produced to show the ruggedness of the terrain in a map of the earth's surface.[5]

The fact that paper in itself is a fibrous mass is important when considering what happens when one erases a signature on a piece of paper. First of all, tiny fragments of ink and eraser fall into the mass of the paper. Therefore, if the paper is illuminated with ultraviolet or infrared radiation, the debris embedded in the paper (which is impossible to remove) gives out an optical signal, which is usually different from the fluorescence given out by the new ink used to create a signature. It is sometimes useful to photograph the system with infrared light, because the paper fibers are virtually transparent to the infrared radiation. Secondly, the physical erasure of writing usually results in the disturbance of the fibers of the surface, which can readily be detected by comparing the structure of the fibers in an area of suspected erasure with those in an area away from the signature. A modern method for erasing writing from a piece of paper involves shining a laser beam onto the surface. The energy of the laser results in the vaporization of the signature. However, it is difficult to balance the power of the laser to erase the signature without also scorching some of the fibers in the vicinity of the erasure.

The structure of the fibrous mat of a paper is often closely controlled for a specific purpose. Thus, the top surface of the paper used to make a postage stamp must take a good image of colored printing without the boundaries running and, at the same time, the undersurface of the stamp must absorb the glue with which the stamp is stuck to a letter. In recent years a lot of paper and envelopes have been made from recycled paper. During the process of recycling, many of the cellulose fibers are broken. Furthermore, the natural bonding between fibers, which made the original paper fibers stick together, is weakened by the recycling process. Sometimes, in the recycled paper a spray of glue must be added to make the fibers stick together. In the early 1990s there were complaints that stamps were falling off envelopes made of recycled paper, because of the poor adhesion between the stamps and the envelopes. I have had a letter returned to me for the addition of postage stamps even though I know that the original letter had the neces-

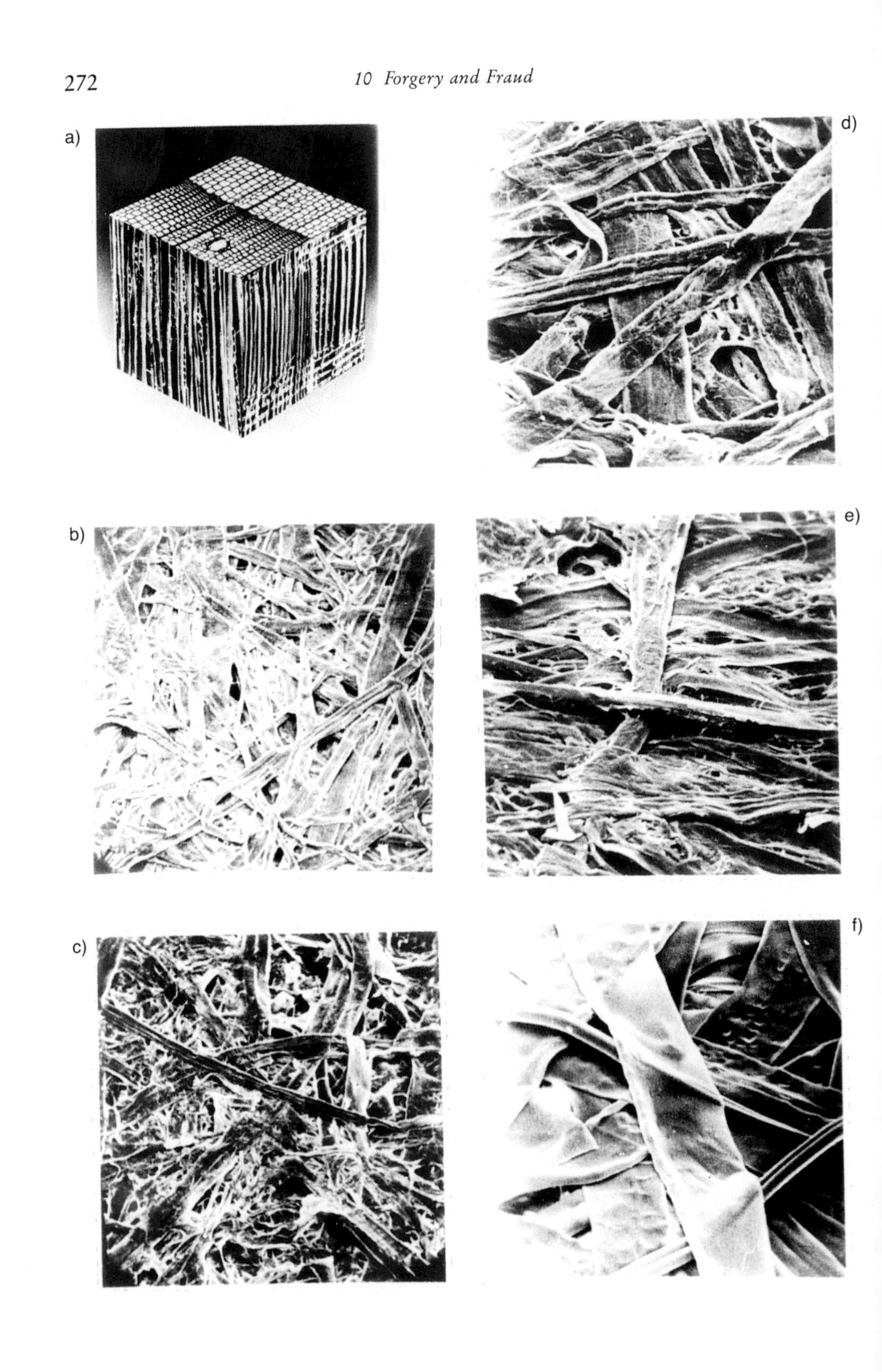
a)
b)
c)
d)
e)
f)

◀ **Figure 10.1.** The type of fibers and the different treatments of the paper during the manufacturing process can easily be identified under the microscope.[4] a) The appearance of black spruce wood. Each side of the cube measures 0.6 mm. The top shows the cross-section, the left face is the tangential section, and the right face shows the radial section. b) The surface of bond paper. c) The surface of newsprint made from ground wood. d) unbeaten kraft paper. e) newsprint containing about 25 % chemical pulp. The center of the field shows the bond between two ground wood fibers. Below this a collapsed chemical fiber crosses the vertical ground wood fiber. f) Sheet of paper made from hollow filament rayon (wide fibers) and normal rayon (narrow fibers). (Images courtesy of Paprican.)

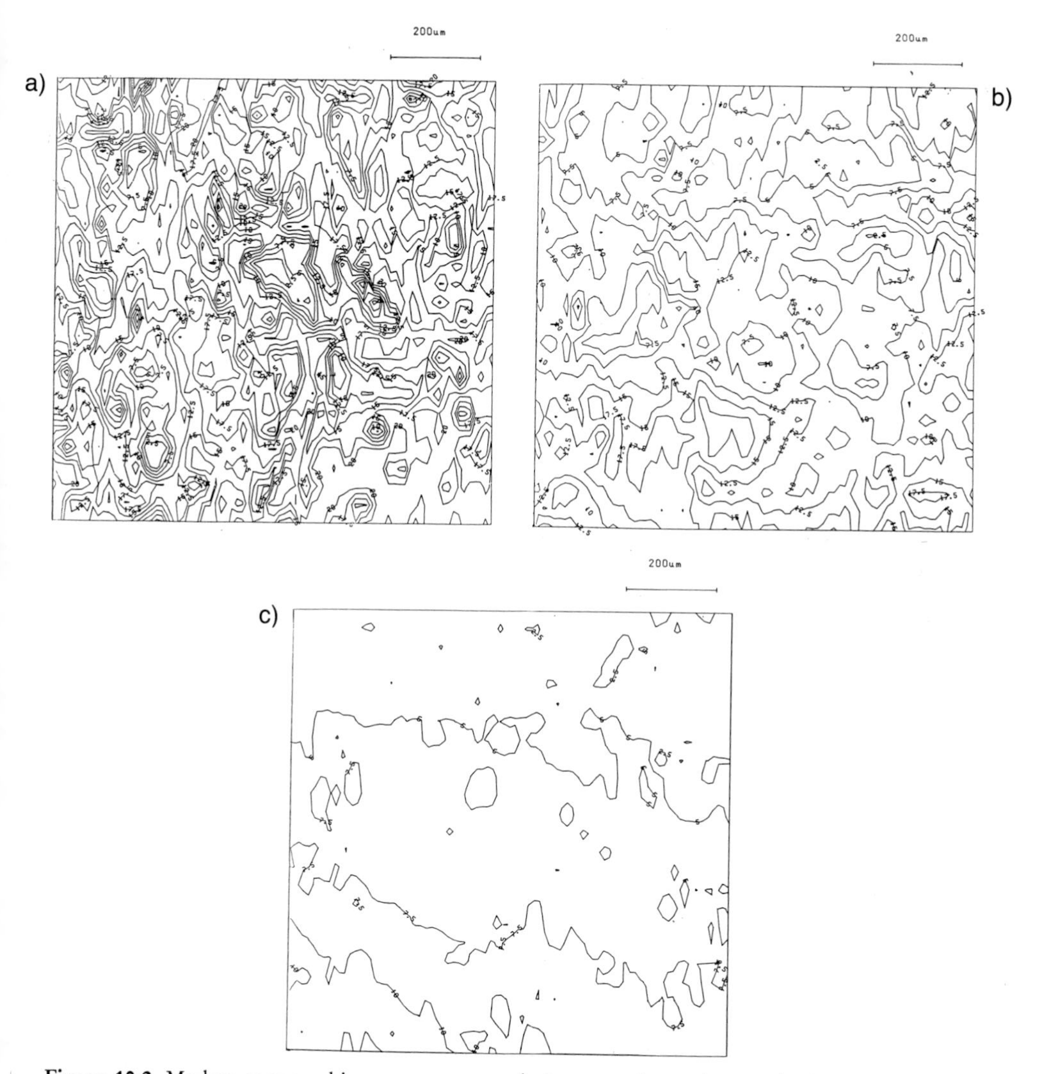

Figure 10.2. Modern topographic measurement techniques can be used to study the surface structure of paper as it is treated to improve its quality. a) surface of paper after the manufacturing process. b) surface of coated paper. c) surface of paper after calendaring. (Images courtesy of ECC, St. Austel, Cornwall.)

sary postage. I glued the second set of stamps on these letters with extra-strength glue!

Because it is porous, paper is very absorbent; recently, saliva absorbed by an envelope licked to seal it was used to identify a suspect in the World Trade Center bombing in late 1992. The letter was claiming responsibility for the bombing from a group calling itself the Liberation Army 5th Battalion and was reviewed by the *New York Times* on March 2nd. When the suspect Nidal Ayad was arrested March the 10th, F. B. I. agents found computer equipment that had previously been linked to the letter. The F. B. I. also claimed that saliva taken from the envelope, when subjected to a DNA test (see Chapter 12) was found to be from the suspect.[6]

The complexity of modern paper manufacture and the complexity of the various processes used to create patterns on paper can be appreciated from the job advertisement shown in Figure 10.3. The series of enlargements of the bird in this picture shows how computers create pictures by using mosaics of black and white squares. This particular job advertisement was seeking people to work on the technology used to make non-carbon copy paper of the type used with modern credit cards and other forms that have to be filled out in triplicate. To create a copy on the lower pieces of paper, the top piece of paper is coated with tiny spheres of invisible ink. These spheres must not be too big or the image created when the spheres are ruptured by the pressure of a ball point pen is blotched. On the other hand, if they are to small, they fall into the mesh of the paper. In practice, 15 micrometers is a good size for the capsules of invisible ink. The paper on which the copy is to appear is coated with a special material that reacts with the ink, when the ink-containing spheres are ruptured, creating an image. The complexity of this process is so high that it is virtually impossible for a forger to reproduce a signature made on non-carbon copy paper. In the next section we will look at the complexity of the world of inks, demonstrating again the type of problems that forgers encounter in their work.

Walls quotes an important fraud, which was supposedly carried out by Thomas Wise, (who died a rich man in 1937) who claimed to have discovered many 19th century first editions of books. Many of these so-called first editions carried dates which placed the printing of the books before 1850. However, when in 1934, two investigators, Carter and Pollard, examined the paper on which the books were printed, they found that the paper contained esparto grass, which was first introduced into Great Britain in 1860 or 1861. Similarly, a whole series of the books dated 1842 to 1873 were printed on paper containing chemical wood pulp first made in 1874.

Butlers Court the Research & Development Centre for the Wiggins Teape Group, based in South Buckinghamshire houses the largest privately owned paper research centre in Europe, where scientists and engineers are continually developing and improving new and existing products. One measure of the importance of this work can be seen in the success of products such as Idem self copying paper – Europe's brand leader.

Some 200 staff work at Butlers Court – many of them recognised internationally as leading authorities in their field.

Currently we are looking for 3 talented scientists to join them.

Process Development Engineer

This is a key role involving the development of coating processes. You'll run a number of long-term projects on our pilot plant: following developments through to their implementation in our production units.

The position would suit a graduate with an Engineering or Physics background but the ability to communicate will be vital – you'll need to 'sell' your ideas to non-technical management.

You'll also need a full driving licence since the troubleshooting of process problems at our units around the country will also be your responsibility.

Development Scientist

We need a Chemist/Materials Scientist with a good BSc or higher degree to work with base papers and mineral coatings. You will be involved in a number of projects concerning cost reductions, quality improvement and new product development. Following these through from test-tube to marketing the finished products, your interpersonal skills will come to the fore as you liaise with personnel in a variety of business functions.

Image Scientist

A self-copying paper is only as good as the quality of the copies. But how do you measure this scientifically?

A good science degree possibly backed by a relevant higher degree or post-graduate experience will help you to develop the necessary physical and psychological tests we need. You'll have to be outward looking – taking ideas from what may seem to be unrelated areas of science – and computer literate.

All of these positions offer exceptional career opportunity through our 'Professional Development Scheme'. You'll have frequent, rigorous appraisal and training courses tailored to your needs.

We offer a starting salary of £11,000 to £13,500, at this level renumeration is performance-related.

Interested? To find out more contact: Miss Dawn Haynes, Personnel Manager, Wiggins Teape R&D Ltd, Butler's Court, Beaconsfield, Bucks HP9 1RT. Tel: (0494) 5652 ext 296/295.

WIGGINS TEAPE

Figure 10.3. For several reasons, including the need to combat forgery, many specially coated papers are being produced. (Reproduced by permission of Arjo Wiggins Appleton).

10.3 Ink and Toner

The original inks used for writing on parchment were made of soot mixed into a gum solution. (Parchment is the dried skin of an animal.) Soot used for this purpose was sometimes called lampblack. Modern drawing inks and printers inks still contain lampblack or carbon black (as it is described scientifically), which is a purified form of soot. The basic constitutions of inks used over different periods of time are shown in Table 10.1.[7]

Table 10.1. Various types of inks and the time periods over which they have been in use.

Ink	Period of Use
Carbon black ink	From very early times to the present
Iron-gall inks	From the 8th century to the present
Logwood inks	From early 1700s to 1800s, still used rarely
Vanadium inks	1890, not extensively used now
Aniline dye inks (with or without iron-gall)	From 1861 to the present
Water-soluble blue dye inks	From 1936 to 1950
Ball-point inks	From 1936 to the present

With regard to the role that specialty papers will play in the business world of tomorrow, Peter Lee, a printing consultant who worked for the bank of England for 30 years, says that the new generation of office copiers has put the skill of a trained printer into the hands of a casual counterfeiter. It is now more effective to add security features to paper at the mill than to the printing press.[8] Nick Ackland of the Portoles in Bathford, England, a company that makes paper for passports and checks, says:

> *It is very hard to counterfeit now unless you are going to make the paper. This is beyond the means of most counterfeiters.*[8]

In the war against forgery of valuable documents, Wiggins Teape has developed a special ink and a coated paper to go with it, which allow officials to verify the nature of a document (see Figure 10.3 and Table 10.2). The ink, called iv, for **Instant Verification**, can be put into felt-tip pens or stamp pads. The paper contains a chemical, which will react with the special ink to make a black mark, but it looks like ordinary paper and can be watermarked and printed on. The company expects the certification ink to be used to check the authenticity of vouchers, record tokens, postal orders, money orders, and other documents that are used once and redeemed. If one tests them with a felt pen or ink pad containing the special ink, only the

Table 10.2. The way in which an ink reacts with various chemicals is characteristic of the type of ink and can possibly identify a fraud.

Spill Reagent	Reaction	Ink
Water	Water is highly colored On blotting with filter paper, the ink is all taken up	Dye ink
Chloroform	Solvent is highly colored Filter paper takes up all color	Ball point ink
	Ink is softened, some color taken up	Printing ink
Dilute hydrochloric acid	Black ink turns blue	Iron ink
	Black ink turns red	Logwood ink
	Slightly bleached	Vanadium ink
	Unchanged	Nigrosine ink
Dilute hydrochloric acid followed by annonium thiocyanate silution	Red color	Iron ink
Dilute hydrochloric acid followed by sodium hypochloride	Bleached	Iron ink
	Bleached	Logwood ink
	Brown	Nigrosine ink
	Unaltered	Vanadium ink
Stannous chloride (10 %), hydrochloric acid (10 %)	Bleached	Iron ink
	Slightly bleached	Vanadium ink
	Red	Logwood ink
	Unaltered	Nigrosine ink

genuine documents will show visible writing. On fraudulent documents, the invisible ink cannot be developed on the paper. Wiggins Teape and the British post office worked for 20 years to develop the ink. A spokesman for Wiggins Teape, who announced the availability of the special paper and ink says that the secret chemicals are added at the start of the paper making and have to be processed carefully to make sure that they are evenly distributed throughout the paper. For such a product the chemicals must be very stable. The company has tested the paper and ink for nine months and is confident that they will still work after several years of storage. The same spokesman said that they had not been able to find any other ink that would react with the chemical in the paper. The effort that is going into the making of new inks for various projects can be appreciated from the information presented in the job advertisement reproduced in Figure 10.4.

The way in which analysis of ink can be used to detect forgery has been reviewed by Brunnelle and Cantu.[9] In their review, the use of thin layer chromatography to analyze the constituents of many different types of ink is described. The physical characteristics of an ink, which can be generated by this analytical technique is summarized in Figure 10.5. The basic piece of equipment is a thin plate of glass covered with silica gel, which consists of

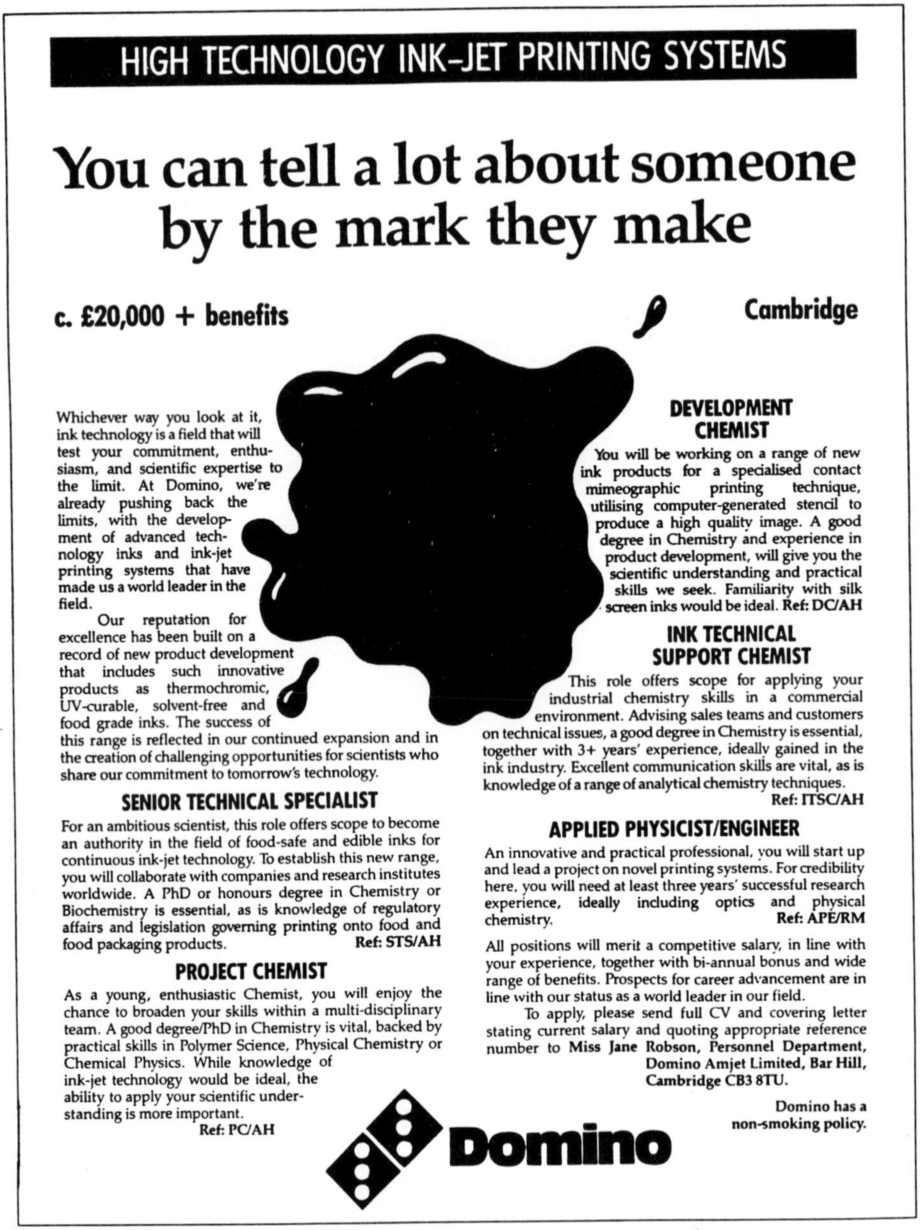

HIGH TECHNOLOGY INK-JET PRINTING SYSTEMS

You can tell a lot about someone by the mark they make

c. £20,000 + benefits **Cambridge**

Whichever way you look at it, ink technology is a field that will test your commitment, enthusiasm, and scientific expertise to the limit. At Domino, we're already pushing back the limits, with the development of advanced technology inks and ink-jet printing systems that have made us a world leader in the field.

Our reputation for excellence has been built on a record of new product development that includes such innovative products as thermochromic, UV-curable, solvent-free and food grade inks. The success of this range is reflected in our continued expansion and in the creation of challenging opportunities for scientists who share our commitment to tomorrow's technology.

SENIOR TECHNICAL SPECIALIST

For an ambitious scientist, this role offers scope to become an authority in the field of food-safe and edible inks for continuous ink-jet technology. To establish this new range, you will collaborate with companies and research institutes worldwide. A PhD or honours degree in Chemistry or Biochemistry is essential, as is knowledge of regulatory affairs and legislation governing printing onto food and food packaging products. **Ref: STS/AH**

PROJECT CHEMIST

As a young, enthusiastic Chemist, you will enjoy the chance to broaden your skills within a multi-disciplinary team. A good degree/PhD in Chemistry is vital, backed by practical skills in Polymer Science, Physical Chemistry or Chemical Physics. While knowledge of ink-jet technology would be ideal, the ability to apply your scientific understanding is more important. **Ref: PC/AH**

DEVELOPMENT CHEMIST

You will be working on a range of new ink products for a specialised contact mimeographic printing technique, utilising computer-generated stencil to produce a high quality image. A good degree in Chemistry and experience in product development, will give you the scientific understanding and practical skills we seek. Familiarity with silk screen inks would be ideal. **Ref: DC/AH**

INK TECHNICAL SUPPORT CHEMIST

This role offers scope for applying your industrial chemistry skills in a commercial environment. Advising sales teams and customers on technical issues, a good degree in Chemistry is essential, together with 3+ years' experience, ideally gained in the ink industry. Excellent communication skills are vital, as is knowledge of a range of analytical chemistry techniques. **Ref: ITSC/AH**

APPLIED PHYSICIST/ENGINEER

An innovative and practical professional, you will start up and lead a project on novel printing systems. For credibility here, you will need at least three years' successful research experience, ideally including optics and physical chemistry. **Ref: APE/RM**

All positions will merit a competitive salary, in line with your experience, together with bi-annual bonus and wide range of benefits. Prospects for career advancement are in line with our status as a world leader in our field.

To apply, please send full CV and covering letter stating current salary and quoting appropriate reference number to **Miss Jane Robson, Personnel Department, Domino Amjet Limited, Bar Hill, Cambridge CB3 8TU.**

Domino has a non-smoking policy.

Domino

Figure 10.4. Special inks are making life difficult for the forger. (Reproduced by permission of Domino Printing Supplies PLC.)

many tiny spheres visible under the microscope. The manufacturers of these modern chromatography systems take great pains to ensure that the spheres are all of the same size. The thin layer of silica fine particles is better than the paper originally used in thin layer chromatography (often referred to as TLC), and gives better separation of the components. To start the analysis of an ink, a drop of the ink is placed on the thin-layer plate, as shown in Figure 10.5(a). Solvent is then made to move up the plate by **capillary action**. The basic physical process is illustrated in Figure 10.5. The word *capillary* in Latin means "a very fine hair." **Capillary tubes** are glass tubes, which have a very thin hollow down the middle about the size of human hair. If one places a capillary tube in a supply of liquid such as water, the

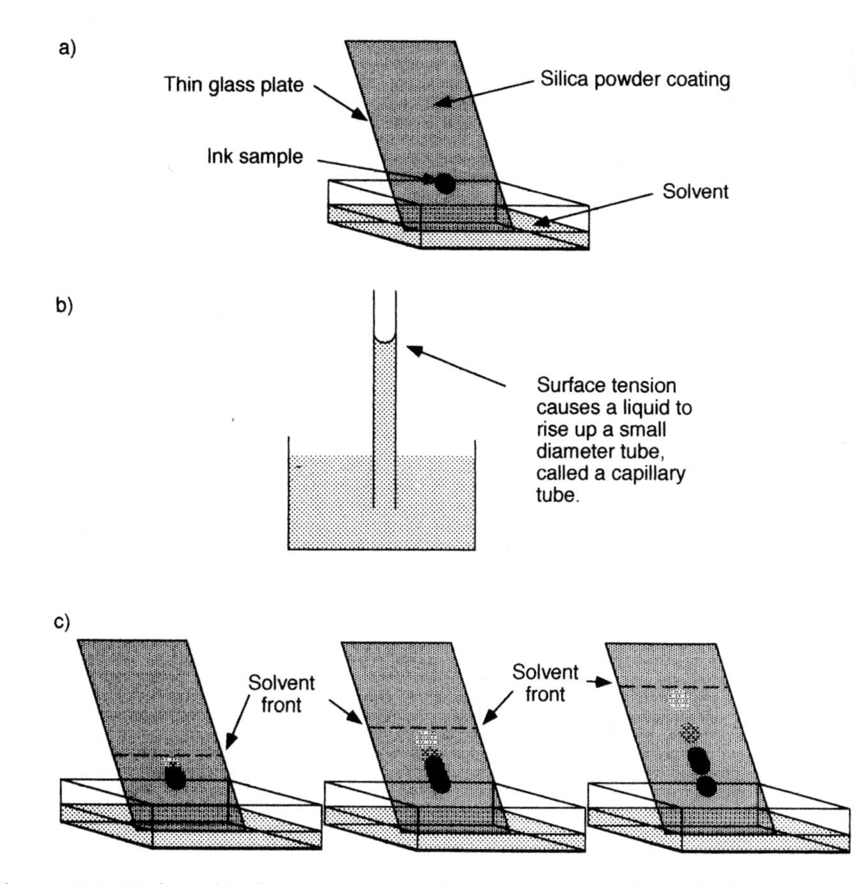

Figure 10.5. Modern thin layer chromatography uses commercially available thin glass plates coated with very fine, uniformly sized, transparent silica spheres.[10] a) Thin layer chromatography equipment. b) The concept of capillary action causes liquid to move up narrow channels. c) In thin layer chromatography, the components of the ink separate as the solvent moves up the plate by capillary action.

surface tension of the liquid causes it to move up the tube. The narrower the core of the tube, the higher the liquid moves up the column. (Note that the curved surface of the rising liquid is known as the **meniscus**, a Latin word which means new moon, because the surface of the looks like the crescent moon.) The tiny gaps between the spheres in the coating of the thin layer of glass act as if they were very tiny capillary tubes. Therefore, the solvent for the ink rises up the thin layer as indicated in Figure 10.5(c). The wick of a lamp draws liquid from the body of the fluid up to its tip by capillary action. For this reason the movement of solvent up a TLC plate is sometimes described as wicking of the solvent.[10]

In Figure 10.6(a) the TLC fingerprint of a series of fountain pen inks are shown (note that these TLC fingerprints are upside down as compared to those of Figure 10.5). In Figure 10.6 (b) the same kind of fingerprint for several blue felt-tip pens are shown. Sometimes it is useful to examine the ink fingerprint under ultraviolet light, to determine whether the critical components of the ink fluoresce.[9]

Brunnelle and Cantu have reviewed some interesting law cases in which identification of ink by their TLC fingerprint played an important role. Thus in a lawsuit in which a large New York bank was accused of illegally awarding loans to small business concerns – a case known as the United States versus Meyers, ink and handwriting analyses showed that many of the loan application forms were prepared by the bank official, rather than by the loan applicant. In a tax-evasion trial called the United States versus Sloan one of the pieces of evidence presented by Sloan was a four page agreement stating that the defendant was to invest sums of money for an anonymous client over a period of time. Also introduced into evidence were a series of notes dating from 1958 to 1966 claiming to prove these investments. An analysis of the ink on each page of the agreement showed that the same ink was used on all of the documents, and that the ink contained a unique dye that was first synthesized by the Ciba-Geigy Chemical Corporation in 1959. The ink formulation containing the dye was not produced until 1960. The analysis of the ink used in the various documents produced by Sloan are shown in Figure 10.6 (c) and it can be seen that, irrespective of the alleged dates of the documents, the ink was identical in all documents.[9]

In another legal case, the United States versus Colasurdo, evidence was presented that a document dated 1965 had been prepared with a ball point ink, which had not been manufactured until 1968. In a similar case, the

Figure 10.6. Modern forensic science laboratories have thin layer chromatography signatures of ink from various types of pens to enable them to detect forgery. (Reproduced from R. L. Brunelle, A. A. Cantu, in G. Davies (ed.), *Forensic Science*, ACS Symposium Series, American Chemical Society, 1975, chapter 14. Used with permission of The American Chemical Society.) a) TLC of blue fountain pen inks. b) TLC of blue felt tip inks photographed in white (above) and ultraviolet light (below). c) TLC of inks used in the U. S. vs. Sloan case. ▶

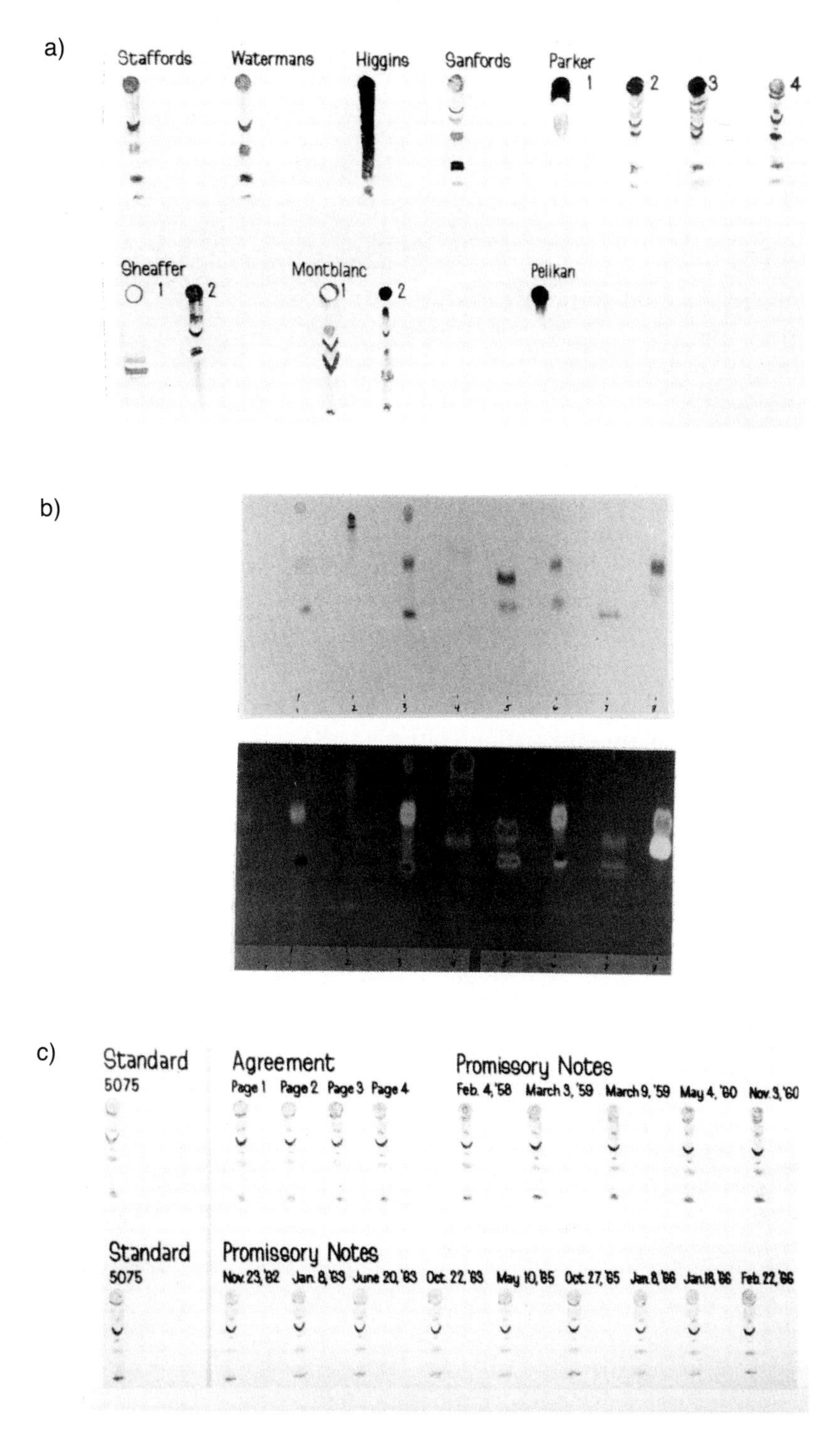
a)
Staffords
Watermans
Higgins
Sanfords
Parker
1
2
3
4
Sheaffer
1
2
Montblanc
1
2
Pelikan
b)
c)
Standard
5075
Agreement
Page 1
Page 2
Page 3
Page 4
Promissory Notes
Feb. 4,'58
March 3,'59
March 9,'59
May 4,'60
Nov. 3,'60
Standard
5075
Promissory Notes
Nov. 23,'62
Jan. 8,'63
June 20,'63
Oct. 22,'63
May 10,'65
Oct. 27,'65
Jan. 8,'66
Jan. 18,'66
Feb. 22,'66

U. S. versus Bruno, the ink used to sign a document dated 1965 was not available commercially until 1967. In this particular case, however, the judge did not accept the evidence because he said there was not an adequate data-bank of inks available in different countries at the time that the document was signed. According to Brunnelle and Cantu, who worked for the Federal Bureau of Alcohol, Tobacco, and Firearms in the U. S. Treasury Department, the bureau continues to build up its library of ink fingerprints. In addition, an increasing number of ink manufacturers are starting to tag their inks with certain products, to enable the ink and its year of manufacture to be identified. This not only helps in the fight against fraud, but it also enables the manufacturers to defend themselves against claims that ink was bought at a certain time or that it has deteriorated with time.

Ink analysis is not only used to look for modern forgeries, but can be used to date historical documents. Thus Richard Schwab and colleagues at the University of California in Davis has been helping Harvard University scientists solve problems associated with the production of the first printed bibles by Gutenberg in 1455. Schwab and colleagues have used an instrument, which is known as PIXE, an acronym which stands for proton induced x-ray emission. In this very specialized instrument, protons are accelerated to high speeds by an instrument known as a cyclotron. The protons are fired at the ink to be investigated. When the protons hit the target, they excite atoms in the paper and ink, which then emit x-rays. By analyzing the x-rays Schwab has been able to look at the ratio of lead and copper in the ink used by Gutenberg and other people busy in the early days of printing. By checking the amount of calcium in each page of the Gutenberg Bible, Schwab worked out that sections of the Bible were made by chopping a single massive sheet into individual pages. His analysis showed that the Gutenberg printers usually composed and printed six distinct sections of the bible simultaneously. By looking at the ink on different pages they have also been able to work out that the printers made up a new batch of ink each day, and that the recipe varied slightly from day to day. Ongoing studies of ink in such documents are concerned with establishing the authenticity of different printed documents. Recipes for the ink are kept secret so that they cannot be copied by would-be forgers.[11]

The title of this Section on fraud includes the word toner. **Toner** is the technical term for powdered dry ink used in electrostatic copying machines such as those manufactured by the Xerox Corporation and Canon. The word for this type of copying is xerography. This word literally means dry writing, from the Greek word *xeros* meaning dry. The word xerography started to make it into the dictionaries in the mid 1960s, thus my 1967 edition of the Chambers Dictionary puts xerography in a two page list of new words placed at the front of the volume. Xerography is defined as a photographic process in which the plate is sensitized electrically. In view of the importance of colored xerography in the fraudulent preparation of documents, it is useful to review the basic physical principles involved in xerography. One has to be very careful when discussing xerography, because the word Xerox is a copyrighted trade name. Once at a scientific meeting I offered to make a xerox copy of my data for another participant, only to be chided by a member of the audience, who reminded me that Xerox was not a generic term for copying and would asked me to please refer to the general process as copying and not xeroxing. Companies with well-known products are very active in defending the copyright name of their products. Currently the Dow Chemical Company is having a vigorous fight to prevent Styrofoam from becoming a generic term, like Kleenex, which has evolved from a protected trade name into a generic term. In Great Britain the term biro, which was originally a trade name for a particular type of pen and is now used to describe any ball-point pen. Similarly, cellophane tape is widely called "Scotch tape," which is in fact a registered trade mark of the 3M corporation.

The basic principles by which one makes a copy of a document in an electrostatic copier are illustrated in Figure 10.7. The copying process begins when a light-sensitive drum inside the copier receives a uniform positive electrostatic charge while in darkness. Next a high intensity light is scanned over and reflected from the original document and then focused on the rotating drum. Due to the properties of the material on the drum, wherever the light strikes, the surface of the drum loses its positive charge. It now bears a reversed image of the original document in the form of a pattern of positive electrostatic charges. Negatively charged toner powder is then put into contact with the drum and it clings to the positively charged areas (these are the dark areas of the original document). Now a piece of positively charged paper is put into contact with the drum and the toner is transferred to the paper. finally, the paper with the toner image is heated to melt the plastic in the toner and fuse the pigments to the page. The toner is made by incorporating carbon black into a plastic. The recipe for the exact mixture of carbon black to plastic varies from manufacturer to manufacturer, but in general a good toner will consist of spheres of the order of 15 micrometers, containing approximately 8 % carbon black. The human eye cannot see details smaller than 30 micrometers, but in the diagrams of Figure 10.7 we have exaggerated the size of the toner spheres to illustrate the physical process. The carbon black is inside the toner beads to act as a pigment. In an electrostatic copying machine the toner beads are not actually sprinkled over the surface of the drum. Large beads known as carrier beads, to which the toner beads cling, roll across the drum, leaving behind fine toner particles where the surface of the drum is charged. You will have noticed that when a copy comes out of a photocopy machine it is hot and that the sheets cling together because of the residual electrostatic charge on the paper passing through the machine. Because in an electrostatic copying machine, the image is created by fused powder beads, the boundary of the copied letters is different than that created by ink moving into the fibers of the paper in a printing process.

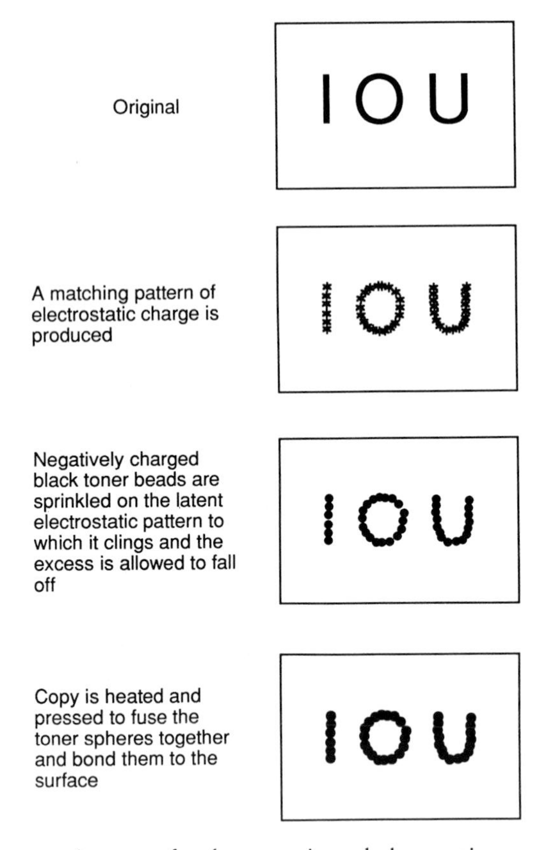

Figure 10.7. An electrostatic copy of a document is made by creating a pattern of electrostatic charge on a light sensitive drum. Toner fine particles, which cling to the electrostatic pattern, are then fused to the surface of a piece of paper.

This is illustrated by the various types of letter boundaries shown in Figure 10.8. The diagrams of Figure 10.8 are taken from a study of the quality of images made by different methods in which the raggedness of the boundary is characterized by the fractal dimension of boundaries. (See discussion of the basic concepts of fractal geometry in Chapter 6). In a modern laser printer, an electrostatic pattern is created by the scanning action of a laser beam controlled by a computer. Like an electrostatic copying machine the laser printer uses toner in a cartridge to create the printouts. Thus, the images are then developed on the paper by essentially the same process that is used in an electrostatic copying machine.

One of the major problems facing law-enforcement agencies is that color copiers can be used to counterfeit currency notes. The basic principles used to create color images with electrostatic-type copiers are illustrated in

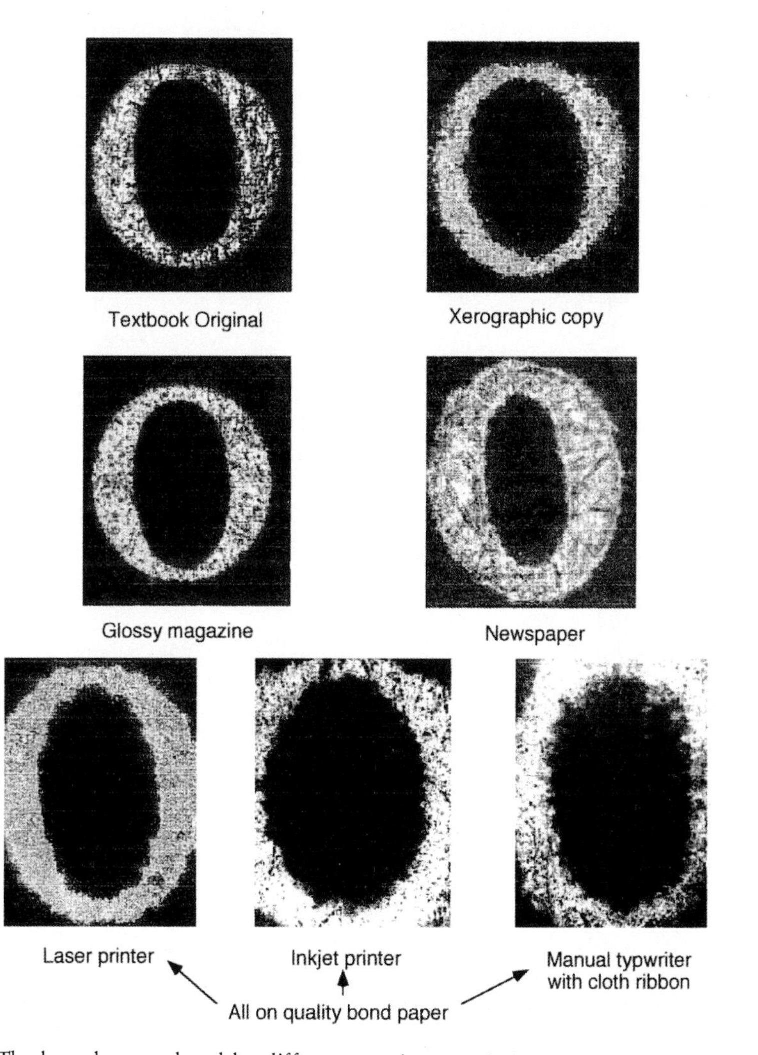

Figure 10.8. The boundary produced by different copying or printing processes on various types of paper has a characteristic ruggedness, which can be used to identify the way in which they were produced.

Figure 10.9. The object is copied three times by using different optical filters. Thus, the picture is first viewed through a blue piece of glass. Only blue light enters the copying process and the pattern is developed using a blue toner powder. Next an image of the document as viewed through a red filter is used to create the second image on top of the first image. The second image is then developed by using a toner powder with red pigment, as illustrated in Figure 10.9. The combination of these two images is shown

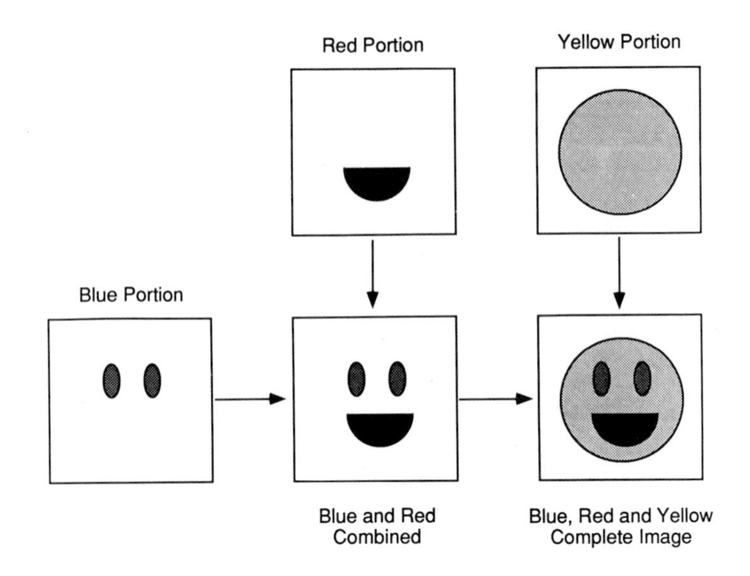

Figure 10.9. To produce a color copy, a series of copies are made on the same sheet using three different colored toner powders.

in the middle part of Figure 10.9. The final stage is to create an image of the original document as viewed through a yellow filter. This image is developed using the yellow pigmented toner to produce a full-color copy as shown in Figure 10.9.[12–14]

You may have noticed that if you place an unprotected electrostatically copied document in a binder with a plastic cover, after a short time the copy will stick to the inside of the binder. When the copy is lifted away, a mirror image of what was on the sheet of paper will be left on the plastic cover. I have not heard of anyone being caught due to a clue left in this way, but it is a possibility. The sticking and transferring of the copy occurs because of the plasticizer incorporated in the cover. This plasticizer also reacts with the toner of the electrostatically copied document, causing the copy to be transferred to the binder. If you wish to keep secret documents in a binder with a plastic cover, make sure there is a clean sheet of paper between the cover and the first copy.

10.4 Counterfeit Currency and Checks

In a district court in Ontario, a man was found guilty of making and possessing counterfeit money. He forged Canadian and American bank notes on a Canon Full Color Laser Copier. In court the prosecutor produced $25,000 worth of the two currencies. Sagus, an expert in counterfeit money serving in the Royal Canadian Mounted Police, gave evidence that the copies would probably have been good enough to escape detection. The counterfeiter, Jose Martins, forged the money at home on the copier, which he bought for $47,000 dollars. However, the check with which he paid for the copier bounced, and the company repossessed the copier. Employees found a few forged notes in the tray of the copier and advised the police of the situation.

In the United States all currency has the same color, but many countries use different colored bank notes to help combat forgery. However, one of the features of modern color copiers controlled by computers, is that the identification numbers on the bank notes can be changed automatically. This helps the forgers who usually produced counterfeit money carrying identical numbers. To help defeat the forger many countries are also introducing advanced technology into the printing of their bank notes. Thus, as shown in Figure 10.10(a), since the early 1990s some Canadian bank notes have been carrying a thin metal tag, which manifests a different color at different angles of viewing. The principles employed in the design of such a metallic tag is based on a familiar phenomenon. Most people have seen that a very thin oil slick on the surface of water produces colored patterns. Some of the energy of an incident light beam is reflected from the top of the oil slick. Part of the beam penetrates the thin film and is reflected from the bottom of the film. If the difference in path length is half a wavelength of light, then the two beams of light leaving the oil slick combine in the eye to produce darkness. The colored patterns result from the fact that white light is made up of many different wavelengths. As one looks at the surface at a particular angle, one color is obliterated, leaving the complementary color to be seen by the eye. As the eye moves, the reflected path lengths vary so that different colors are observed. The metallic tag mounted on Canadian bank notes is very thin, roughly one tenth of the wavelength of light, so that as one varies the angle of inspection, the interference effects makes the metallic tag change from gold to green. The metallic tag, developed by J. Dobrowolski of the Canadian National Research Council, is made by depositing layers of zirconium and silicon oxide in a vacuum onto a thin plastic webbing. Next the patches of the thin film are transferred under pressure to the newly printed bank notes. An adhesive secures them to the paper before the plastic backing is removed. It is estimated that the procedure raises the cost of producing the notes by 2.5 Canadian cents per note. John Rolf, the Bank of Canada's scientific advisor, supervised tests indicating that the optical effects can still

a)

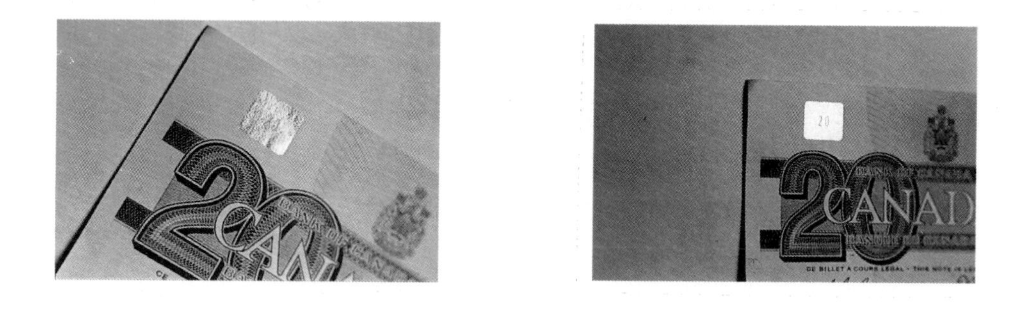

b)

Silver thread

Watermark

Figure 10.10. To fight forgery of modern bank notes, different strategies based on advanced technologies are used. a) Canadian bank notes have started to carry a forgery fighting, variable color metallic tag, which is colored by optical interference effects. b) British bank notes have an interwoven silver thread and a watermark to help fight forgery.

be seen after the bills have been crumpled, scratched, washed, dry cleaned, and treated with chemicals. The first bill to carry the new metallic tag was the Canadian $50.00 note. It is estimated that a bank note has an average life in circulation of 3.5 years. Experiments show that the metallic tag will keep its color variability throughout this period. The metallic tag is superior to holograms, which apparently do not survive crumpling and washing.[15]

The British have adopted different strategies to help fight forgery of their currency. Two of these strategies are illustrated in Figure 10.10(b). A silver thread is woven through the paper from which the notes are made. This is very difficult to forge. Second, the area outlined on the note contains a watermark showing Queen Elizabeth's head, which can be seen when the note is held up to the light. This is another feature of the bank note, which is virtually impossible to forge.

The United States authorities are well aware that U. S. money is one of the more easily copied currencies. There are indications that they will start issuing new dollar bills with improved anti-forgery devices. Apparently, however, there is considerable public resistance in the United States to the introduction of different colored money, because of the traditional greenback of U. S. society.[16]

If you notice that someone has passed you a forged bank note you should be aware that in theory the loss is yours unless the forgery is detected immediately as the bill is acquired. In Sudbury a woman was given some hundred dollar bills before she started her vacation. When she reached the United States she found that some of the bills were counterfeit. It would have been difficult for her to prove she had actually obtained the hundred-dollar bills from the bank. Theoretically she would have been forced to sustain the loss herself, but the bank, as an exercise in customer relations, made up the loss.

If you use a bank note in a store and are told that it is counterfeit, ask to speak to the police immediately. In 1993 a Canadian vacationing in the United States was paying for gas at a service station when the attendant advised him that one of the 20 dollar bills he was using was counterfeit. Somewhat irritated by this, the tourist paid with some other bills and left the gas station only to be arrested 5 minutes later by the state police. The fact that he left the gas station without reporting the existence of the forged bills to the state police made it look as if he knew that the money was counterfeit, and was trying to pass it for personal gain. The tourist was arrested and later had to return to the United States to fight the lawsuit. Although acquitted of the charge, the episode cost over $500 in legal fees to clear his name.

Australian scientists have developed various strategies for preventing forgery. In 1987 the Australians issued a plastic bank note, which is believed to be the first of its kind in the world. The bill is described as being:

> *... packed with anti counterfeiting devices which were developed by the Commonwealth Science and Industrial Research Organization (the central government research labs of Australia).*[17]

The cost of developing this type of note is estimated at twenty million Australian dollars. The main security feature of the plastic bank note is an **optical variable device (OVD)**. An image of Captain Cook in the top corner of the notes is essentially a diffraction grating, the color of which changes as the bank note is tilted – the same way that the colors seen from the surface of the compact disk change with the angle of viewing. (See discussion of diffraction grating in Chapter 2). Because of this, the color of Captain Cook's picture changes as you hand the note over to someone. The image can be viewed from both sides of the bank note. The note also has an aluminum coating, which forms a reflective surface but cannot be felt with the fingers. If the note is held up to the light, diamond shaped patterns appear. To test the durability of this new plastic bank note tests were carried out, which included boiling the note, burying it in soil for nine months and placing it for hours in a washing machine. A more advanced version of the same strategy for defeating forgery has recently been announced by the Australian organization that developed the diffraction grating.[17,18]

Checks can be copied and passed fraudulently unless certain precautions are taken. Two of the ways in which check forgery is being combatted are illustrated in Figure 10.11. The top check is from Britain. You will notice that there are two vertical lines halfway along the check, accompanied by the statement "not negotiable." When this notation is added to a check in Great Britain, the document is called a crossed check. They can only be placed in the banking account of the person named on the check. Thus, if I had signed the check and given it to my wife, she would not have been able to place it in an account in her own name. However, when I cashed this check in Canada the procedure was unknown to the bank and my wife was able to deposit it in her account without anybody checking up. In defence of the bank however I should point out that both my wife and I were well-known at the branch involved in the transaction.

The lower part of Figure 10.11 shows a copy of another check that I received, which was protected from fraud in three different ways. First of all you will notice the sentence written across the top of the check "The face of this document has a colored background, not a white background." (Of course the check copy shown in the diagram has a white background.) You will also see the words "void" and "faux." These two words were printed on the check in ink that is not visible in daylight but which fluoresces during the photocopying process. Therefore the writing appeared only on the copy. Finally, the bottom of the check reads "the back of this document contains an artificial water mark, hold at an angle to view." It is not possible to reproduce this artificial water mark.

a)

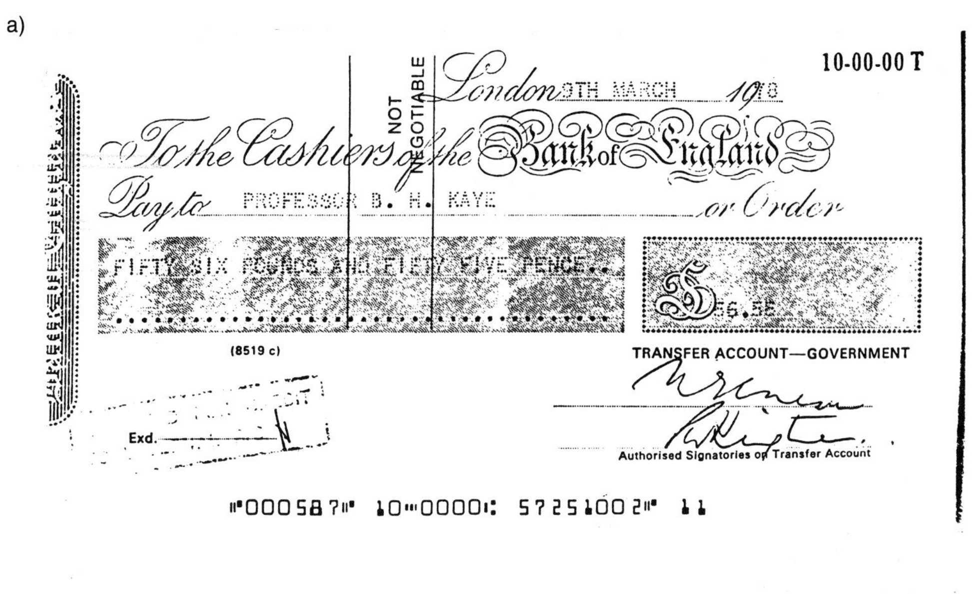

10-00-00 T

NOT NEGOTIABLE

London 9TH MARCH 1978

To the Cashiers of the Bank of England

Pay to PROFESSOR B. H. KAYE or Order

FIFTY SIX POUNDS AND FIFTY FIVE PENCE..

£56.55

(8519 c)

TRANSFER ACCOUNT—GOVERNMENT

Exd.

Authorised Signatories of Transfer Account

⑈000587⑈ 10⑆0000⑆ 57251002⑈ 11

b)

THE FACE OF THIS DOCUMENT HAS A COLORED BACKGROUND – NOT A WHITE BACKGROUND

0010890

CTV

CTV Television Network Ltd.
42 Charles Street East
Toronto, Ontario M4Y 1T5

Canadian Imperial Bank of Commerce
Main Branch Commerce Court, Toronto, Ontario

Pay

3dol's00cts

Cheque Number 07506

Amount

*****4

To the order of KAYE, B H DR

CTV Television Network Ltd.

P3E 3P3

VOID FAX

THE BACK OF THIS DOCUMENT CONTAINS AN ARTIFICIAL WATERMARK – HOLD AT AN ANGLE TO VIEW

⑈0010890⑈ ⑆00002⑈010⑆ 21⑈00118⑈

Figure 10.11. Banks, in a constant battle against forgery and theft, can adopt several strategies as illustrated above. a) A British check which is "crossed" to make it non-negotiable. b) A Canadian check protected against forgery by three different strategies.

A rather curious case of counterfeiting of checks, bank notes and other documents was reported from Australia in 1992. Apparently, in the Australian state of Victoria, there is a prison in which the prisoners are allowed to have color televisions and computers. The government discovered that there was a counterfeiting group, which was operating from this prison, with connections to all the other prisons in the state of Victoria. Checks were being forged with the computers and color printers and sent to relatives who cashed them in Melbourne. Paper for the printing was either brought by visitors or sent through the mail! Apparently prisoners were also forging Thomas Cook and American Express travellers checks, prison visitor passes, and fifty dollar bills.[19]

Once a document has been declared fraudulent one can use the various techniques outlined in the forgoing sections of this book to study the source of the ink and to identify possible printers who have participated in the forgery.

10.5 False Stamps and Postal Fraud

The adhesive, prepaid postage stamp, was invented in 1840 by the British postmaster Sir Roland Hill. The very first British stamp was the famous penny black shown in Figure 10.12 (a). Figure 10.12 (b) shows the listing in Stanley Gibbons' concise stamp catalogue for the stamp issued to commemorate the centennial of Sir Roland Hill. Each stamp from the Roland Hill special issue, in which some of the color was left out during printing, was worth $6,000 dollars in 1990. If a mistake has been made in the printing of a stamp, the prices that people are prepared to pay for one square inch of paper is astronomical. Postage stamps have been known to sell for prices of the order of four million dollars. Such huge prices for small pieces of paper attracts forgers like bees to a honey pot. Throughout the history of stamp issuing and collecting, there have been spectacular frauds. One example resulted in a famous trial, which took place at the **Old Bailey** in London, England in 1892 (the Old Bailey is a famous law court of Great Britain). A group of three men known as the London Gang, Benjamin, Sarpi, and Jeffreys were convicted of stamp fraud. Benjamin and Jeffreys were sentenced to six months of hard labour and Sarpi was given four months of hard labour. The three worked together in the following manner: Benjamin ran a shop in London. Sarpi positioned himself in a back room with the door ajar so that he could listen unseen to the conversations that Benjamin was having with customers. When he heard what stamps the customers wanted, he advised Jeffreys, who made fake stamps to meet the commercial demand. Stamp collecting was a very active hobby in the 1880s and large sums of money could be made by defrauding rich customers. At the trial of the three men, particular attention was focused on the way in which Count Ferrary was duped. This eccentric millionaire went to city stamp shops each day returning home laden with purchases. He employed two clerks to sort out and file his new acquisitions. The Count became one of the gangs most frequent victims, with Jeffrey's faking stamps virtually to order.

a)

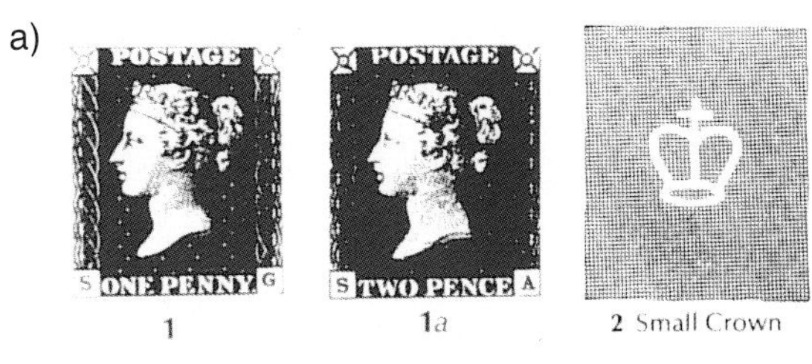

1840 (6-8 May)
Letters in lower corners Wmk Small Crown W 2. Imperf.

No.	Type		Un	Used	Used on Cover
1	1	1d. intense black	£3250	£190	
2		1d. black	£2750	£140	£225
		Wi. Wmk inverted	£3750	£350	
3		1d grey-black	£3000	£190	
4	1a	2d. dp full blue	£7000	£375	
5		2d. blue	£5500	£300	£650
		Wi. Wmk inverted	£7000	£650	
6		2d. pale blue	£7000	£375	

b)

573 Sir Rowland Hill

574 Postman, circa 1839

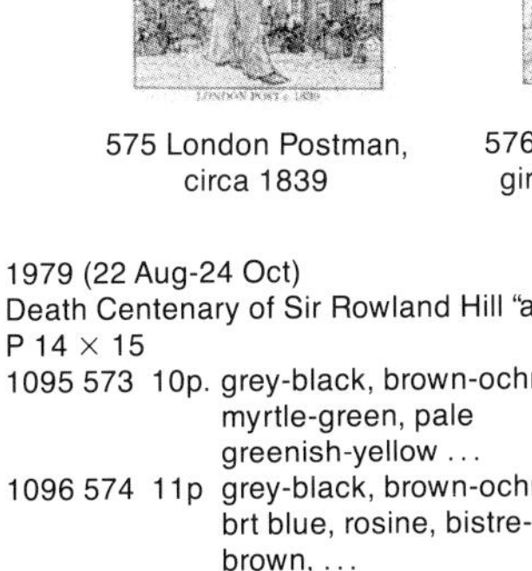

575 London Postman, circa 1839

576 Woman and young girl with letters, 1840

1979 (22 Aug-24 Oct)
Death Centenary of Sir Rowland Hill "all-over" phosphor.
P 14 × 15

1095 573	10p.	grey-black, brown-ochre myrtle-green, pale greenish-yellow ...	25	25
1096 574	11p	grey-black, brown-ochre brt blue, rosine, bistre-brown, ...	30	35
1097 575	13p	grey-black, brown-ochre, brt. blue, rosine, bistre-brown, ...	35	40
1098 576	15p	grey-black, broen-ochre, myrtle-green, bistre-brown, rosine, ...	40	40
		Set of 4	1.10	1.25
		Set of 4 Gutter Pairs	2.40	
		Set of 4 Traffic Light G. Pr.	3.75	
		First Day Cover		1.25
		Presentation Pack	1.60	
		P.H.Q. Cards (set of 4)	1.25	2.25

c)

d)

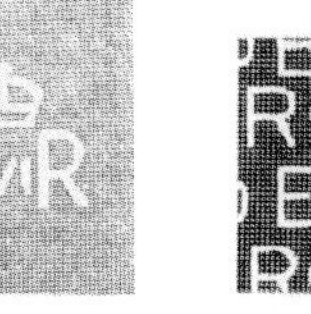

Figure 10.12. Forgers stand to make substantial amounts of money by making fake postage stamps. a) The Penny Black, the world's first self-adhesive postage stamp. b) Used postage stamps can be worth very large sums of money. c) A forgery attempt carried out in Canada in the early 1990s involved monochrome roll stamps, a genuine example is shown. (Note the two straight, unperforated edges.) d) Typical watermarks, which have been used in the making of stamps. (Reprinted with permission of Stanley Gibbons, Ltd.)

Forgers have understandably avoided duplicating the more exotic stamps concentrating instead on the middle value stamps. However, there have also been forgeries of current postage stamps in an attempt to defraud the post office, rather than stamp collectors. In Canada in recent years there have been two cases of forgery involving common postage stamps. Both of them involved monochrome lithographed coil stamps (lithography is a type of printing, see reference 29). The current 43-cent stamp of the kind forged is shown in Figure 10.12 (c).

The appearance of the stamps under ultraviolet light gave the forgery away. Phosphors are added to stamps in modern technology to enable robots to sort and cancel the letters going through the mail system. Phosphors, which glow in ultraviolet light, are used to add the postal codes to the envelopes. Also, the class of mail, that is, first or second class, is marked with different stripes of phosphor on the stamps. When viewed under ultraviolet light, the robots see only the printed bars, which enable them to sort the mail. All other information on the envelope is not visible to the robot eyes. Absence of phosphors and their associated glow in ultraviolet light may reveal a forged stamp.

In a 1990 counterfeiting episode, when the postal rate in Canada was 39 cents, forgers printed 6.5 million dollars worth of counterfeit 39-cent stamps. The crooks planned to unload the booty on the market by swapping them for cigarettes and similar items at convenience stores, because these stores stock stamps, and from time to time clerks do take stamps as currency for small purchases. The police were alerted to the counterfeit operation by a store owner, who noticed that when he was offered some of the stamps, the perforation looked irregular. In fact, the thieves had tried to create the perforations in the stamp with sewing machines. It is very difficult to obtain perforation machines, which simulate the edges of genuine stamps. When they were tipped off, the Montreal police, accompanied by Canada post investigators, raided a printing shop in Montreal. They found a printer preparing to print more counterfeit stamps. When the building was searched, police found cartons, each containing 44,000 phony stamps in sheets of 440, hidden in the ceiling.[20]

It is interesting to note that, although forgers take great care with their printing, they sometimes trip themselves up with small errors. Thus, the crooks who were trying to create the forged coil stamps in 1990 made the mistake of perforating the stamps on all four sides, whereas coil stamps, as can be seen from the example shown in Figure 10.12(c), have two straight edges.

The reader will not be surprised to find out that analysis of the ink of suspected stamps forgeries can indicate whether or not a valuable stamp is a forgery. In general, if a stamp is on an envelope, the combination of cancellation and stamps still mounted on the envelope often increases the value of the stamp itself. Billmeyer and Priess at Rensselaer Polytechnic Institute in

New York State, were asked to look at an envelope with a green postmark that was said to have been issued around the time of the Civil War. However, the investigators determined that the dyes used in the green ink had not been invented until 1920.[21]

Another well-known forger in the history of stamp faking was a French man called Sperati. It is said that he became wealthy during the German occupation of France in the Second World War by selling fake stamps to German soldiers. Near the end of his life, when his eyesight was failing, he actually contacted the British philatelic association, offering not only his stock of forged stamps, but the things that he needed to produce them along with the promise that he would not make forgeries anymore. The association raised about 20,000 dollars to obtain the stamps and the equipment. The money was raised by selling the indelibly marked bad stamps, to collectors and dealers, for use as reference to protect themselves from counterfeit. However, Sperati made some rather startling mistakes in his forgeries. For example, he forged a U. S. Civil War stamp showing a Confederate stamp with a convincing Richmond postmark carrying the date of May 10th, 1865. Had Sperati studied his history lessons, he would have known that the post office had ceased to exist before that date because the Northerners had captured the city. Another Confederate stamp forgery that he made showed a postmark of a city in Vermont, also of May 10th, 1865. The Confederates never reached that far north! He also issued some unbelievable Southern Confederate forgeries, with post marks of places such as Cincinnati, Ohio and Detroit, Michigan.[22]

Because stamps can be so valuable, collectors should protect themselves against theft by having their stamps photographed at high magnification. Robertson tells us of a case where photography was able to identify a stolen stamp. The stamp in question was a Canadian five-cent stamp issued in 1959 to celebrate the opening of the St. Lawrence Seaway. The Canadian banknote company printed forty million of these stamps, but two sheets of them had one color printed upside down. The stamps of this type that survived became very valuable. Thus in 1979, the value of a single unused stamp was placed at around 15,000 dollars. In 1977 Canada post sent Queen Elizabeth of England a corner block of four five-cent Seaway inverts as a gift on the 25th anniversary of her succession to the throne. This corner block is currently in the Royal philatelic collection. It is believed that approximately 150 of the stamps have survived in various conditions from unused to damaged.[23]

Val GreenLeaf, a prominent Belleville, Ontario, collector of stamps was able to include a Canadian invert stamp in his collection. After his death, while his widow was in Toronto visiting relatives, a burglar entered their house. Word of the GreenLeaf robbery was circulated to police departments across Canada and stamp dealers were alerted to the robbery. A youth entered a Toronto stamp dealer shop shortly after the theft; he said he wanted to sell his family's stamp collection. When the dealer saw the collection contained an inverted Seaway stamp, and finding that the youth knew little about the value of this stamp the dealer became suspicious and contacted the police. The youth was questioned by the Toronto police and, at a subsequent trial, the original vender of the stamp to the GreenLeaf collection, Kasimir Belinski, was subpoenaed. He had to travel from Vancouver to participate. (He had to come to the trial because the word *subpoena* comes from two Latin root words meaning under penalty, and if he failed to appear in court, he would have faced a heavy penalty.) When giving evidence in court, Belinski made the following statement:

> *Identifying individual inverted Seaways is very simple. They printed them in two colors. The alignment never matches right on. It is very easy to identify a particular stamp from photographs of the stamps.*[23]

Rather than place a tiny mark on the back, Mr. Belinski said that when he sold the stamp he always took magnified close-up photographs of the stamp. The prints of the stamp were stored with the records of the purchasers. In his presentation Mr. Belinski said:

> *It is absolutely impossible to have two the same. A perforation may be nibbled, it could be off center. Like a fingerprint, which cannot be duplicated by another person, the broken paper fibers on one stamp can be easily matched with its photographic image. If it is a used stamp, this an absolute fool proof record. No cancellation on a stamp is the same.*

The thief of the stamp collection apparently thought it was impossible to identify individual stamps. It is recorded that after his conviction the thief had learned his lesson. He is now married with two children and sells vegetables for a living![23]

Lane has reviewed the way in which the British post office is trying to develop better paper and printing techniques for making stamps to prevent forgery.[24] In Figure 10.13(b), two new strategies are illustrated. As already pointed out, forgers have difficulty obtaining perforation machines. However, in the mid 1980s the Irish post office had to withdraw a one pound stamp because of high quality perforated phonies. The one pound fifty stamps in Figure 10.13(b) have an unusual set of perforations with two elliptical cuts, one on each side of the stamp. Apparently this is extremely difficult to forge. Even if they obtain the perforation machine, as can be seen from the forty-one pence stamp shown in the figure, it is very difficult to reproduce the position of the elongated cutaway, which varies from one stamp to the other. In an article on this new perforation, Ian Robertson points out that collectors were not particularly pleased to see this new development, because it makes it difficult to tear the stamp from the sheet without damaging it. However, in the long run this means that undamaged, used specimens of the stamp will have a higher value.

Figure 10.13. Automation in the post office creates new opportunities for fraud; scientists constantly try to change the technology of stamp making to protect postal revenue. a) Robot failures invite "theft" by leaving unfranked stamps that are easily re-used. b) New perforations and images which change color are the latest technical innovations in the fight against forgery.

Earlier versions of these stamps showed a detailed photographic profile of the Queen instead of a silhouette. Britain breaks two general rules in the printing of their stamps. First, as a rule, one cannot reproduce the likeness of a living person on a stamp! You have to be dead before you become famous on a stamp. Second, Great Britain does not print the Country of origin on the stamps. The British claim that since they were the first to use pre-paid stamps, they have the right to remain anonymous. The silhouette of the Queen on the new stamps is printed in a special ink. The ink is called **optically variable ink**. It is manufactured by Sicpa of Switzerland. Depending on the angle of viewing, it changes color. Sometimes it appears to be gold, sometimes a glowing green. The recipe for this special ink, for obvious reasons, has remained a secret. I suspect that it contains pigment grains of a specified fraction of the wavelength of light. If so, the change in color is again an optical interference effect.

In 1993, the British post office introduced a new ten pound (twenty dollar), stamp. This stamp again will have optically variable ink as a security feature, but it will also be the first British stamp to feature Braille embossing. Braille is a system of raised dots, which can also be read by blind people. The dots have different configurations for different letters of the alphabet, and by moving one's fingers over the dots, one can decipher the letters. Although this will be a convenience for blind people, the main reason for adding the embossed writing is to make it even more difficult for forgers to reproduce the stamp. The value of such a stamp when used (lightly canceled) is such that if you ever receive one, don't throw it in the wastebasket; save it for your stamp collection or give it to a friend.

In his article on the technology being developed by the British post office, Lane tells us that the British post office believes it may have lost several million pounds in recent years through stamps being reused, either because the cancellation ink had been washed off, or because it had not been cancelled.[24] Stamp washing for criminal reuse is not a new crime. The British penny-black stamp was changed to a red-brown version one year after its initial issue, because the red canceling ink that the post office was using with the black stamp was easy to wash off. When the red-brown penny stamps went on sale, the canceling ink was changed to a black ink, which was more difficult to wash off. Robertson tells us that in England a major crime operation that had removed cancels from millions of stamps was broken by police and postal inspectors. He further comments that in the United States, similar stamp washing scamps prompted the post office at one time to ask charities to stop selling used stamps in bulk to collectors because they suspected that this was the crooks' source of used stamps.[25]

Until 1990, the traditional stamp cancellation ink used in Great Britain was a combination of carbon black and mineral oil with no drying agents added. The post office relied on the cancellation to become permanent when the ink penetrated and was absorbed into the paper. However, with the type of printing, which is now becoming popular when making stamps, the old carbon-black based ink tends to smear and does not penetrate the stamp, making it easier for stamp washers. Therefore in the 1990s, the British authorities introduced a new ink, which relied on resin binders for adhesion and which is very difficult to remove from the printed surface. There is a basic conflict of interest between the post office and stamp collectors. The post office likes to see a hefty cancellation mark on the stamp, whereas the stamp collector likes to see the type of cancellation shown on the one pound fifty stamp of Figure 10.13. Because I am a stamp collector, my son took this envelope along to the post office and asked them to stamp it lightly and in the position shown, because this is the style of cancellation which enhances the value of a used stamp. Such a light cancellation is obviously easier to remove than the heavy cancellation favored by members of the post office, who are known as stamp butchers to those of us who collect stamps. The post office has to strike a balance, since the sale of franked stamps to stamp collectors is a good source of income.

In 1993, the Canadian post office issued a stamp with a hologram of a satellite circulating the earth. The face value of the stamp was 42 cents. Some of these stamps were found to have the hologram missing. Within six months of the mistake's discovery, collectors were willing to pay $3,000 for one of the faulty stamps. However, dealers of stamps are warning that it is very easy to remove the hologram from an ordinary stamp to simulate this error. Belinski has however said that:

> *You can check under a microscope, and you can detect a forgery, i. e. a stamp which has had the hologram removed, because of the broken fibers you will see and also because when the holograms are added to the stamp, they are put there under pressure and one can see an indentation in the paper if a hologram has been removed from a normal stamp.*[25]

10.6 Jurassic Adventure in the Land of Fraudulent Archaeological Specimens

In 1993, many people who had previously been unaware of Paleontology became interested in this scientific discipline, because of a movie entitled "**Jurassic Park**." The term **Jurassic** is used by geologists to describe a period in time when dinosaurs roamed the earth. (The name is taken from the name of the Jura mountains in Switzerland). The basic theme of the movie is that scientists recover genetic material from a biological specimen entombed in amber, and from this they recreate dinosaurs. At about the same time as this movie became well-known, the scientific world was surprised to hear about a forgery involving a fly entombed in amber.[26]

Amber is defined in the dictionary as a yellowish fossil resin used for ornaments and jewelry. As mentioned briefly in Chapter 2, the Greeks name for amber, which was usually found floating in the sea, was *electra*. They thought amber was pieces of the sun's rays that broke off when they hit the water. We now know that amber is the fossil resin that oozed out of pine trees and became preserved when the ancient forests were buried. Amber washes up on the coast, because some of these forests are now under the sea. As the resin oozed out of the bark of the tree, objects such as flies became trapped in the material to become a permanent part of the piece of amber. The natural history museum of Great Britain has over 2,500 specimens of insects trapped in amber. One of the specimens was an entombed fly known as the latrine fly. The specimen is treated in textbooks because it was described by world famous insect specialist Hennig as an example of a living creature which had not changed in forty million years of evolution. Andrew Ross, a student of ancient insects, was examining this entombed fly under the microscope. When one is looking at this kind of specimen under a microscope, one has to be careful that the heat from the lamp of the microscope does not damage it. As Ross looked at the specimen he became alarmed when suddenly two cracks appeared on either side of the fly. He quickly removed the specimen from the microscope, fearing that he may damage the fly and destroy vital anatomical detail in the specimen. However, when he turned the amber piece around to check the extent of the damage, he noticed that a clear line ran through the amber. He was amazed to note that the specimen seemed to be sitting in a hemispherical depression. The line on the hollow could be seen from both sides and he deduced that a genuine piece of Baltic amber had been carefully cut in half and a small excavation made in one surface. A forger had then taken a fly and buried it in the cavity with a resin-type medium. He then glued the two pieces together to form the complete specimen. No one knows the origin of this particular fossil. (The term in Latin *fossere* means to dig and the word fossil originally meant something dug out of the ground. In modern English it has come to mean the remains of a prehistoric plant or other living creature which has been preserved in a recognizable form through encapsulation within some material). It is known that the specimen was listed in the collection of H. F. Loew, a German expert on flies. The Natural History Museum bought 300 of Loew's amber specimens in 1922. In the words of Di Palma, who described the discovery of the forgery:

> *the criminal's identity is still a mystery, but the motive was probably monetary. In Victorian times, collectors like Loew paid good prices for interesting specimens, creating a lucrative market for forgers. The forger is unlikely to have chosen the fly deliberately to confuse future students, it was probably the nearest insect at hand.*

It is interesting to note that after the news of the fake fossil was described in the magazine *New Scientist*, Glynn Owens wrote a letter to the journal pointing out that in the Sherlock Holmes detective story "The Yellow Face," published in 1893, Sherlock Holmes remarked about pipes with amber mouthpieces:

> *I wonder how many real amber mouthpieces there are in London. Some people think a fly in it is a sign (of verity). Why, it is quite a branch of trade that, putting of flies into the amber.*[27]

Owens also points out that although this remark was in the original story published in the *Strand* magazine it is often omitted in later editions of the story.[27]

Incidentally, scientists believe that the art of reconstructing creatures such as dinosaurs from fossilized genetic material in amber is still a long way from reality! So the readers need not loose any sleep over fears that they will be eaten by dinosaurs.

Perhaps the most famous forgery from the world of paleontology is known as the Piltdown skull. The saga of this famous skull began in late 1912 when a British Lawyer Charles Dawson (1864–1916), who was also an amateur paleontologist, claimed to have found fragments of an early human skull in a gravel bed near the village of Piltdown, Sussex. Famous people and well-known scientists were involved in the study of the Piltdown skull and its supposed importance for the history of human evolution. The group included Arthur Connan Doyle, the writer of the Sherlock Holmes mysteries, Pierre Teilhard de Chardin, a famous Jesuit priest who at the time was a student in the seminary, Sir Arthur Keith, a famous anthropologist, and Arthur Smith-Woodward, a geologist. Both Smith-Woodward and Keith recreated skulls from the fragments as shown in Figure 10.14(a).[28] We now know that the entire skull was a fake. It had been created out of the jaw of an orangutan and part of a human cranium, which was a few hundred years old. The teeth were filed and some of the bones were stained with the chemical potassium dichromate to make them look old. The skull was exposed as a forgery in 1953 by Joseph Weiner of the University of Oxford, who carried out chemical tests on the bone. Until the exposure of the skull as a forgery, it was reluctantly accepted; many people did not know how it fit into the evolutionary scheme. Several books have been written on the mystery of the Piltdown skull and various individuals have been blamed for creating the fake skull. Dawson, the original collector, has been suspected of making the skull in order to make himself famous. The most recent book on the subject blames Arthur Keith as the creator of the fraud. Yet others suspect that Teilhard de Chardin created the skull as a joke and that he was unable to take the heat of exposing his own forgery when eminent individuals accepted the forgery as a genuine paleontological specimen.

Other experts, including a very famous British scientists, Lord Solly Zuckerman, suspect a person called Martin Hinton.[28–33] At the time of the discovery of the Piltdown skull fragments, Hinton was a lawyer's clerk and an amateur paleontologist. For two years he had been working with Smith-Woodward, whom he had come to dislike because he felt he was a pompous person. Some people believe that he deliberately planted the bones in the gravel bed to make a fool out of Smith-Woodward. He found himself in a difficult position, when experts pronounced the fragments as genuine. According to one theory he was so disgusted with the gullibility of the experts over the Piltdown man, that he forged a further fossil by taking a leg bone from an extinct species of elephant and carving it to look like a cricket bat. On the jocular assumption that if the Piltdown man had been British he would have loved to play cricket, Hinton planted the "bat" in the gravel

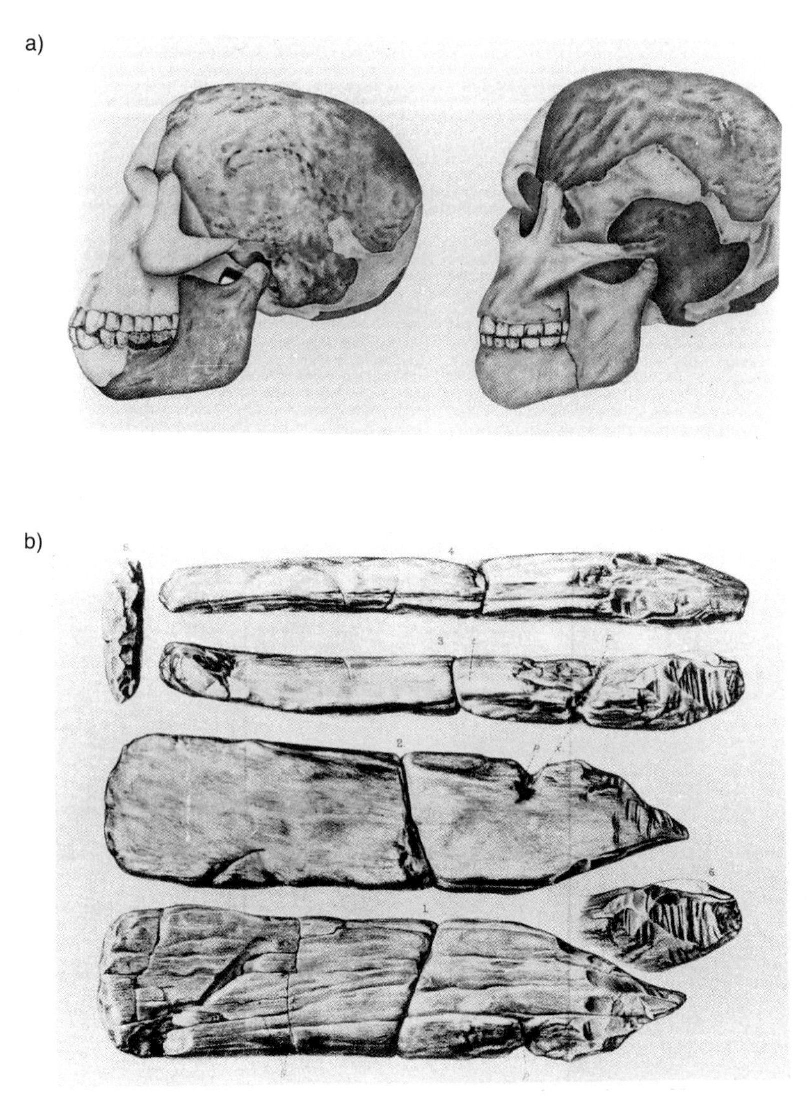

Figure 10.14. The saga of the Piltdown skull is probably the most famous example of a scientific forgery. a) Two reconstructions of the fragments of the Piltdown skull. The left was assembled by Arthur Smith-Woodward and the right by Arthur Keith. b) "Batty Bones" carved from elephant bones and planted in a gravel pit to "fool" archaeologists. (Reproduced from *American Scientist*.[28])

bed. The cricket-bat forgery is shown in Figure 10.14(b). However, once again the forgery misfired because Smith-Woodward pronounced the artifact a superbly important example of the work of Paleolithic mind. Apparently he went to great lengths in describing its details and even interpreted the remains of a hole pierced through the bone. According to his theory, a string was threaded through the hole and used to attach the bat to Piltdown man's waist belt. Ralph Estling points out that at this point Hinton gave up the unequal struggle with the experts and the deception continued.[32] I will leave the reader to study the saga of the Piltdown skull in detail for himself, but I would like to point out that if someone tried to do the same thing today, carbon dating of the bone would immediately reveal a forgery.

We will now review the basic physics of carbon dating by studying the experiments carried out to test the authenticity of a religious relic known as the Shroud of Turin.

In 1978 the shroud of Turin was placed on public display in Turin. In an article entitled "Christ Under the Microscope," Hanlon, a well-known scientific writer describes his impression when looking at the shroud.[34]

> *When I saw the shroud on a Saturday night in Turin Cathedral my first impression was surprise that this burnt, water stained, and wrinkled piece of cloth should be so venerated. Yet, the image of a long hair bearded man with his hands crossed in front of him is unmistakably and strikingly there. The picture is anatomically correct for a human corpse, and has considerable detail such as hair and beard. The reddish brown image is faint and seems clearer if you stand farther away from it.*
>
> *Proponents claim to see evidence of all that the bible says happened to Jesus: scourge marks on his back, swelling below the right eye from being struck on the face, bleeding from the crown of thorns, and so on. I viewed the shroud for some minutes and to me many of these details seemed more like water stains and burn marks. Are the faithful simply reading their expectations into ink blots? These details have the backing of experts such as J. Malcolm Cameron, who is a home office pathologist and a London University professor of forensic medicine. Photo interpretation experts find remarkable things hidden in the pictures.*[34]

The comments by Hanlon hint at the fact that the authenticity of the shroud was the subject of controversy, but also that the shroud has received some very high-powered examinations from a whole range of forensic experts. As we will discuss later in this section, these experts have finally reached a conclusion as to the authenticity of the shroud. The range of investigations is worth reviewing because of the wealth of information on the sophisticated studies that can be carried out when faced with a mystery involving cloth and/or a corpse.[34–43] A panoramic view of the shroud of Turin is shown in Figure 10.15(a). A news story describes the Shroud of Turin as follows:

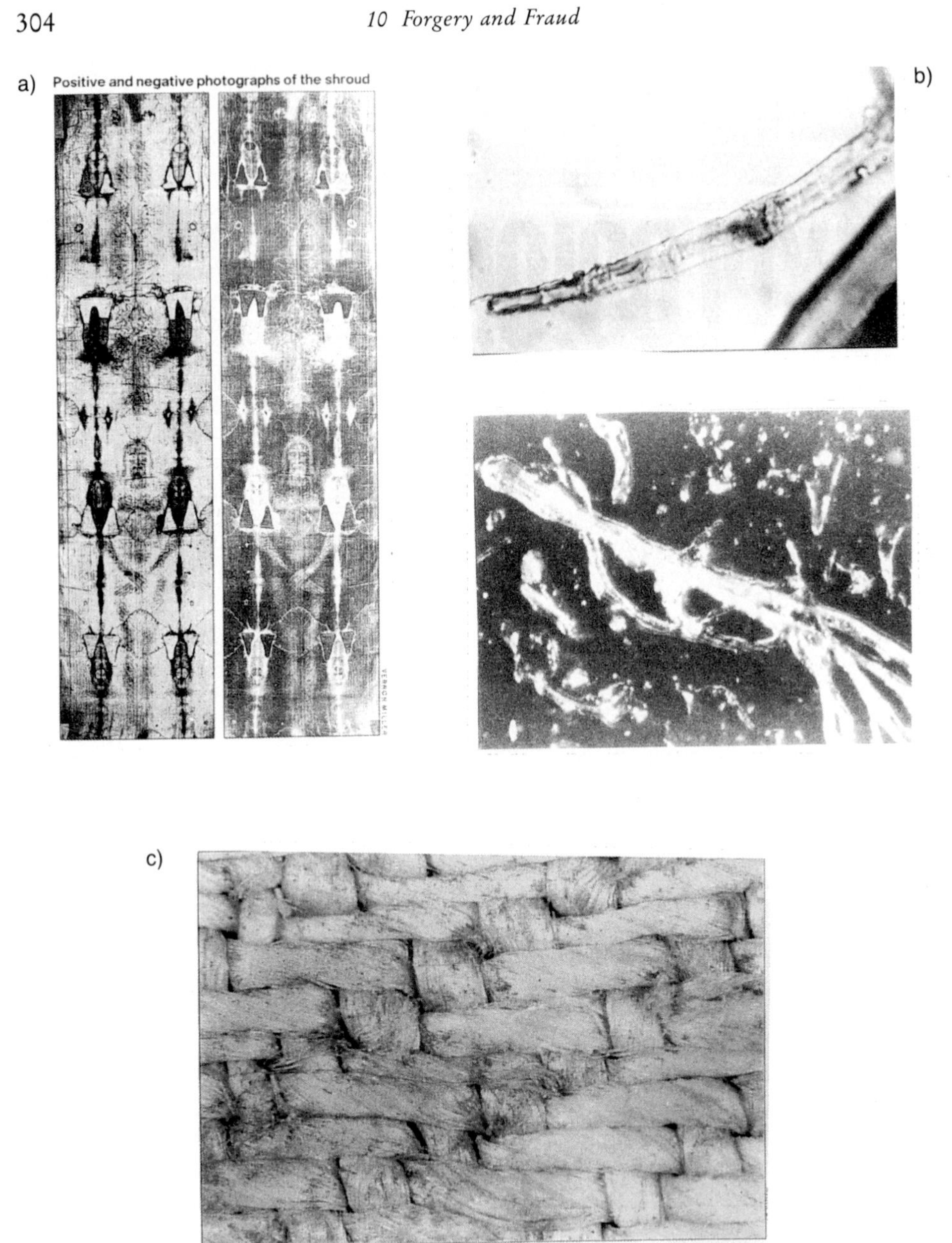

Figure 10.15. The Shroud of Turin has been studied by a very wide range of techniques. a) Panoramic view of the Shroud of Turin as both positive and negative images. b) Pollen and fibres from the weave of the cloth of the Shroud. c) McCrone found that red blotches on the surface of the cloth were artists pigment. (Reprinted with permission from *Discover*.[36])

The shroud is a 4.3 meter piece of linen with the impression of a man lying with his hands crossed on his stomach. The impression is remarkably clear with the hair and beard clearly distinguishable and the face is strikingly similar to that traditionally chosen by artists who represent Christ. The image is like a photographic negative. The shroud's history can be traced only back to 1357. It has been frequently branded as a 14th century forgery.

In medieval times many claims were made for different objects as being from the time of Christ. Thus, in an article on the shroud Angier states:

Around the 16th century... so many pieces of the true cross had been unearthed that if they were gathered together, the great skeptic Charles MacKay said in 1852, there would be almost sufficient to build a cathedral.[36]

During the middle ages, more than 40 authentic burial shrouds of Christ were being peddled throughout Europe. Most of these shrouds have long since joined the ash heap of history but for six centuries millions of people, Christians and non-Christian alike, have continued to believe that one of the cloths, the Shroud of Turin, is genuine.

The first authentic historical record of the existence of the Shroud of Turin is to be found in a document dated from 1357. Already in that year Bishop Henri De Poitiers pronounced the shroud a forgery.[36] The shroud survived a chapel fire in 1532 with only a few scorch marks (caused by drops of molten silver from the casket that contained it, and water stains. The movement of the shroud has been traced from Jerusalem to Turkey and then to Europe. It was given to the Church in Turin by the Royal Family of Italy. One of the modern studies of the Shroud was carried out by Max Frei, Director of the Zürich Criminal Police Scientific Service. He placed tape on the cloth and lifted pollen caught in the weave of the cloth. Through the microscope he identified pollen from 50 different plants. Some of the pollen was from types of plants, which grow in Turin, and others were from the south of France where the shroud had been exhibited. He stated at a conference that:

It was possible to identify pollen of a significant number of plants from the Palestine desert including some which do not grow in the surrounding deserts and came from the Anatolian steppes of Turkey. My findings must be interpreted as valid proof that during its history the Shroud was exposed to the air in Palestine and Anatolia.[37]

Another scientist who examined the fine particles on the Shroud was Walter McCrone.[42] Some of the items studied by McCrone are shown in Figure 10.15 (b) and (c). After he had studied the supposed blood stains shown in Figure 10.15 (c), McCrone summarized his conclusions in a scientific article as follows:

Physical microanalytical instrumentation has confirmed the earlier conclusion based on polarized light microscopy that the red component of the Shroud image contains a paint pigment iron oxide. It further strengthens the earlier conclusion that the images were painted by finding substantial amounts of a second red pigment vermilion. Hopefully these results will initiate an objective reappraisal of all of the data and lead to the only remaining test capable of resolving the question of authenticity, carbon dating. Although the shroud may have been painted at any time during or before the 1350s or the cloth may even be first century, it seems most reasonable that both painting and cloth are the products of the 14th century.[42]

This comment was published in 1981 and finally in 1988 three laboratories were allowed to carry out the **radio-carbon dating** (also known as **carbon dating**) referred to by McCrone in his discussion of the pigments. Previous to 1988, part of the problem of allowing radio-carbon dating was that the analytical method would have required relatively large samples of the cloth, which would be destroyed in the test. However, by 1988 the technique had been developed to the point where carbon dating could be carried out on a single, inch-long fiber. Because radio-carbon dating is so important, not only in the study of art forgery, but also in the study of dating bones, we will discuss the technique at some length.

We have already mentioned that many atoms in the universe exist with the same outer electron structure, but different central nuclei, the different forms being called isotopes. It was pointed out that the most common form of carbon in the atmosphere (in the form of carbon dioxide) is an isotope with six neutrons and six protons, known as Carbon 12. The earth is being constantly bombarded by cosmic rays, which are energetic fragments of an atom. Some of these cosmic rays, traveling from the sun or other stellar bodies with high energy, can collide with nitrogen atoms, which have seven neutrons and seven protons, and change them into an isotope of carbon, which has six protons and eight neutrons, known as Carbon 14. This isotope of carbon is radioactive and a given fraction of any carbon 14 atoms in the atmosphere will spontaneously decay into nitrogen atoms. Scientists have studied how fast carbon 14 atoms break down to form the original nitrogen atoms and they have discovered that half of a population of carbon 14 changes back into nitrogen in a period of 5,730 years. The population of carbon 14 continues to disappear at this rate so that a quarter of the original population will be left after 11,460 years and the population would be down to one eighth after 17,190 years.[39,40] In the study of radioactive materials, the time required for half of an original population of radioactive atoms to disappear is called the **half-life** of the isotope. In a review of carbon dating given by Hedges, the following facts are given

Every year circa seven kilograms of carbon 14 are produced in the earth's upper atmosphere. This newly formed atmospheric radioactive carbon is rapidly mixed throughout the atmosphere and is then incorporated by plants and eventually animals. The plants take up carbon dioxide to form sugars and proteins and then the plants are eaten by animals, and other animals eat these animals so that all living forms eventually incorporate proportional amounts of carbon 14 into their body structures.[39]

At death the amount of carbon 14 trapped in a bone or a plant is frozen and from that time on the ratio of carbon 14 to carbon 12 continuously decays. Therefore, by measuring the ratio of carbon 14 to carbon 12, one can discover how old an object is. The accuracy of the method depends upon the sophistication of the actual analytical method used; the most sensitive method in current use is mass spectrometry. At the time of writing, objects up to 70 thousand years old could be dated with an accuracy of plus or minus about two hundred years.

In the examination of the linen from the Shroud (linen is 20% carbon), samples were sent to three laboratories, one at the University of Arizona in Tucson, the other to the University of Oxford in Britain, and the third to the University of Zürich in Switzerland. The Arch Bishop of Turin, who announced the results of these studies, reported that the laboratories dated this shroud between 1260 and 1390 A.D. In order to check up on the accuracy of the three laboratories, they were also asked to date a sample of the mummy cloth of Cleopatra, which is dated between 9 B.C. and 78 A.D. They also dated a sample from a Vestment of St. Louis d'Angou, which is dated between 1263 and 1283 A.D. It seems fairly certain that the Turin Shroud is a medieval fake, and now those interested in the shroud are turning their attention to the fact that nobody really knows how the image was made. One intriguing suggestion is that it might have been the work of Leonardo Da Vinci, who died in 1519.[44]

In 1956 the British archaeologist James Mellaart made some important discoveries concerning Neolithic culture at a place called Haciliar in Anatolia, which is part of modern Turkey. After Mellaart's discoveries, the antique market was flooded with Haciliar artifacts, such as pottery. So many objects became available that archaeologists thought that someone was digging illegally at the location where the first discoveries were made. However, the suspicion that some of the pottery available was being forged, increased and peaked in 1965, when the Ashmolean Museum at Oxford bought a two-headed ceramic vase on the London Antique market. Very similar vases had been sold around that time for over $7,000 each to collectors in Europe and America.[45,46] As a result of the doubts of the museum officials, the Haciliar pottery was tested by a technique known as **thermoluminescence testing**. (The word thermoluminescence literally means light given out by heated objects.) Pottery, heated to a temperature of 900 °C, gives out light, which can be measured by means of sensitive photoelectric devices. The amount of light emitted by the clay is related to the age of the object.[47] The test showed that the Ashmolean vase had been made in the 20th century.

The reason that the light given out is related to the age of the object arises from the way in which the light is generated by heating. Clay contains very small amounts of radioactive material, such as thorium, uranium, and the isotope potassium 40. When radiation given out by these decaying atoms passes through the clay it interacts with atoms in the quartz crystals (quartz is one of the main constituents of clay) knocking out some of their electrons and leaving behind positively charged ions. Most of the electrons recombine with the ions immediately, but a few become trapped and stay in the clay for up to millions of years. When clay is first made into pottery, the heat of the firing frees all previously trapped electrons produced by the radioactivity, in fact resetting the process and producing a reference point for dating the pottery. The half-lives of the radioactive materials in the clay are so long that one can assume that the level radioactivity in the clay is constant and the creation of trapped electrons proceeds at a constant rate. If the pottery is now heated to about 900 °C, enough energy is transferred to the trapped electrons that they are able to escape from their crystal prisons and once they are free, they recombine with ions or atoms and the released energy results in the emission of light, in an amount related to the age of the object. Hence the name of the technique.

There are several factors that can alter the level of thermoluminescence in certain clays, and for this reason one must always compare a piece of pottery to be dated with other samples of known antiquity. Thus, in the case of the Haciliar pottery the light given out by the heated vase was compared to authentic pottery pieces known to be from 700 B. C. and sent to the British museum by the Turkish government.

The technique of thermoluminescence has been used to detect some significant frauds. For example, the British Museum issued a publication entitled "Greek and Roman Pottery Lamps."[47] On the front of the book was a Roman oil lamp shaped like a Gladiators helmet. After the book was published, thermoluminescence testing showed that the lamp was made in the 1920s.

Interestingly, a Mexican official stated that he thinks there are five times as many forged pieces of pottery supposedly made by ancient Indian tribes than there are genuine pieces. As always, when purchasing such artifacts, one should remember *caveat emptor* a Latin phrase meaning "let the buyer beware."

Francis Drake, who plundered the Spaniards for Queen Elizabeth I of England, visited what we now call San Francisco Bay. In 1628. Francis Fletcher, who was the Chaplain aboard Sir Francis Drake's ship, the Golden Hind, wrote that 50 years earlier Sir Francis Drake had installed *a plate of brass whereon is engraven her grace's name and the day and year of our arrival all there and of the free giving VP of the province and kingdom, both by the King and people, into her majesties hands.*[48]

One can imagine the excitement when in 1936 a brass plate bearing the name of Sir Francis Drake and this inscription was found near San Francisco Bay.

It is not surprising that the claimed discovery of such an interesting historical item would attract the attention of those who doubted its authenticity. Early studies, carried out by two metallurgists at Columbia University in New York, seemed to indicate that the plate might be the one left by Sir Francis Drake in 1579. Other experts, however, criticized both the letter forms and spelling used on the plate and they also speculate that the quality of the engraving was below that expected of workers known to be on Drake's ship. Other metallurgists at Oxford University used x-ray fluorescence to determine the constituents of the brass plate. They found that it contained 34.8 % zinc, 61.2 % copper, and impurities such as silver (120 parts per million), tin (500 parts per million) and lead (0.05 %). From historical records it is known that brass used at the time of Drake had a zinc content of less than 30 % and much higher amounts of lead. Neutron activation analysis was carried out at the University of California at Berkeley and scientists pointed out that the ratio of the different components of the brass practically matched those in a modern sheet of brass alloy registered in the records of alloys as number 268. When Professor Emeritus Cyril S. Smith, a world famous expert of the Massachusetts Institute of Technology, was asked to look at the plate he made the following comments, which he felt suggested that the plate was a forgery. Firstly, he noted that the corrosion seemed superficial and inconsistent with a four hundred year exposure. Secondly, the plate seemed to be more uniform in thickness than could have been achieved by 16th century workers and the edge of the plate seemed to have been sheared and then hammered to make it look genuine. In the 16th century, one would have cut the plate by using a chisel and filed the edge. Finally, Smith suggested that x-ray diffraction studies would show whether the plate had been produced by hammering or by rolling. When this final test was carried out by a Dr. Hart, sure enough, the metal showed aligned metal grains in the brass, indicating that it had been produced by a modern process in which the brass is squeezed between rollers in a press.[48]

One of the points raised by Dr. Smith was the odd appearance of the corrosion. Forgers go to great lengths to make pottery, bones, and metal objects look as if they have been corroded and/or exposed to the atmosphere for a long time. Old copper and bronze vessels are usually covered with a greenish coating known as patina (a word which is literally the Latin word for a dish; presumably because the coatings were first noted on old bronze dishes.) Bishop tells us that anyone trying to make a modern bronze look old needs a lot of skill to imitate the patina. She also tells us that when some students at Camberwell College of Art in London, England, wrote a book on how to produce attractive surface effects, they were rather taken aback when a "bunch of heavies" approached them wanting to create some interesting effects, which make new objects look old. A true patina forms very slowly, but fakers can accelerate things by dipping the objects in a solution of chemicals. However, this type of false patina can usually be detected by analyzing the chemical composition of the coating. Sometimes forgers mix the appropriate chemicals together as powders and stick them to the object This type of fake patina can be readily detected by viewing the object under ultraviolet light. Steps taken to artificially age forgeries have been described in some detail by various workers.[47–50] For example, clay pieces are sometimes brushed with a thin layer of lime before they are fired. This will remove any imprints left by tools. Depending on the style and age of the item intended to be reproduced, the forger may also add an extra layer of different earth to produce various colors on the surface, or grains of sugar, which turn black in the heat, to simulate the attack of fungal growths on the pottery. After removing a ceramic piece from the firing furnace, the forger can saturate the object with water and age it underground. Sometimes such buried objects are first sprinkled with urine to produce ochre, brown, and green shades. The forger even sometimes pastes small roots on the outside. As one writer points out, for a final touch of authenticity some craftsmen break their wares and then carefully repair them.

As in the case of the faked plate attributed to Drake, analysis of the alloy in question and comparison with typical alloys of known age can help to detect a forgery. Thus, Fleming has discussed alloy composition in the detection of forged historical coins. He has used x-ray fluorescence to look at forgeries. He also records an episode in which some individuals were punished for forging coins when in fact they were innocent. Apparently in the reign of Henry I of England, English coin makers were accused of diluting the silver used to make coins to pay the King's troops in Normandy, by adding tin. When King Henry heard of the alleged forgery he summoned all the money makers to Winchester and quickly deprived them of their testicles and right hands. However, modern examination of the coins involved in this episode shows that the coin makers were innocent.

Cyril Smith, in his examination of the Drake plate, also pointed out that a knowledge of the way in which workmen made objects can help to detect a forgery. For instance, in the modern manufacture of bracelets, a jeweler makes gold wire by pushing and pulling a thin rod of gold through a series of holes with decreasing diameter. The round edges of such holes are never perfect, and slight imperfections around the edges leave lines on the gold wire called striations. In the time of the Pharaohs of Egypt, gold was first beaten into very thin plates and then wire was made by twisting the plates, which gives a totally different appearance to the surface of the wire as compared to modern gold wire. Thus, when the gold wire of a supposedly Egyptian bracelet is examined under a microscope, striations would indicate that the wire had been made in the modern way and prove that the bracelet was a forgery. Similarly, when a collection of seventeen gold beakers believed to have been made in 15th century B. C. was examined with a microscope the inscriptions in the gold surface were seen to have been engraved rather than traced, as was the custom in ancient times. Tracing is shallower than engraving and the microscope examination also detected tiny hammer marks on each of the vessels, which were traditionally smoothed away by ancient craftsmen. It is now believed that the gold beakers were made in the period from 1919 to 1922.[51]

X-ray images of a supposedly valuable object can also yield some surprising information. Thus, when an apparently perfect Chinese vase was examined with x-rays, scientists were able to show that the object had obvious cracks and consisted of several pieces stuck together. Subsequent analysis of the composition of the vase showed that the base had actually been part of a different vase. Moreover, the restorer had doctored the surface with ground malamute, a copper bearing chemical mixed with colored plaster of Paris to give it the required patina. (For an extensive scientific study of real and forged patina see reference 50).

The fossil of a bird has been the subject of much controversy recently. Those who believe in its authenticity claim that it is the fossil of an ancient beast which bridges the evolutionary gap between reptiles and birds. It is known as an **archaeopteryx**. The name comes from two Greek words, *archaeo* meaning ancient, and *pteryx* meaning wing. Thus, the name of the fossil means ancient wing. Unfortunately people break the word wrongly in modern English and pronounce the word archae · opteryx. This hides the meaning of the word in the same way that the meaning of helicopter, which means spiral wing, is hidden by the splitting of the word into heli·copter instead of helico·pter. The Funk and Wagnall's Encyclopedia has the following entry with regard to this fossil:

Archeopteryx is the oldest known bird, the remains of which have been obtained in the Solenhofen lithographic stone, a limestone of Jurassic age (which is quarried near Eichstatt in Bavaria). Only two specimens of the Archaeopteryx are known; one in the British Museum, London, and the other in Berlin. The bird appears to have been about the size of a crow. Its vertebral column extended so as to form a long tail from each vertebrae of which grew quilled feathers. The entire body was covered with feathers. Archeopteryx was once thought to be a link in the evolutionary transition from reptiles to birds.[52]

In 1985 a well-known British Astronomer Fred Hoyle and his colleagues published an article in the *British Journal of Photography* in which they claimed that a study of pictures of the fossil indicate that the archaeopteryx fossil is really a fossil of a reptile and that someone had coated parts of the fossil with cement and glue and pressed feathers of a bird into the wet glue. The British Museum claims to have carried out tests, which established that the fossil is genuine, but in correspondence published in *Scientific American* Willis states that:

> *Before we close the book on Fred Hoyle's charge of fraud against archaeopteryx, it should be remembered that from as early as 1800, it has been well known that one can obtain exquisitely detailed relief patterns on Solenhofen limestone with a greasy ink, and a nitric acid etchant. Certainly, by the time the first feathered specimens were disclosed, any lithographer might have been able to make excellent impressions of feathers in the limestone without using any tell-tale binding material. A properly executed etch would probably leave the effected material virtually indistinguishable from the surrounding stone, except under the scanning electron microscope.*[53]

(The term lithograph literally means drawing with stone, the first lithography was actually carried out using printing plates of a type of limestone in which a grease pattern prevents the uptake of ink when the plate is inked. The whole stone can then be used to make printed copies.) The reader may find it interesting to review the scientific evidence for and against the authenticity of this famous fossil.[54]

10.7 Artistic Frauds

A joke amongst art dealers, which illustrates the extent of forgery in the art world, is that Camille Corot painted 800 pictures in his lifetime, of which 4,000 ended up in American collections.[55] Whole books have been written on art forgery. All we can hope to do in this section is to tell the reader where to look for extensive literature on this subject and to highlight a few interesting stories of how forgeries were detected.

Fleming, who is an expert on art forgery, tells us that the first time that scientific evidence was used in the conviction of an art forger was in 1932, when Otto Wacker was convicted of painting several fake Van Gogh's. At the trial x-ray photographic data was presented by a chemist, Helmuth Runemann, from the State Museum of Berlin. These x-ray pictures showed that the underlying technique was not that of Van Gogh, even though the final painting looked like a Van Gogh. Fleming said that until that time:

> *a scientist's evidence of fraud would be swiftly overwritten by the weighty opinion of an expert.*

At another famous trial the experts and a scientist disagreed on a forgery created by Van Meegeren. The trial took place in 1947. One of his fake paintings is shown in Figure 10.16. It is known that Van Meegeren baked his painting in an oven to simulate various paint surfaces. Experts differ as to whether Van Meegeren (1889–1947) was really a very good painter or a "talentless and paranoid hack" who felt that the Dutch art world had joined in a conspiracy of silence against him (for this latter type of opinion see Reference 55, for the opposite viewpoint see 56). In a review of the Van Meegeren's forgeries Hughes remarks the following:

> *Although the baked surface of the forged veneer was almost perfect, Van Meegeren's bodies were boneless, the faces coarse and thick lipped, their expression stereotyped.*

Yet Dutch experts led by the art historian Abraham Bradius were very impressed:

> *what a picture! What we have here is, I am inclined to say, the masterpiece of Johannes Vermeer of Delft.*[55]

Van Meegeren produced a stream of Vermeers over several years. During the German occupation of the Netherlands he sold them to the German

Figure 10.16. Some art dealers suspect that there may be more faked oil paintings, such as those illustrated above, in private collections than genuine works of art. Shown here is a Vermeer created by Hans Van Meegeren. (Reprinted from *Time*.[55])

occupational forces. Hermann Göring bought one of them. As a consequence of this sale of a supposed Vermeer, Van Meegeren was charged after the war with treason for selling national treasures to the Nazis. This was a crime that carried the death penalty. At the trial Bradius and all his colleagues testified that the Vermeer was genuine. However, Van Meergen convinced the court that he had made it himself and was sentenced to a year in prison for fraud. At his trial, a Belgium chemist, B. J. Cormans, gave evidence that a scientific study showed that the pictures were indeed forgeries. Hughes points out that, to everyone's intense relief, Van Meergen died before the sentence of one year was finished. Hughes states that:

> *not only should the buyer beware but he should remember another Latin phrase* "perites nec crede," *which means put not thy trust in experts.*

Experts are understandably put off when an item they have verified before a museum or that an art gallery buys, is exposed as a fraud. Yet one does not always need scientific evidence to prove that a painting is a fraud. Thus, Warren Sanderson tells the story of how he examined what was claimed to be a painting by the Flemish artist Rubens, which was displayed in a museum in the mid-western United States. He tells us that he suspected it to be a forgery, because the brush work was too cold and rigid. To confirm his suspicions, he took an expert to see the painting. The expert pointed out that Rubens must have been clairvoyant, because not only was the painting very fresh, but the subject of the painting was sitting in a chair the design of which dated from the Napoleonic era, two centuries after Rubens had supposedly painted it. When Sanderson suggested to the museum staff that the painting was a forgery, they refused to discuss it.

Just as the creator of fake money trips up because he is unaware of the subtlety of paper making, some very good forgers have made mistakes by painting good facsimiles on wood that gave the game away. Thus, Nathan Stollow exposed two oil-on-pine panels, which were allegedly painted by Maurice Cullen and James Wilson Morrice, as forgeries. He noted that both had been cut from the same piece of pine, and that the pattern of the grain in the wood fitted together perfectly.[57] By studying the patterns of thick and thin rings on the edge of the piece of wood bearing a painting, one can date the piece by comparing the ring structure with that of trees grown in the supposed region of the painting's origin.[58] Similarly, Stollow exposed a fake 1920 painting by R. Lisimer, which had been painted on Japanese plywood of a type that was not imported into Canada until 1949. Another painting supposedly by Tom Thomson was painted on modern fir-ply panels artificially aged with umber or ochre.

Stollow also played a major part in breaking up group creating forgeries of the paintings of the Canadian Group of Seven. One of the paintings was exposed because in a routine examination, infrared light showed that the painting had a copying grid under the paint surface. As a result of these studies, two Toronto dealers were jailed.

It should be noted that Van Meegeren, when he was painting his forgeries, obtained low value paintings from the time of Vermeer and scraped the paint off before he went to work on his forgeries.

Another way to study the pigments of a painting is to shine a beam of x-rays through tiny pieces of paint. Just as a light beam is scattered by a fingerprint, the x-rays (which are a much smaller-wavelength radiation than visible light) are scattered by the crystal planes in the pigment. One can examine resulting the x-ray diffraction pattern to detect different substances in the pigment.[49] Fleming points out that if the pigment in a painting is very fine and uniform in size, this indicates that a modern paint was used. Older pigments were pulverized by hand with a pestle and mortar, and contained many angular grains in a range of sizes. Fleming has given an interesting account of the way in which pigment technology has changed over the last several centuries.[49,50] Closely allied to the question of fraudulent paintings is the production of photographs, which can be used as false evidence in the law courts. *New Scientist* contains a very informative article on how modern technology can produce fraudulent pictures. In his article on forged photographs Robert Mathews claims:

> *By and large people believe what they see in the newspaper because of their belief that the camera never lies. But not anymore. In the world of digital imaging, seeing is definitely not believing. Once a picture has been turned into a series of digital picture elements, or pixels, its can be manipulated with a computer; people, faces or entire locations can be added or taken away. With little effort it is now possible to have the British Prime Minister being led away from the British parliament and bring Einstein back from the grave. Modern digital technology does not do anything that cannot have been done before with a much larger amount of time and care. But its increasing availability means the potential for images to be easily and undetectably cleaned up or enhanced or simply invented is now greater than ever. Eventually digital technology may simply make people realize that visual images can no more be trusted than the written word.*[59]

10.8 Fraud in the Marketplace

The subject of commercial fraud is so large that, again, we can only highlight a few areas where modern technology is helping to combat fraud in the marketplace. One important area is the protection of trade names for designer clothes. In 1981 it was estimated that the sale of jeans with name-tags such as "Levi Strauss" fraudulently added to them was costing the textile industry $40 million. In 1980 it was estimated that the recording industry lost $400 million in the sale of counterfeit recordings. In 1980 three companies, Levi Strauss and Company, Mattel, and Warner Communications came together to try to combat the counterfeiting of trade names. In response to their research, the Polaroid Corporation introduced a device called Polarproof. This is a thin transparent polyester film embossed with 1,500 linear parallel cylindrical lenses on one side, and parallel color lines on the other. Designers' or manufacturers' logos can be easily incorporated into the film and it is neither economically nor technically feasible for crooks to duplicate the film. In its final form, Polarproof is circa 1 millimeter thick, pliable, and usable for labels, stickers, or hang tags. The Polarproof label is machine readable for high-speed verification. Viewed under normal light, the material produces an ever-changing array of visual effects.[60]

Another strategy for combatting fraud with designer labels of clothing was developed by scientists at the Department of Energy laboratories in the United States. This approach uses small computer-chip identity tags. A small, cheap computer chip can be sewed into clothes complete with a message saying , for instance, "Authentic Levi Strauss Jeans." The news story concerning the release of this computer-chip label indicates that U. S. companies may be losing six billion dollars a year overall to the fraudulent sale of designer-label clothes overseas.

Another area of fraud in the clothing industry is the mislabeling of goods. Thus, many garments which are labeled, for instance, 70% cashmere and 30% wool contain no more than 5% cashmere. (Cashmere is a very fine form of wool obtained from goats, which originally were to be found only in Kashmir. Today the best cashmere cloth is woven from the hair of goats that are raised in China and Mongolia.) The rest can be recycled rag, human hair, acrylic or asbestos fibres, rabbit fur, and even newspaper fiber. Spilhaus is the Director and Chief Detective of the Boston-based Camel Hair and Cashmere Institute of America. He claims that nearly 30% of the cashmere products sold in the United States are mislabeled and he points out that the wave of counterfeiting of cashmere clothing is partly due to the surge in demand – cashmere cloth can cost up to $200 a yard. Lower grades of cashmere wool from Iran and Afghanistan goes for $100 a yard. This problem was reviewed in *Time Magazine*.[61]

Fraud is also rampant in the drink industry. Thus, a French company found sugar that had been added to ten of 40 brands of orange juice, which were described as pure and unsweetened. Seventeen of twenty-three samples of Holland's gin from Belgium and the Netherlands, contain sugar from sugar-beet alcohol, when the product should be made entirely from grain alcohol.[62] The company analyzed the gin and orange juice using a method developed at the University of Nantes. The technique, which is very sophisticated, is called site specific natural isotope fractionation by nuclear magnetic resonance (SSNIF!). This technique can reveal whether natural or artificial perfumes and flavors, such as lavender and vanilla, are being used in commercial products. The problem is not so much that the fraudulent goods taste any different to the average consumer, but that the claims on the brand label are not true. In further studies the investigators found that some wines had up to one half of their alcohol derived from added sugars. A similar fraudulent sale of allegedly genuine olive oil has been studied by mass spectrometry. The innovation in this case was the use of a neural network.[63]

Another major area of commercial fraud is that of false insurance claims. A conundrum widely circulated amongst insurance agents indicates the type of problem. The riddle asks: "When does a Timex change into a Rolex?" The answer is: "on an insurance claim." In 1992, it was estimated that fraudulent insurance claims were costing Canadian Insurance Companies one to two billion dollars a year. Similar figures describe the situation in other western countries. In Canada in 1993, the insurance industry set up a special bureau to track down cheaters, particularly with regard to automotive insurance. An incident, which illustrates the problem, occurred in Vancouver in 1980. A youth called the "bicycle bandit" made a series of successful hit-and-run assaults with his bike, including a number of purse snatches. To quote the news story describing the crimes:

> *The youth didn't pawn any of his ill gotten gains but instead hid what he had stolen, keeping a meticulous list of what he had taken and from whom. There were 40 insurance claims made after these robberies and in 39 of them it turned out the value of the stolen items was exaggerated.*[64]

Investigating insurance fraud is likely to be a major area of forensic activity in the coming years.

In the last section we discussed the problem of experts differing on the authenticity of an object. Sometimes in the marketplace experts will indicate that an object is not authentic, while a scientific study will show that the object is, indeed, what it was claimed to be. Palenik gives an example of this type of situation involving jade sculptures. A client with a large personal collection of carved jade objects came to McCrone laboratories because an expert had said the objects were not jade. The dealer who had been asked to buy the Jade said that visual examination and the feel of the objects indicated that they were not genuine. Scientists at McCrone laboratories removed minute particles from the crevices at the bottom of the statuettes and examined them under the microscope. Thus, they were able to establish that the objects were indeed jade and a successful sale of the objects was completed.[65]

10.9 Checking up on Suspicious Signatures and Documents

Experts can sometimes examine signatures under a microscope and compare the sequence of lines in the signature to those in a suspected forgery. We have already discussed how infrared light can be used to see if a signature has been altered or if the amount of money on a check has been changed. In this section we will be more concerned with the way in which modern science can be used to check on the structure of a signature. Banks and other organizations are often anxious to determine the authenticity of a signature being written on a check. As a precaution many banks require that on a cash withdrawal the signature be made in front of the teller or other officials of the bank. When using a credit card to pay a bill, the salesperson compares the signature on the back of the card with the signature written in his presence. Individuals can learn to copy a signature, but a new technique can be used for checking on the authenticity of the signature, which involves studying the dynamics of the way the signature is written. A device known as an **accelerometer** is used for this purpose. This is a scientific device, which measures acceleration and deceleration of an object. The basic system of a pen incorporating an accelerometer is shown in Figure 10.17.[66–69] This type of pen is also called a ballistic pen. The principle employed in the ballistic pen is the same as that used in the design of a seismograph. (For a discussion of the basic principles of a seismograph see Reference 70). In the main cylinder of the pen there is a lead slug attached by a spring to the top of the cylinder. When the pen is moved suddenly the lead slug is slow to follow the movement of the pen, because of its inertia. Thus as illustrated in Figure 10.17 (a) if the pen is moved upwards suddenly, the slug tries to stay in its position so that the spring extends as the body moves. To an external spectator the slug appears to move down the pen when in fact the slug is staying in the same position as the pen moves upwards. (It is rather like the problem of observing relative motion from inside a railway carriage. When an adjacent carriage moves, the observer is often quite convinced that his own carriage is moving; it is the relative motion that makes the stationary carriage appear to move.) On the internal wall of the pen there is a conducting strip of metal and the electrical resistance of this strip depends on the position of the slug with respect to the connecting electrode. When the pen is moved up, the conducting portion of the strip becomes longer and as the spring pulls up, the slug returns to its initial position. The change in resistance of the wall strip as the slug moves creates a pulse of electricity in an external circuit. As the pen is taken through a series of movements to generate the written signature, the pattern of electrical pulses recorded is called a **ballistic signature**, as shown in Figure 10.17 (b). Note a popular term for this trajectory signature is a **spasm signature**. One of the advantages of the ballistic signature is that it is possible to store the signature directly in the

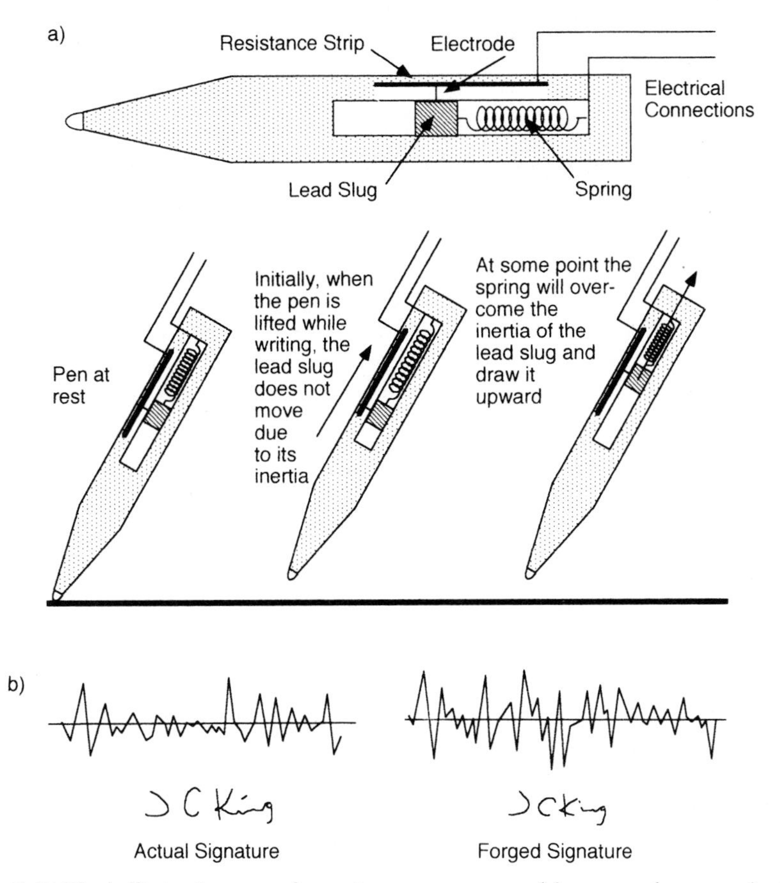

Figure 10.17. The ballistic signature of a pen's movement created by an accelerometer in the body of the pen, makes it very difficult for someone trying to forge a signature. a) Basic concepts employed in a ballistic signature pen. b) Ballistic signatures for the same name written by two different individuals are clearly different.

computer. A signature is usually written several times, so that the computer will recognize the inherent variation in the structure of a given signature. One could also store a series of signatures as the person ages or suffers from a serious disease, so that authentic changes in a signature can be established over a period of time.

Many banks are beginning to look into the possibility of reducing fraud involving cash machines by installing machines that can recognize an individual by his speech pattern. Thus, in a news story entitled "How Cash Machines Will Recognize You When You are Drunk," Irwin discussed the use of speech patterns linked to neural networks in voice recognition of individuals using automatic tellers.[71] For a discussion of severe fraud problems with bank credit cards in the early 1990s in Great Britain, see reference 72.

Laser-diffraction pattern analysis has opened up a whole new technology for characterizing signatures and writing. Fournier, of the Holland Institute in Massachusetts, pioneered the application of such techniques for recognizing the authenticity of ancient documents and also re-examined a very famous case in which forged writing played a key role – the Dreyfus case in France.[73–75] One of the problems tackled by Fournier involved the recognition of the writing style used by various scribes in the preparation of ancient Hebrew documents.[76] Thus, Figure 10.18(a) shows the optical transforms of several Hebrew letters; Figure 10.18(b) shows variations in the structure of a Hebrew letter generated in a line of script, as studied by Fournier and Venoit.[76] Note that the study of ancient writing is called **paleography**. Not only can one look at the individual letters and signatures using optical information processing, one can also process a whole page of writing to take into account the overall structure of a handwritten piece of information. Fournier used this technique to look at the documents involved in the Dreyfus case. In 1894, a cleaning lady working for the French Intelligence Service intercepted a handwritten memo listing a number of documents pertaining to the national defence and intended for the enemy. (She found the writing in a wastebasket kept at the German embassy.) This document became known as the bordereau. A replica is shown in Figure 10.18 (c) (i). Captain Dreyfus was accused of writing this letter and in 1895 he was sentenced to jail and sent to Devil's Island. But many suspected that the real culprit was another officer – Esterhazy. The whole affair was charged with anti-Semitism and it proved very difficult to get at the truth. Fournier made diffraction patterns of the bordereau document, as well as of the handwriting of Dreyfus and Esterhazy. These specimens of handwriting are shown in Figure 10.18 (c) (ii) and (iii). In his discussion of the case, Dr. Fournier uses guarded language, but it appears without doubt that Dreyfus did not write the bordereau. Also in Figure 10.18 (c), the computer-generated, two-dimensional Fourier transforms of the various handwriting specimens of Dreyfus are shown.

Today modern computer technology is making it possible to use two-dimensional Fourier transforms, which are similar to the Diffraction patterns discussed earlier in fingerprint recognition (Chapter 2). At Laurentian University, we are currently attempting to teach a neural network to distinguish and recognize the handwriting specimens via their two dimensional Fourier transforms. In Figure 10.19 two specimens of a signature as analyzed in our laboratories, along with the two-dimensional Fourier transforms are shown. In scientific terms the two dimensional Fourier transforms can be split up into several regions.[77] The energy region in the center of the pattern shows low-frequency information. Smudges on the paper would appear in this part of the energy-pattern. The outer areas contain so-called high-frequency information; this is information generated by items such as crossed T's, dotted i's and capital letter E's. The energy in the various areas can be registered as a list in the computer and the computer can then check any signature by

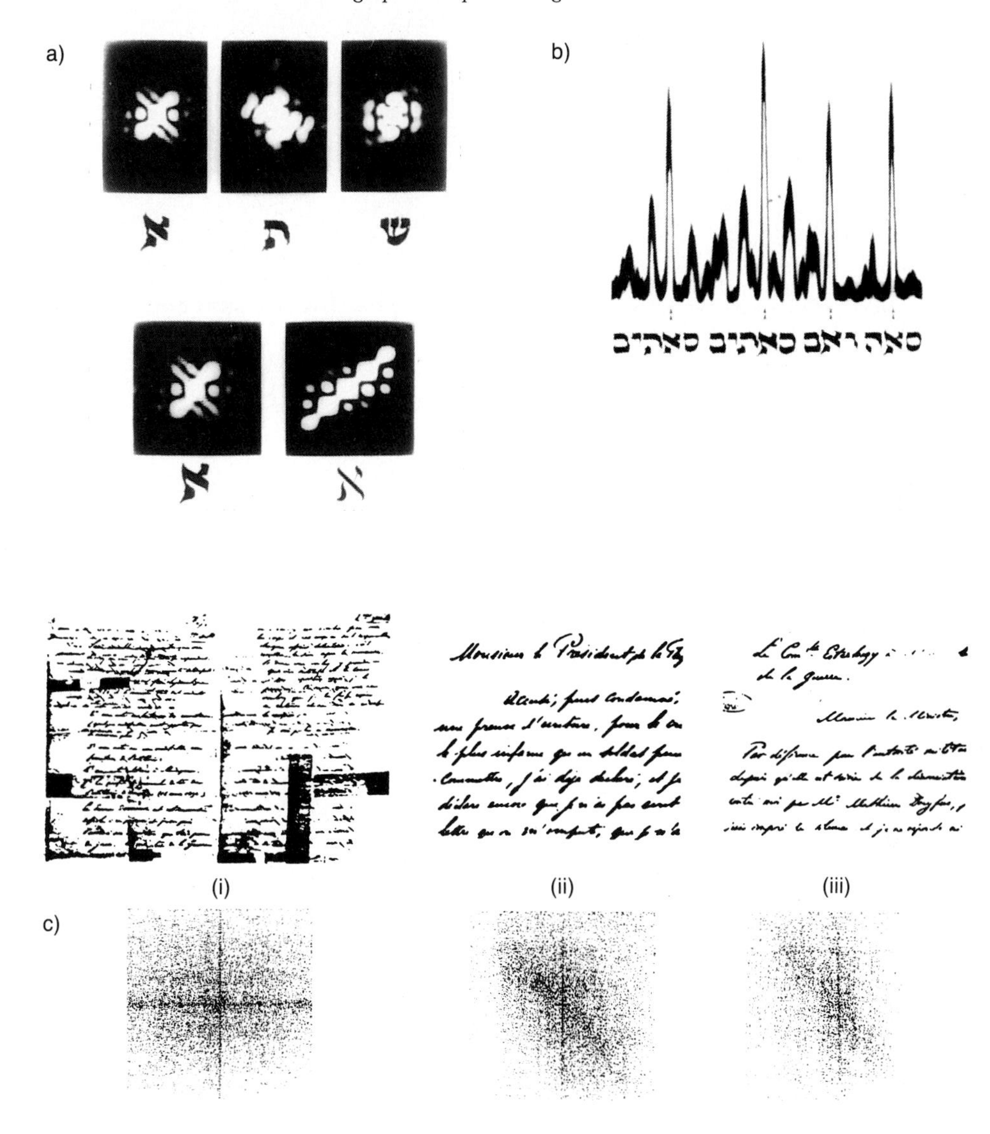

Figure 10.18. Optical image processing, similar to that used in fingerprint recognition, can be used to study ancient handwriting styles and to compare handwriting used in documents of suspect authenticity. a) Diffraction patterns of different Hebrew letters. b) Computer recognition of the variation in a letter written by the same scribe. c) Various documents associated with the Dreyfus case and their diffraction patterns. (i) The "bordereau" found in a waste basket. (ii) Authentic Dreyfus writing. (iii) Esterhazy's writing. (Reproduced with permission of J. M. Fournier.[76])

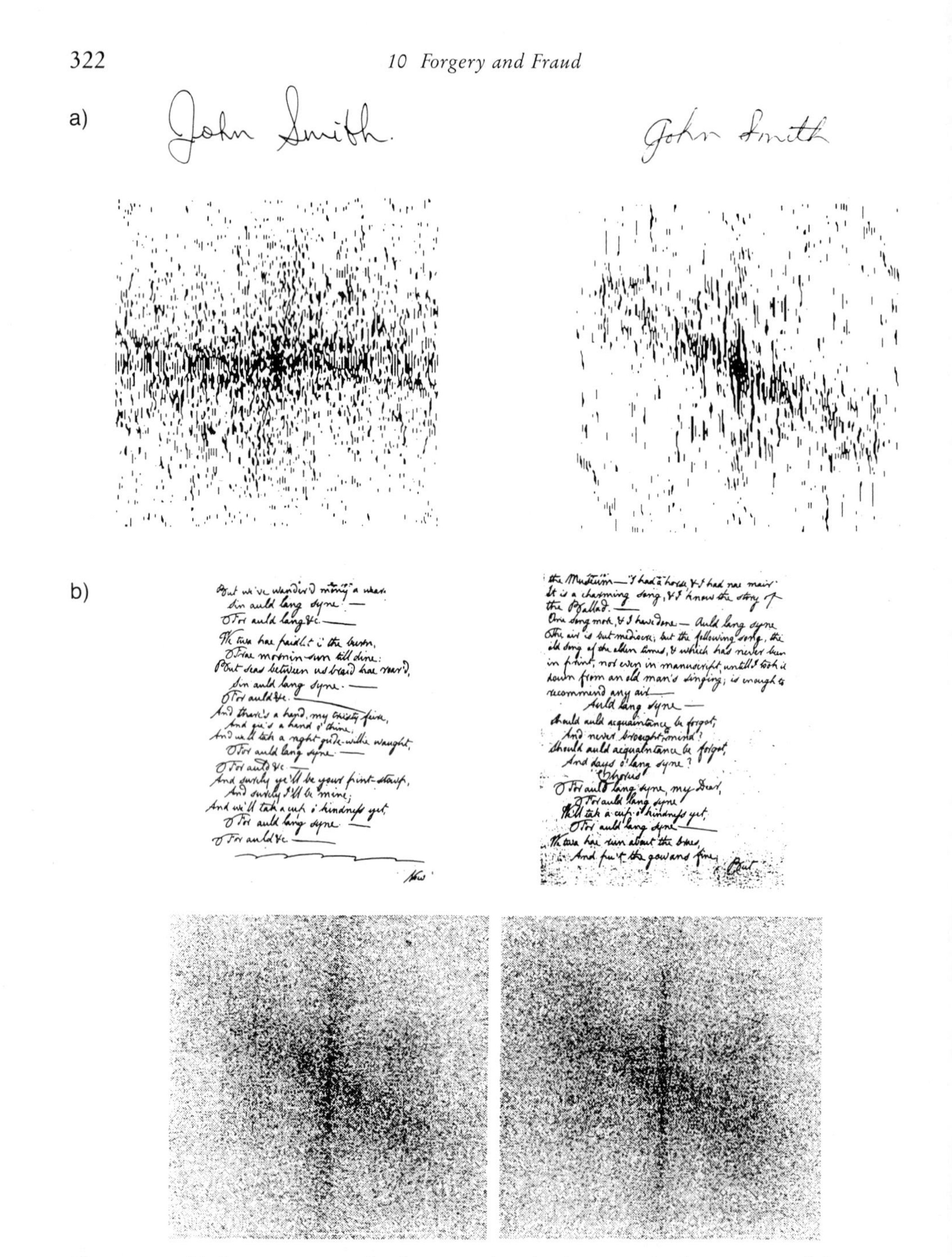

Figure 10.19. Modern computer technology permits scientists to study the structure of signatures and pages of handwriting. a) Two-dimensional Fourier transforms can represent the structure of a signature so that a new and a stored signature can be matched by a computer. b) Pages of handwriting can be stored so that the overall veracity of the document can be checked against the style of a given author.

transforming it into a list of frequencies and checking this against the information stored about the signature being verified. (For an introductory discussion on how the different areas in the two-dimensional Fourier transform can yield information on low and high frequency elements of a signature, see reference 78.) Again, as in the case of optical transforms, the two-dimensional Fourier transform can be used to process whole pages of script and Figure 10.19(b) shows two pages of a manuscript, written by Robbie Burns, along with the two-dimensional Fourier transforms.[77] The structure of these transforms can be stored in memory and if a new manuscript supposedly by the same author is discovered in a library archive, the computer could check to see if the document is authentic.

When it comes to determining the authenticity of an entire written document, one can also use literary style analysis. This involves analyzing the way in which various authors construct sentences, and studying the frequency with which they use given words. One form of style analysis looks at the frequency of common words used by a writer, as compared to the overall frequency of words in a written language. Thus, in Figure 10.20 (a) the 20 most frequently used words in written and spoken English, along with the same information for French and German, is shown. This list is probably somewhat out of date since inclusive language has tended to lower the frequency of he and his in this list and also, in the German list, DDR refers to the East German Republic, which has now disappeared. As an example of how a difference in the frequency of the words can hint at the background of a writer, you could analyze the English in this book. You would find that the word "the" is not the most frequent word in my writing, although editors struggle to add "the" to my sentences. The reason is that in the East Yorkshire dialect, which is spoken where I grew up, the word "the" is seldom used. Writing style can also be analyzed by determining the frequency with which a given writer uses certain nouns that are not particularly common in the general written literature of the language. For example, I know that I use the word manifest more often than most people.

The first attempt to use the frequency of words to analyze the authorship of a document was carried out by T. C. Mendenhall (1841–1924) who investigated the authorship of several plays, attributed to Shakespeare, but which some scholars suspect were written by Marlow or Bacon.[79,80] The frequency with which words occur in the English language has been extensively studied by Zipf, who has enunciated a law describing the frequency of words in any language.[81] Recently, computers have made it possible to carry out much more extensive studies of word frequencies in written English and the question of the authorship of Shakespearean plays has again received considerable attention in both the English literature publications and in computer journals.[79] Another aspect of style that can be described by mathematical relationships, the average length of sentences by different authors, is illustrated by the data of Figure 10.20 (b). It is interesting to note that the very first

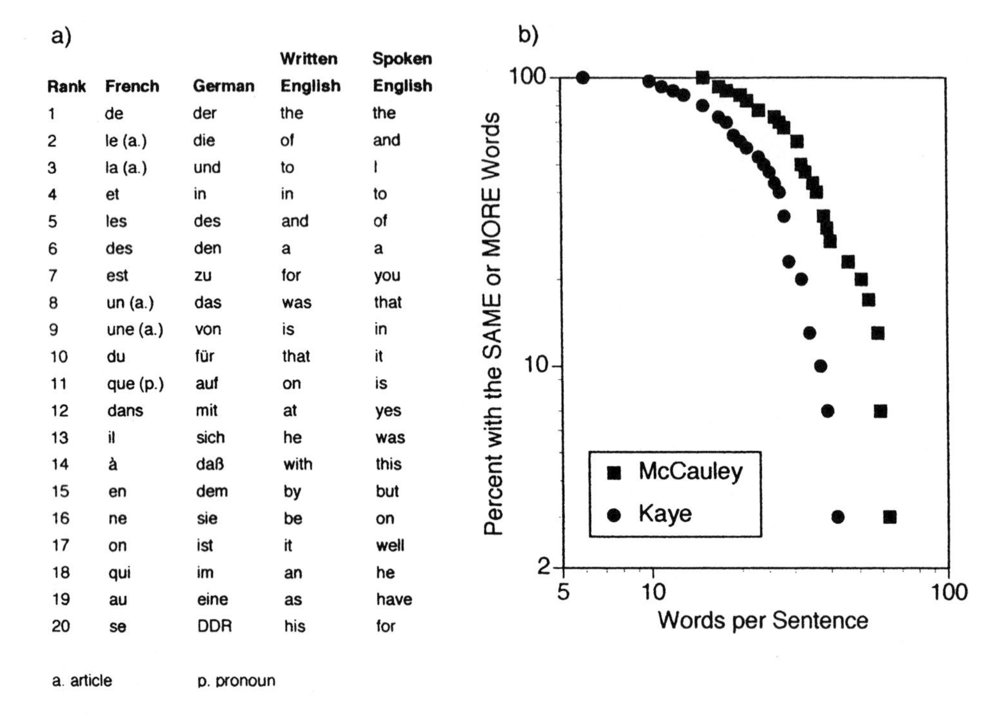
a)

Rank	French	German	Written English	Spoken English
1	de	der	the	the
2	le (a.)	die	of	and
3	la (a.)	und	to	I
4	et	in	in	to
5	les	des	and	of
6	des	den	a	a
7	est	zu	for	you
8	un (a.)	das	was	that
9	une (a.)	von	is	in
10	du	für	that	it
11	que (p.)	auf	on	is
12	dans	mit	at	yes
13	il	sich	he	was
14	à	daß	with	this
15	en	dem	by	but
16	ne	sie	be	on
17	on	ist	it	well
18	qui	im	an	he
19	au	eine	as	have
20	se	DDR	his	for

a. article p. pronoun

Figure 10.20. Various strategies can be used to quantify the style of different writers. a) Ranking of the most frequently used words of various Indo-European languages. b) Sentence structure of two different authors.

time that I submitted a sample chapter for a book to a publisher, when I was in my late 20s, one of the criticisms of the chapter was that I used some archaic English and some Germanic grammar. In fact, the source of these two trends in my writing lies in my grammar-school education. During my formative years in school I lived in rural East Yorkshire, where the dialect is still quite Germanic and words and grammatical structures that have disappeared from standard English are still used.

Different types of stylistic analyses have been applied to written criminal confessions. The technique for analyzing confessions is known as **forensic linguistics**. At the time of writing this book, the technique was the subject of considerable controversy as a general technique for the analysis of short confessions. However, in one particular instance style analysis strongly indicated that an innocent man had been put to death. The case occurred in 1950, when a Timothy Evans was hanged for the murder of his wife and child at a house in London. Three years later the police discovered several bodies at the house and the person who still lived at the house, a lodger, John Christie, was charged with these murders and hanged. After considerable discussion of the case, a public inquiry was held, which led to Evans being granted a posthumous pardon in 1966. (The word posthumous means after death.) The way in which style analysis was applied to the confession supposedly made by Timothy Evans has been reviewed by Crystal, who states:

The central piece of evidence against Evans was the statement he made to the police in London on the 2nd of December 1949 in which he confessed to the murders. Evans was largely illiterate so the statement was made orally and written down by the police. At the trial he denied having anything to do with the murders, claiming that he was so upset that he did not know what he was saying and that he feared the police would beat him up if he did not confess.[82]

In an analysis of the Evans statements, amounting to nearly 5,000 words, it was shown that the language contained many conflicting stylistic features. Thus Crystal quotes two statements allegedly made by Evans: "I done my days work and then had an argument with the governor and when I left the job he gave me my wages before I went home." and, "She was incurring one debt after another and I could not stand it any longer so I strangled her with a piece of rope and took her down to the flat below, the same night whilst the old man was in hospital." Analysis of these statements showed that words such as done and governor are compatible with the type of language used by an illiterate person, whereas it is most unlikely that such a person would have used words such as incurring and whilst. A complete study of the stylistic analysis of Evans' statement is beyond the scope of this book and the interested reader is referred to Crystal and to the works of Matthews quoted in references 79, 80 and 82.

It is a sobering thought to realize that an innocent man was so terrified by the process of interrogation that he made a false confession to avoid what he thought was a possible beating, and that no one had the sense to realize that the vocabulary used in the confession looked suspiciously more like the work of a far more literate person. The fact that innocent people have been subjected to the death penalty is a powerful argument against such drastic methods of criminal punishment. I am sure that posthumous pardon was not of much satisfaction to Timothy Evans. Ubelaker and Scammell give some telling statistics which should make supporters of the death penalty think very carefully. Thus they state:

America is one of only four industrialized nations still taking the lives of their citizens (the others are Russia, South Africa and China) and the process here is nowhere efficient enough to significantly increase space available in penal institutions. Time between initial proceedings and execution often run ten years or more. Some appeal courts like those in California spend half their time on capital cases. Executions don't save money. On the contrary, tax payers in Texas pay an average of seven million per legal killing and in

Florida the cost is over three million. In Massachusetts, where the death penalty has not yet been restored, the cost of life imprisonment without parole averages a comparatively trifling $900,000.[83]

Another statistic worth citing from Ubelaker and Scammell for those not persuaded by economics:

Such monitoring organizations as Amnesty International, The National Association for the Advancement of Colored People (NAACP), and the American Civil Liberties Union (ACLU) say that approximately 10% of those executed in the United States in this century were subsequently proved innocent for the crime for which they died.[83]

10.10 Fraud in Science

Scientist are usually viewed as individuals who are out to seek the truth and would never be party to fraud. However, as in athletics, the prizes to be won for eminence in science are now so high that there is considerable temptation for people to manipulate the data of their research. Perhaps the most spectacular case of alleged fraud in science in recent years occurred at the Massachusetts Institute of Technology. This college is one of the leading technical universities in the United States. The case involved three major players. A post doctoral worker Margot O'Toole, an internationally known immunologist, Tereza Imanishi Kari, and David Baltimore, a Nobel Prize winning biologist. As reviewed in an article by E. Masa, the story began when O'Toole, who was working with Imanishi Kari, came across laboratory notes written by Kari, which were not in accord with some published results.[84] O'Toole took her concerns to Kari and eventually to Baltimore, who curtly dismissed her claims, calling O'Toole "a discontented post-doctoral fellow." (When one attacks the person rather than looking at the evidence, one is using a strategy known as *ad hominem*. This is a Latin phrase meaning "for the needs of a person" – the idea being that you act out of personal motives.) O'Toole was fired from her laboratory position and no one else would hire her to do scientific work. She was reduced to taking a job answering phones for her brother's removals company. However, the story reached the public press and the *New York Times* wrote a scathing editorial stating the initial investigation of Dr. O'Toole's complaints smacked of an old-boy network drawing up the wagons to protect scientific reputations. Two National Institute of Health scientists, Walter Stewart and Ned Feder, looked into the case and persuaded forensic experts at the U.S. Secret Service to analyze Imanishi Kari's laboratory notes. They are said to have proved that her data entries were falsified at a later date. Under pressure, Baltimore retracted his support for the publication by Imanishi Kari to which his named had been appended. Later, when the prosecutors decided not to proceed in the case, he stated that Imanishi Kari had wrongfully been put through six years of hell.[85] However, the decision by the public prosecutor not to pursue the case was based on the fact that he would have had difficulties proving Kari's intent to defraud. It was not an absolute vindication or clearance of Imanishi Kari. The office of scientific integrity of the National Institute of Health issued a draft report in 1991, presenting evidence that Imanishi Kari had reacted to the doubts about the paper by fabricating whole sets of data and publishing some of them as corrections to the original paper. Evidence against Imanishi Kari was assembled by forensic experts from the U. S. Secret Service. They examined the paper and ink of computer print-outs of raw data that Imanishi Kari had pasted into her notebooks and compared them to similar tapes from notebooks kept by the other researchers in the laboratories. The experts said that some of the tapes pasted in Imanishi Kari's notebooks could not have been produced in 1985, the year she said she had performed the experiments. The paper and ink from these tapes allegedly matched those from other research laboratories notes that were written in 1981 and 1982. The evidence collected by the secret service was turned over to Imanishi Kari's lawyers, who employed their own forensics expert. Their expert, Albert Lyter, claimed that the Secret Service analysis was flawed.[84] In a curious twist, the prosecutor decided not to lay criminal charges against Imanishi Kari, partly on the grounds that a case would have required delving into the details of science and "many of these are practically incomprehensible even to many other scientists." Korinther, the government prosecutor, stated that he believed the secret service report was convincing, but that Lyter's competing reports still helped persuade him not to file charges. He said:

Anytime you have countervailing expert testimony you have got trouble.

Korinther says that he would have had to convince a jury that Imanishi Kari's notebook represented corrupting intent and not just sloppiness. The news story describing the decision not to file charges in the Imanishi Kari case had the headline "Abandoned Fraud Case Too Baffling for Jury!"[85] While this may not be the real reason why charges were not pressed, it is an interesting comment on the complexity of considering modern scientific evidence in lawsuits. This story stresses the need for a wider education of lawyers, and citizens who are potential jurors, in forensic science.

In the language of science, O'Toole's attempt to call the attention of the scientific community to what she believed was fraud, is called **whistle blowing**. Examining the history of would-be whistle blowers seems to indicate that the attitude of the general community to those who try to expose fraud is not favorable and that the attitude is summed up in the children's rhyme: "Tell tale fib your tongue shall be slit and all the little dogs shall have a little bit." Apparently, this attitude does not change when school yard children become adults.

John Dingel, Chairman of the Energy and the Commerce Committees in the U. S. House of Representatives, who has played the leading role in bringing scientific fraud to public attention, especially if the misuse of public funding is involved, stated that whistle blowers are at best left unprotected and they are often subject to harassment.[86,87]

In the case of the O'Toole and Imanishi Kari, members of the general public were not harmed by the situation. However, in another case dealing with the effect of drugs, forged results presented by a scientist named Glueck could have had serious effects on the health of children. Glueck had been studying a controversial treatment for children at risk of developing heart disease. Since the drugs under investigation absorb not only cholesterol but also nutrients there was, and continues to be, real concern that children's growth could be stunted by the drug. Glueck published a scientific paper dealing with the long-term effects of diet and the drug. He claimed that the treatment was perfectly safe. Shortly before his study appeared in print, the National Institute of Health received two phone call from sources, who requested anonymity, warning that the pending paper was "grievously flawed". Essentially Glueck claimed to have studied children and given them regular examinations when some of them had not been checked for up to six years. As a result of the subsequent investigation, Glueck was banned from receiving National Health Institute grants for five years. He was asked to retract the article containing the false data in a letter to the *Journal of Pediatrics*. He also resigned his university position. The reader interested in this case can read the details in a review of the case by Roman.[88]

LaFollette, in a book on scientific fraud, draws a distinction between a whistle blower (one who brings attention to misconduct from the inside) and a nemesis (one who does it from the outside). As mentioned earlier, *nemesis* is the name of the Greek goddess of retribution. (A dictionary describes retribution as punishing suitably, the root-word means to give something back to somebody). Probably the most recent case of a scientific nemesis active in exposing fraud is the case of Robert Sprague, a psychologist at the University of Illinois. He exposed fraud in the writings of Steven Breuning, who had been studying the side effects of powerful tranquillizers known as neuroleptics. Though the drugs have been used for years to help calm violent mentally retarded patients, Breuning suggested that the treatment might do more harm than good, and he claimed to have carried out studies that showed, amongst other things, that intelligence quotients miraculously doubled when patients were taken off the drug. Details of some of the studies that Breuning claimed to have carried out made Sprague uneasy. When he reviewed Breuning's data in depth, Sprague came to the conclusion that Breuning could not possibly have carried out the tests. There was physically not enough time for him to have completed the studies in the period that he claimed covered the work. In a review of this case Roman states that, in the case of Breuning, the work was not just bad, it was potentially deadly:

> *A researcher who fudges his drug findings is playing with more than the moon and stars, he is playing with life.*[88]

Sprague finally presented his findings in a six page letter to the National Institute of Health. Again, quoting Roman:

> *With his accusation in the mail, Sprague thought the job was over and that he could sit back and relax and let the federal agency go to work.*

Sprague is quoted as saying, "Boy was I naive." The detailed controversy of Sprague versus Breuning is beyond the scope of this book, but the interested reader can read Roman's review of the case.[88]

The person acting as whistle blower or nemesis may also be at risk. After 17 years of continuous, steady funding Sprague found his projects suddenly receiving less funding than usual. After acting as Breuning's nemesis, the people awarding the funds vehemently denied that the new reduced funding was connected to Sprague's activity as a nemesis, but it would seem to be at the very least a strange coincidence.

Another scientific type of crime is known as plagiarism. **Plagiarism** involves taking somebody else's work and putting your own name to it. (The term *plagiarism* comes from a Latin word meaning "to kidnap." This word in turn is taken from a Latin word meaning a "net.") Thus capturing somebody else's work in a net and putting your own name on it is kidnapping intellectual merit. Perhaps the most famous case of known plagiarism in present time is that of Elius Al Sabati. In the words of Roman:

> *operating without a medical degree, Al Sabati fooled U.S. medical schools and hospitals for several years by amassing a long list of scientific publications. He would retype articles from obscure journals, put his name at the top, sometimes honoring himself with a Ph.D., then submitting the paper as original research. Between 1977 and 1980 Al Sabati published 60 phoney articles in little known publications like Neo Plasma in Czechoslovakia and Tumor Research in Japan. Toting his lengthy list of publications, Al Sabati conned his way onto the research staff at a number of institutions including South West Memorial Hospital in Houston. Shortly after researchers began spotting their work under Al Sabati's by-line, Al Sabati disappeared. To this day no one knows where he is.*

Sometimes scientific fraud involves deliberate fabrication, but sometimes it is just a little imaginative polishing of data to convince skeptics of the validity of a theory. Thus Millikan, a physicist awarded the Nobel Prize for his work on electric charge, is alleged to have withheld unfavorable results, while claiming to report all results. Mendel, whose work formed the basis of modern genetics, is believed to have fudged his results to make them very convincing. Even Newton is supposed to have introduced fudge factors to increase the apparent power of prediction of his equations. It has even been suggested that Galileo did not actually conduct all of the experiments he wrote about, because a few were so difficult to reproduce.[89,90] Although these possible historic examples of fraud are not of particular interest to general forensic science, the reader deserves a warning that science is not above fraud. The current general opinion amongst scientists is that as the competition for scientific research grants gets stiffer, and as the pressure to publish scientific papers increases, there is likely to be more fraud in the scientific world – some of which may have serious consequences for the general public.

References

1. A. M. MacDonald (Ed.), *Chambers Etymological Dictionary*, Chambers, Edinburgh, 1971.
2. The story of Archemide's scientific discoveries in his bath tub is told in many physics textbooks, e. g., P. A. Tipler, *Physics*, Third Edition, Worth Publishers, New York, 1991, p. 340.
3. O. Schisgall, "The Man all Forgers Fear," *Readers Digest*, April 1969, 142.
4. Scanning Electron Microscope pictures of various types of paper provided by and used by the permission of the Pulp and Paper Research Institute, Montreal.
5. Pictures of the change in the surface of specialty paper as it is produced provided by and used with the permission of English China clays, St. Austel, Cornwall, England.
6. "Envelope Led Police to Bombing Suspect," *The Sudbury Star*, 20 June, 1993, A8.
7. (a) A good discussion of the role of classical ink studies in the detection of fraudulent or altered documents is found in H. J. Walls, *Forensic Science. An Introduction to the Science of Crime Detection*, Frederick A. Praeger, New York, 1968; (b) a good, brief history of ink constitution is found in L. C. Nickolls, *The Scientific Investigation of Crime*, Butterworth, London, 1956.
8. E. Geake, "Invisible Ink that Foils Forgers," *New Scientist*, 22 June 1991, 21.
9. R. L. Brunnelle, A. A. Cantu "Ink Analysis – A Weapon against Crime by Detection of Fraud" in G. Davies (Eds.), *Forensic Science. ACS Symposium Series 13*, American Chemical Society, Washington D. C., 1975, chapter 14.
10. For a discussion of the physical principles of modern thin-layer chromatography see the standard text books on Analytical Chemistry C. W. Keenan, J. H. Wood, D. C. Kleinfalter, *General College Chemistry*, Fifth Edition, Harper and Row, New York, 1976, p. 288.
11. J. Beard, "How Medieval Printers put Bibles Together," *New Scientist*, 18 April 1987, 5.
12. D. A. Hays, I. K. Morrison, L. S. Smith, "Role of Particles and Dispersions in Electrophotography," *Particulate Science and Technology 5* (1987), 39.
13. M. Scharfe, *Electrophotography Principles and Optimization*, Wiley, New York, 1984.
14. D. Owen, "Copies in Seconds," *the Atlantic Monthly*, February 1986, 65, 73.
15. Canada Brings in Thin Films to Foil the Counterfeiters, *New Scientists* 16 December 1969, 22.
16. M. Skoler, "The Old Buck Stops Here," *Discover*, November 1985, 92.
17. "Plastic Money Gains New Currency," *New Scientist*, 7 January 1988, 43.
18. G. O'Neill, "Flipping Images Fool the Fakers," *New Scientist*, 10 March, 1990, 40.
19. "Computer Counterfeiters Profit in Prison," *New Scientist*, 30 May 1992, 5.
20. I. Robertson, "They Almost Pulled it off – Sharp eyed Storekeeper Blew Whistle on Counterfeiters," *Canadian Stamp News*, 7 April 1992, 5.
21. "RPI Applies Chemistry to Stamp Collection," *Optical Spectra*, April 1979, 30.
22. H. Herst, "The Master Fallen was not a Master," *Canadian Stamp News*, 12 April 1988, 6.

23. I. Robertson, "Care of the Identified Invert," *Canadian Stamp News*, 22 September 1987, 15.
24. H. Lane, "Closely Observed Stamps," *New Scientist*, 21 December 1991, 34.
25. I. Robertson, "Fakes not Easy to Spot," *Canadian Stamp News*, 23 March 1993, 6.
26. D. Palmer, "Fatal Flaw Fingers Fossil Fly," *New Scientist*, 13 November 1993.
27. R. G. Owens, "Elementary Amber," *New Scientist*, 4 December 1993, 57.
28. K. S. Thompson, "Piltdown Man: The Great English Mystery Story," *American Scientist 29* (1991), 194.
29. L. Keith, "Piltdown Plot," *New Scientist*, 20 October 1990, 59.
30. S. Zukerman, "A New Clue to the Real Piltdown Forgery," *New Scientist*, 3 November 1990, 16.
31. F. Spencer, "Piltdown: A Scientific Forgery," Oxford University Press, London, 1990.
32. R. Estling, "Leg Before Cricket," *New Scientist*, 24 November 1990, 67.
33. P. Shopman, "On the Trail of the Piltdown Fraudsters," *New Scientist*, 6 October 1990, 52.
34. J. Hanlon, "Christ Under the Microscope," *New Scientist*, 12 October 1978, 96.
35. K. F. Weaver, "The Mystery of the Shroud," *National Geographic*, June 1980, 730.
36. N. Angier, "Unraveling the Shroud of Turin," *Discover*, October 1982, 54.
37. "Shrouded in Mystery," *New Scientist*, 22 September 1977, 720.
38. R. I. Johnson, "Scientists Examine the Shroud of Turin," *Industrial Research Development*, February 1980, 145.
39. R. Hedges, "New Directions in Carbon 14 Dating," *New Scientist*, 2 March 1978, 599.
40. For Review of the concepts involved in quantifying the rates of decay of radioactive material in terms of the quantity "half life," see the discussion of simulated radioactive decay in B. H. Kaye, *Chaos and Complexity. Discovering the Surprising Patterns of Science and Technology*, VCH Verlagsgesellschaft, Weinheim, 1993, chapter 4.
41. I. Anderson, "Teams Agree on Medieval Origins of the Shroud," *New Scientist*, 22 October 1988, 25.
42. W. C. McCrone, "Microscopical Study of the Turin Shroud III". *Microscope 29* (1981), 19.
43. J. Nickell, *Inquest on the Shroud of Turin*, Prometheus, Amherst, NY, 1982; see a review of the book by R. Hedges, *New Scientist*, 2 June 1983, 640.
44. M. White, "While Others Seek the Forger with Flair," *New Scientist*, 22 October 1988, 25.
45. "The Fakes of Haciliar," *Time*, 6 September 1971, 50.
46. "Time and the Forgers," *Newsweek*, 23 April 1973.
47. C. Bishop, "There's no Fraud Like an Old Fraud," *New Scientist*, 7 January 1988, 52.
48. "Plate of Brass, a Hoax," *Technology Review*, January 1978, 27.
49. S. Fleming, "Science Detects the Forgeries," *New Scientist*, 4 December 1975, 567.
50. S. Fleming, *Authenticity in Art*, Institute of Physics, London, 1975.
51. These objects can be viewed in the Natural History Museum, London.
52. R. S. Phillips, *Funk and Wagnall's New Encyclopedia Vol. 2*, Funk and Waganall Inc., New York, 1979, 201.
53. See Correspondence between C. I. Willis and P. Wellinhofer in *Scientific American*, November 1990, 10.
54. S. Connor, "Riddle of the Missing Rock Resurrects Archeopteryx Controversy," *New Scientist*, 13 August 1987, 27.
55. R. Hughes, "Brilliant, but not for Real," *Time*, 7 May 1990.
56. "Art Fraud Rampant, says Expert," (news story), *Sudbury Star*, 8 January 1981.
57. "The Lab Detectives," *Time*, 31 January 1972, 14.
58. S. Bunney, "A Firmer Footing for tree Ring Dating of Old Masters," *New Scientist*, 1 May 1986, 29.
59. R. Matthews, "When Seeing is not Believing," *New Scientist*, 16 October 1993, 13.
60. "An Optical Weapon in the War Against Counterfeiting," *Optical Spectra*, June 1981, 20.
61. M. Smilgis, "A Crackdown By Cashmere Cops," *Time*, 14 March 1988, 38.
62. T. Patel, "Pour Yourself A Gin and Sugar," *New Scientist*, 19 December 6.
63. "Olive Oil Scam Caught in the Net," *New Scientist*, 12 December 1992, 19.
64. "Industry to Crack Down Hard on Insurance Fraud." (news story), *Sudbury Star*, 11 August 1993, A5.

65. J. G. Delly, S. Palenik, "The Microscope in Art and Archeology. Is it Jade?," *Industrial Research*, May 1976.
66. N. M. Herbst, C. N. Liu. "Automatic Signature Verification based on Accelerometry," *I. B. M. J. Res. Develop. 21* (1977), 245.
67. J. S. Lew, "An Inspired Regional Correlation Algorithm for Signature Verification which Permits Small Speed Changes between Hand Writing Segments," *I. B. M. J. Res. Develop. 27* (1983), 181.
68. J. Vredenbregt, W. G. Kosler. "Analysis and Synthesis of Handwriting," *Philips Tech. Rev. 32* (1971), 73.
69. "IBM Forges Ahead on Verification," *New Scientist*, 20 May 1976, 148.
70. (a) F. Press, R. Sicker, *Earth*, W. H. Freeman, San Francisco, 1974; (b) For an introductory discussion see B. H. Kaye, *Golf Balls and Other Missiles Interesting*, VCH Verlagsgesellschaft, Weinheim, chapter 8, to be published.
71. A. Irwin., "How Cash Machines Will Recognize You When You're Drunk," *New Scientist*, 21 August 1993, 18.
72. B. Fox, "Phantom Card Tricks," *New Scientist*, 7 August 1993, 45.
73. J. M. Fournier, "New Progress in Optical Writing Appraisal Applied to the Dreyfus Affair." Available from the author at Laboratoire de Physique Generale de Optique, Université de Franche Comte, 25030, Vesancon Cedex, France.
74. J. M. Fournier. "Approache Analogique d'une Expertise en Ecriture: Un Example Concemant l'affaire Dreyfus," *Le Courrier du CNRS 16* (1975), 23.
75. See article on Dreyfus in *Funk and Wagnall's Encyclopedia Vol. 8*, Funk and Waganall Inc., New York, 1979, p. 172.
76. J. M. Fournier, J. C. Vienot., "Fourier Transform Holograms used as Matched Fillers in Hebraic Paleography," Symposium of Engineers Application of Lasers, Tel-Aviv, Israel, June 29 – July 1, 1970.
77. Two-dimensional Fourier transforms of the signatures and pages of handwriting generated by G. G. Clark. Hand written manuscripts of poem by Robbie Burns reproduced from *Canadian Stamp News*.
78. For an introductory discussion of how sharp edges generate high frequency information in the Fraunhofer diffraction pattern and two dimensional Fourier Transforms see B. H. Kaye, *A Random Walk Through Fractal Dimensions*, Second Edition, VCH Verlagsgesellschaft, Weinheim, 1994, p. 84.
79. D. Crystal, *The Cambridge Encyclopedia of Language*, Cambridge Press, Cambridge, 1987, pp. 68, 86.
80. R. Matthews, T. Merrian, "A Card by Any Other Name," *New Scientist*, 22 January 1994, 22.
81. For an introductory discussion of Zipf's Law see B. H. Kaye, *Chaos and Complexity. Discovering the Surprising Patterns of Science and Technology*, VCH Verlagsgesellschaft, Weinheim, 1993, chapter 11.
82. R. Matthews, "Linguistics on Trial," *New Scientist*, 21 August 1993, 12.
83. D. Ubelaker, H. Scammell *Bones, A Forensic Detective Casebook*, Edward Burlingame Books, New York, 1992.
84. "Thin Skins and fraud at MIT," *Time*, 1 April 1991.
85. D. Charles, "Abandoned Fraud Case Too Baffling for Jury!" *New Scientist*, 25 July 1992, 4.
86. J. Marks, "Scientific Misconduct: Where Just Saying No Fails," *American Scientist 81* (1993) 380.
87. R. Lewin., "Pressure to Publish Leads to Increase in Fraud," *New Scientist*, 4 April 1992, 7.
88. M. B. Roman, "When Good Scientists Turn Bad," *Discover*, April 1988, 50.
89. T. Marsa., "Scientific Fraud," *Funk and Wagnall's Year Book 1993*, Funk and Wagnall Inc., New York, 1993.
90. A cartoon summary of possible famous historic scientific fraud has been presented by Kate Charlesworth, *New Scientist*, 31 August 1991, 45.

Chapter 11

Bodies, Bones and Blood

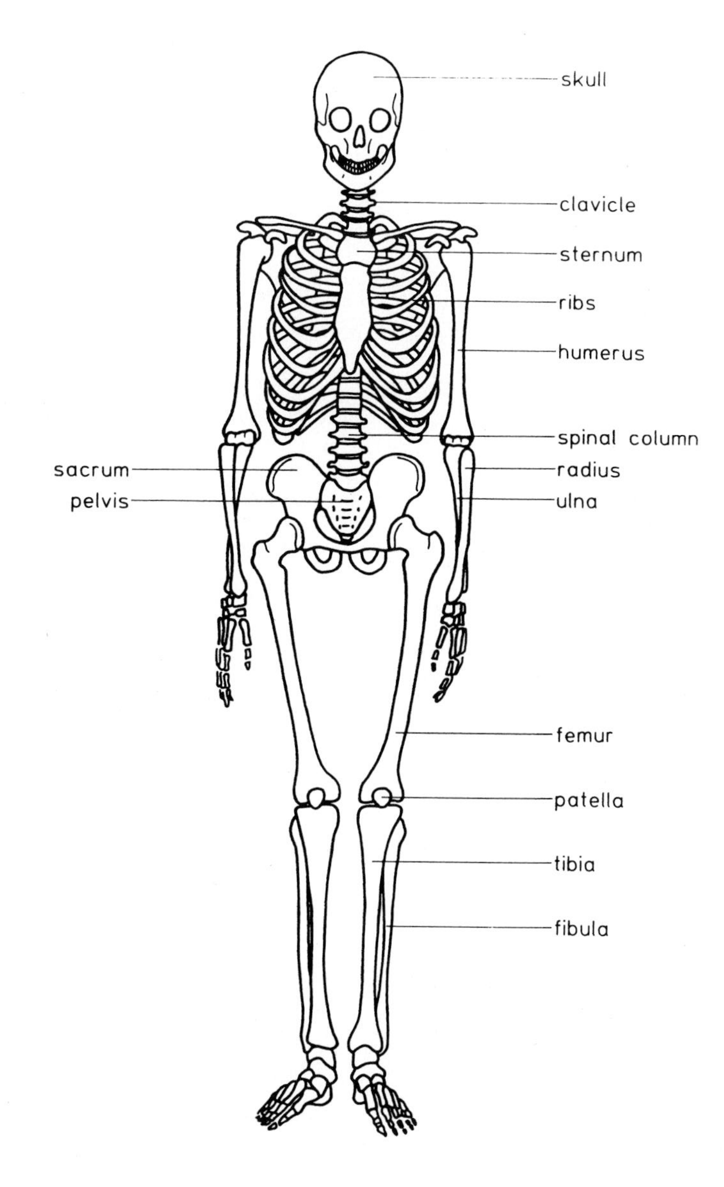

Chapter 11

Bodies, Bones and Blood

11.1 Talkative Bodies

The title of this section was inspired by an article that appeared in the Canadian edition of *Reader's Digest* at the time that this chapter was written. It dealt with the work of Michel Evenot, a forensic ondontologist working in France. The main idea of the article, which was entitled "Corpses Can Have Talkative Teeth,"[1] was that by examining the teeth of a corpse or skeleton, one can gain tremendous information regarding the lifestyle and manner of death of the person. A **forensic ondontologist** is one of the specialists that we shall look at briefly. Whole books have been written on deducing evidence from bodies, bones and blood. Some of the detailed case histories require a strong stomach.

This chapter treats the more physical side of the interpretation of evidence and we refer the reader to a more detailed account of forensic studies of the bodies of crime victims.[2] Bernard Knight is a forensic pathologist who has written about deducing information from the examination of corpses.[3,4] A **pathologist** is an expert who studies the origin and cause of disease. In popular television shows on crime, the pathologist is usually the person who shows up at the scene of a crime to study the corpse and who is asked to determine the time of death of the victim. In general, however, pathologists also examine tissues and fluids taken from sick people to determine the origin of a disease. By extension, the word **pathological** is used to describe a criminal whose crimes are horrendous and by inference these must be the act of a sick person rather than a sane person. Knight tells us that the pathologist at the scene of the crime must both minutely describe the fatal injuries and offer an opinion as to their cause, and, where appropriate, determine if they relate to any actual disease. With regard to determining the time of death when examining a corpse, Dr. Knight makes the following statement:

> *The sight of a doctor in a television drama laying a hand on the head of a cadaver and gravely pronouncing "she died at 3 o'clock last Tuesday" is guaranteed to reduce a real forensic pathologist to hysterical laughter.*

Knight points out that forensic scientists have published a multitude of papers on how to increase the accuracy of "time of death estimates" during the first day after death. He tells us that the way in which body temperature declines after death holds the greatest promise for improved estimation of the time of death but that, although people have studied body cooling rates for a century and a half, the many variables that can affect the rate of cooling still preclude precise estimation of the time of death. A popular twist in television murder mysteries is turning up the thermostat in the room where a body is lying to delay its cooling, thereby suggesting that the victim died later than the time at which the crime was actually carried out.

Factors that can affect how quickly a body loses heat include the amount of clothing on the body; the temperature of the surroundings; the humidity; the posture of the corpse; the body's initial temperature, which may not be 37 degrees; how fat or thin the person was; and the ratio of the body surface area to its weight. Recent developments in computing hold some promise. However, these programs depend on measuring a series of temperatures at intervals and on different sites on the body.[3]

Ubelaker describes in some detail the work of Bill Bass of the University of Tennessee, who studied the cooling rates of bodies exposed to different weather conditions and lying on different types of ground. Ubelaker tells us that Bass' work is not popular with the local residents, who do not like to think of dead bodies lying around for the purpose of scientific experiments. Knight has discussed in some detail the popular ideas about **rigor mortis** (the stiffening of the body after death) and comes to the conclusion that:

many people, especially crime writers and police officers, seem to have a touching faith in the ability of doctors to estimate the time of death from rigor mortis, that is the development of stiffness in the body.[4]

In an article written in 1986 Knight states:

standard forensic textbooks perpetuate the overly optimistic view of rigor mortis interpretation. A typical quotation would be that rigor mortis takes twelve hours to come on, lasts for twelve hours and takes another twelve hours to pass off. All that is consistent in this statement is that it is consistently wrong.[4]

If the body of a victim in a murder case is found more than two or three days after the crime, the physician is no longer needed to interpret the significance of temperature changes and the task of deducing the time of death becomes the territory of another specialist – the forensic entomologist. For an extensive discussion of the forensic entomologist see references 5 and 2. The **entomologist** is the name of a specialist in biology who studies the life cycle and behavior of insects. The forensic entomologist can look at the different insects that are living on the remains of a victim and, from the known life cycle of the maggots, determine the time elapsed since the victim was killed. Ubelaker has this to say about the work of the forensic entomologist:

Maggots are repulsive to almost everyone. The most seasoned medical examiners have learned to tolerate the associated odor but less experienced examiners instinctively arm themselves with a few deep breaths and anoint their upper lips with oil of camphor or tiger balm before approaching a death scene at which maggots are still at work. No matter how many such cases they see or how fascinated they may become by the process of translating the mysteries to which the maggots may provide the key, there is still an element of shock in seeing one of the lowest forms of life feast on the remains of the highest. Because of that shock and because it is very difficult to see below the surface state when the soft tissue is dry or in advanced decomposition, some medical examiners turn away sooner than they should and murder goes unnoticed.[2]

Knight has described a case in which the entomologist provided useful information:

In 1964 police called the forensic pathologist Keith Simpson to a body buried in a Berkshire wood. On the body were fat, third instar maggots of bluebottles (a type of carnivorous fly), indicating a minimal interval since death of about 10 days. There were beech leaves on the body, but no beech trees in the wood. The prosecution claimed that the man, Peter Thomas, had been murdered by William Brittle in Gloucestershire and carried in the trunk of the car to Berkshire (England). Three witnesses were produced who claimed to have seen Thomas alive some days after the accused had left Gloucestershire. The man was convicted on the basis of the entomological evidence.[3]

Erzinclioglu reports another interesting case in which entomological evidence proved the innocence of a man who had spent eight years in jail for a murder he did not commit.[5] The case involved the discovery of the body of a man found murdered on a ferry that had docked at six o'clock in the evening one day in September. A postmortem was carried out the following day at two p. m. at which time large numbers of blow-fly eggs and larvae were found on the body. This entomological evidence was ignored at the trial and the captain of the ferry was sent to prison for life. Eight years later however, the case was reopened. At the new trial Dr. F. Mihali of the Budapest National History Museum said that no carrion flies (blow flies) are active in Hungary at six o'clock on a September evening. He also pointed out that the presence of larvae on the corpse meant that the eggs could not have been laid on the day that the postmortem was carried out. They would not have had time to hatch, so the eggs must have been laid before six p. m. on the day before the discovery of the body, that is, before the Captain arrived on the ferry. The captain was acquitted of murder and released. The reader interested in this aspect of forensic science will find several other cases discussed in the article by Erzinclioglu.[5]

Knight has also given us information on how examining water from the lungs of a victim of drowning can yield interesting information. In water there are very tiny creatures called diatoms. There are many thousands of types of **diatoms**, some characteristic of sea water and others of freshwater. So it may be possible to decide whether a body found in an estuary, where a freshwater river meets the salt water of the sea, was washed in from the sea with the flow of the tide or came down the river. In a murder, which took place on the yacht Christine in the English Channel, forensic pathologists were able to show that a body found on the Belgium coast had gone into the water off the Isle of White, because of the local variety of diatoms that was discovered in the lungs at the postmortem of the victim.

Botanists may assist the police with an opinion regarding how long it takes for undergrowth to lose it chlorophyll beneath the corpse or how long grass or weeds need to grow through the ribs of a decaying body. They may even be asked to estimate the rate of growth of tree rootlets in a buried skull or bone marrow cavity. Details on this type of examination have been discussed at length by Ubelaker.[2]

11.2 What Bones can Tell Us

When all of the flesh has disappeared from the body and we are left only with the skeleton, specialists, including the forensic ondontologist mentioned earlier, and the forensic anthropologist, can give us information about the victim by studying its bones and teeth. It is amazing how much information the forensic specialists can deduce from even a few fragments of bone. Again, since this topic has been discussed at length in other books, we will just briefly look at different types of information that can be extracted from studying bones. One of the first things that one must do is to determine the age of the bone. If one suspects that the bone is ancient, radio-carbon dating can be used to deduce its age (see discussion of radio-carbon dating in Chapter 10). Knight tells an interesting story of the use of carbon dating when studying a skull. In 1983 following the discovery of a woman's head in a peat bog, a man called Peter Reyn-Bardt confessed to the murder of his wife 20 years earlier. He was convicted at Chester Crown Court but when scientists used radio-carbon dating to test his supposed wife's skull they found that the skull had been in the bog since 210 A.D.![3]

Deciding whether a bone is human or animal is quite a difficult task. This is discussed in detail by Ubelaker, who recounts a case which had most of the experts baffled. A break in a bone discovered in Alaska had been fixed with the aid of a surgical plate. Everybody presumed that the bone must be human, since it was presumed that no one would go to the trouble of fixing an animal break with a surgical plate. However, it turned out to be the bone of a dog. The origin of the bone was determined by examining it under a microscope. In describing his work, Ubelaker states:

We removed a sample of the bone from one side of the bone fragment from the fracture. We then prepared a ground thin section from the sample for microscope analysis. In particular we were looking for the pattern of osteons in the section. ***Osteons*** *are small circular bone structures that form continuously throughout life. In humans they are randomly scattered throughout the cortex of the bone while in many animals they frequently line up in rows.*[2]

Note that the term *cortex* comes from a Latin word meaning bark; the cortex of a system is its outer layer. Thus, one talks about the cortex of the brain as well as the cortex of a bone. In Figure 11.1 (a) various femural (upper leg) bones from different animals are shown. The structure of bone specimens from humans and animals as viewed through the microscope are shown in Figure 11.1 (b). The study of osteons showed that the Alaskan bone came from a dog.

Whether or not an ancient bone is from a fossil hominid (The term hominid means man-like) or an animal can be the source of great debate in anthropological circles. Thus, the nature of a particular piece of bone was the source of a fierce debate between two anthropologist, Dr. Tim White of the University of California at Berkeley and Dr. Noel Boaz from New York University. Dr. Boaz maintains that the bone in question, found at Sahaibi in Libya, was from the human family Hominidea, whereas Dr. White maintains that it is the bone of a dolphin. Anderson states that the problem with anthropologists is that they want so much to find a hominid bone that any scrap of bone becomes a hominid bone.[6]

In a previous chapter we mentioned the case of Evans, who was wrongfully hanged for a crime committed by John Christie. Dr. Knight tells us that John Christie killed at least six women and hid the bodies under the floorboards and in a cupboard, which was then papered over. He buried two more bodies in the garden. Apparently his garden fence was propped up with a human thigh bone! In the examination of the remains found in the Christie case, a forensic ondontologist played a key role in identifying one of the victims. Investigators recognized that one of the victim's molars had a metal crown of central European workmanship. As a result, they were able to identify one of the victims as an Austrian woman called Ruth Fuerst.

One of the cases cracked by Evenot required an amazing eye for detail.[1] In September 1982 he was asked by he police to investigate the possible origin of two teeth found under a carpet during an arrest of a suspect identified by the victim of a vicious assault. The suspect, who was a Moslem, claimed that they were sheep's teeth from a sacrificial rite. Evenot said that even a layman could see that they were human teeth. Using fiber optics and electron microscopy, Evenot studied the teeth and spotted cracks, a bone chip, and traces of steel that pointed to brutal extraction of the teeth. He reported:

these teeth were ripped out by a metallic instrument.

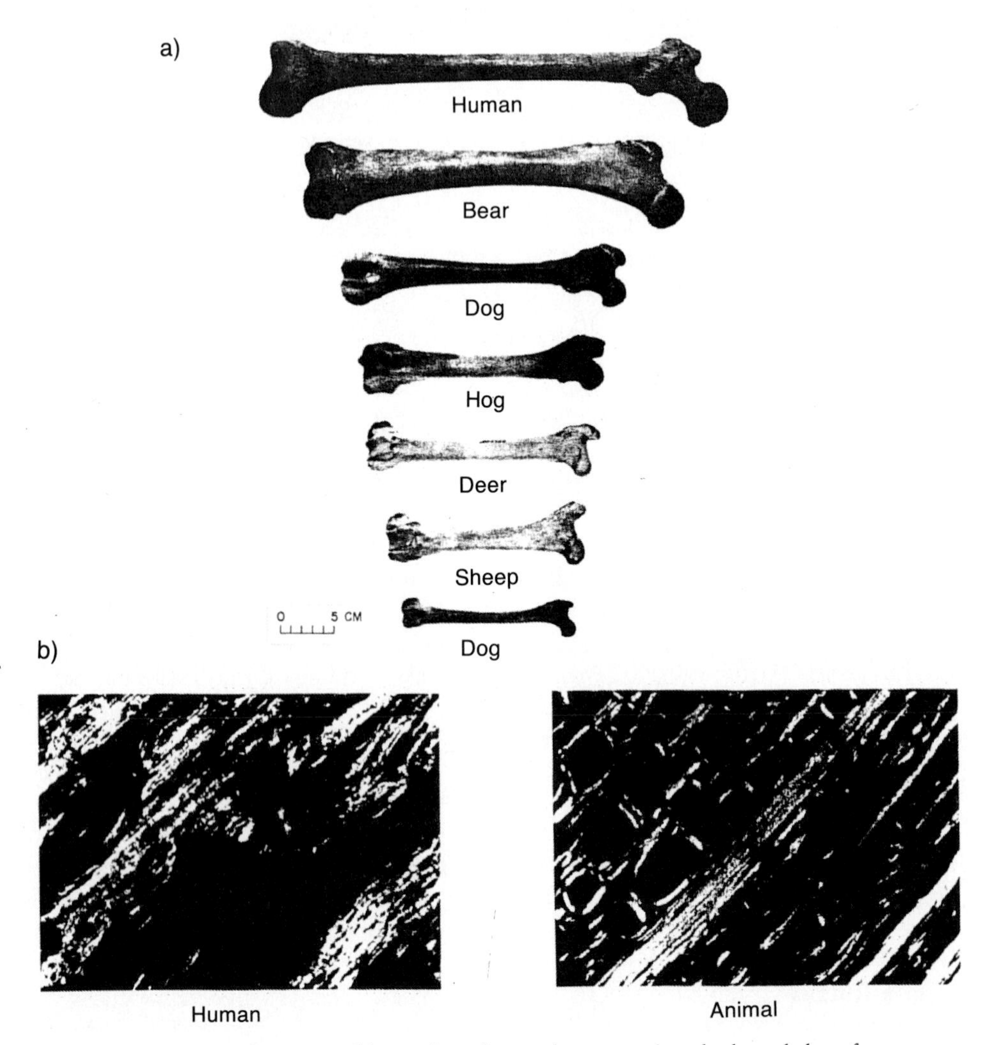

Figure 11.1. Distinguishing animal bones from human bones requires the knowledge of an expert. a) Femur bones from human and animal remains. b) The structure of human bones differs from animal bones when viewed under a microscope. The small structures visible are know as osteons. (Reprinted by permission of HarperCollins Publishers, Inc.[2])

From the fact that the teeth had small brown scratches on the enamel, he was able to say that as a child the victim had lived in a region where the drinking water was high in fluorine. However, the really interesting observation was a tiny notch on the side of one tooth, suggesting that the person had been a seamstress who was in the habit of biting off thread. Putting all these details together, the police deduced that the teeth belonged to a thirty-year-old Moroccan woman, who had grown up in a region where the water was rich in fluorine. Her body had been found earlier and had been sent back to Morocco. After two years of negotiation, the justice officials exhumed the corpse in Morocco and Evenot was able to show that the two teeth fitted exactly into the Moroccan woman's jaw.

Sometimes what is known as a jaw print can be matched with dental x-rays to pinpoint a corps' identity, as illustrated by the photograph in Figure 11.2 (a). In Figure 11.2 (b) an enlarged photograph showing the distinct markings and irregularities present on individual teeth is shown.[7] Sometimes, if the victim has been bitten, the forensic ondontologist can make a pattern of the suspect's teeth and match them with the bite marks, as illustrated in Figure 11.2 (c). The skills of the forensic ondontologist are very important when attempting to identify the victims of an air crash, where many of the bodies are burnt beyond recognition.

An example of how a tooth can be used to identify a badly burnt body is described by Evenot, who examined a charred body discovered in Normandy. The remains were suspected to be those of a young girl who had disappeared from a girls' camp. Evenot was able to identify the body after he had been given a baby tooth by the girl's mother. On the baby tooth he found microscopic cracks caused by a cavity that had dislodged the tooth prematurely. This meant that the fragments of the tooth must still have been in her jaw. Therefore, he matched the tooth to the jaw bone found in the woods, and it was a perfect fit. In another case he was able to identify a body and link it to fragments of teeth found in an automobile used by a suspect in the murder case.

Dental records as a possible means of identification have been used in the study of bones claimed to be those of Hitler and of one of his infamous henchman Joseph Mengele. Mengele was also known as the "Angel of Death" because of the inhuman experiments he carried out when he was a doctor at the concentration camp at Auschwitz in the Second World War.[8,9] In 1992 the Russians displayed what they claimed were fragments of Adolf Hitler's skull. Adolf Hitler and his wife Eva Brown committed suicide in an underground bunker as the Russian army closed in on them in 1945. It was claimed that Russian troops had taken these fragments from the site of the cremation and that they had been in Russian archives until they were displayed in 1992. A French team of forensic ondontologists from the Institute for Forensic and Social Medicine in Lille contested the claim that the bone fragments were from Adolf Hitler's body. The case has been reviewed by Tara Patel.[8]

After the Second World War many notorious Nazis wanted for war crimes, fled from Germany and some of them took refuge in South America. One of the most wanted criminals was Joseph Mengele. In June 1985, the body of a drowning victim buried in 1979 was exhumed because two Austrians who lived in Brazil claimed that it was the body of Mengele.[9] Three teams of forensic scientists from international groups converged in Sao Paulo, Brazil,

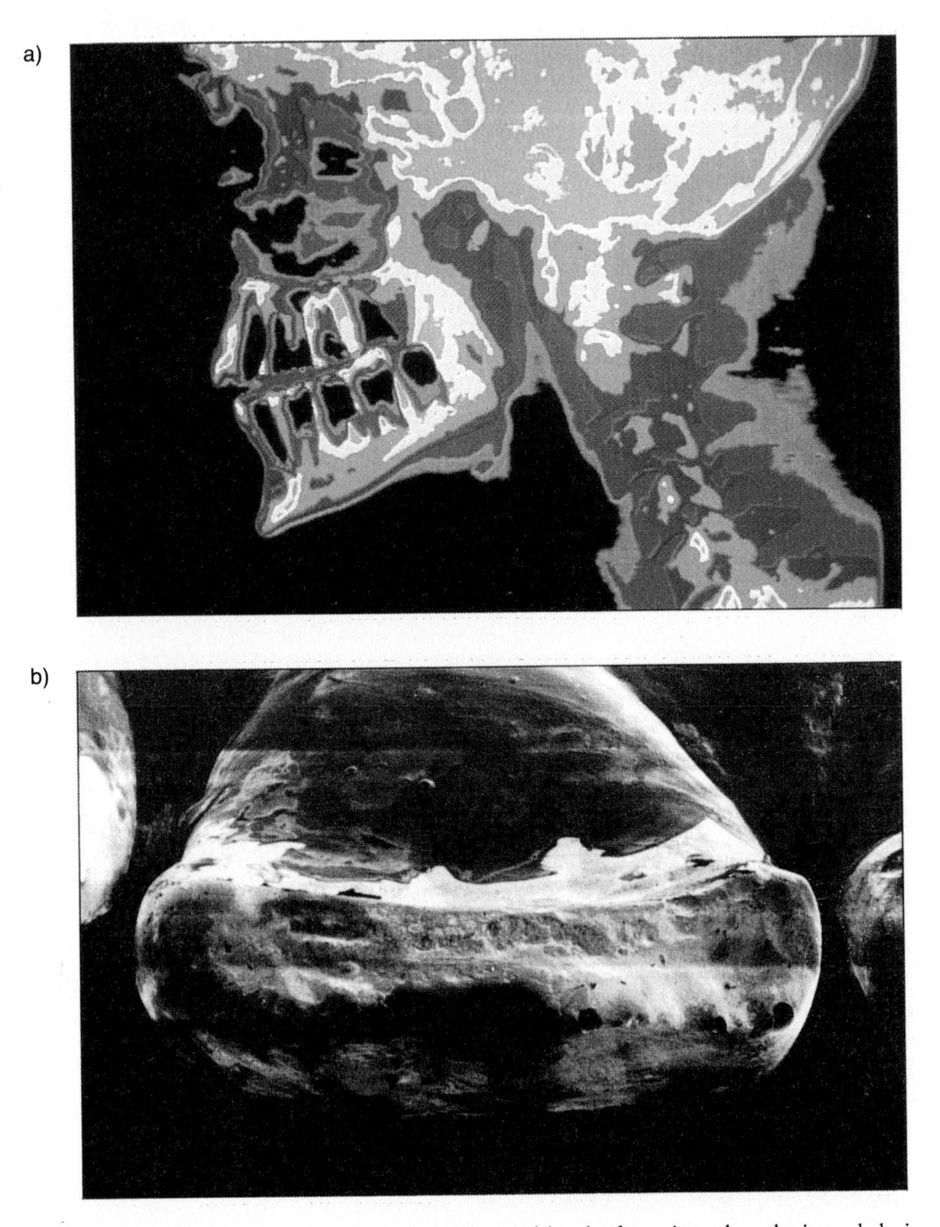

Figure 11.2. Various types of information can be used by the forensic ondontologist to help investigate the identity of a victim. a) The position and condition of the teeth can be matched to dental x-rays. b) Distinctive markings on individual teeth can indicate a profession or habitual activity. c) Impression of bite marks on a victim can be matched to a suspect's teeth. (Reproduced with permission from *Science Digest*.[7])

c)

to carry out extensive tests on the remains. It is worth studying their techniques and conclusions in detail, since, as pointed out by Dr. Leslie Lokash, Chief Medical Examiner of Nassau County, New York, the case probably represents one of the most exhaustive studies of a set of remains carried out by forensic experts. In his account of the examination Dr. Lokash describes how the height of the person was estimated. One of the leg bones indicated that the body stood 174 centimeters, which is the exact height of Mengele as listed in records from the German S. S. Other evidence was that the body had a **diastema**, a technical term for a gap between the upper two front teeth. Evidence of a diastema in the Brazilian remains was found in the hard pallet, the part of the skull just above the top teeth. Three molars remained in the skull, two on the upper right and one on the upper left side. All three contained fillings that matched Mengele's dental records.

Dr. Ellis Kerely, a forensic anthropologist at the University of Maryland, determined the probable age of the Brazilian remains by studying a thin section of bone from the left femur. The section of bone was ground and polished until it was 100 micrometers thick and then examined under the microscope. The older the bone, the higher percentage of osteons that are fragmented or broken (see discussion of the determination of animal versus human above). The official medical report of the forensic team states that there was an 86.7 % probability that the age of the individual at death was 64.25 to 74.25 years of age, and that most probably he was 69 years old. It was determined that the bones were those of a male by looking at the standard skeletal check points; the mastoid, a bone behind the ear, the sciatic notch in the pelvis and the configuration of the sacrum. (See the labelled skeleton on the front page of this chapter to locate the latter bone). Visual examination and x-rays of the bones showed injuries consistent with those that might have occurred in a motorcycle accident. It is known that Mengele had a motorcycle accident, which took place during the Second World War. The details of the accident were recorded in the Auschwitz documents.

Another study carried out by members of a West German team, led by Richard Helmer, involved a photographic study of the skull. The team of West German scientists marked the skull with pins at 14 characteristic points often used for identification, such as the brow and chin. Helmer filmed the skull and cast its image onto a video screen. Then he did the same with a photo of the young Mengele from S. S. files. By merging the two, Helmer superimposed the features of the face onto the skull and found 24 points of comparison that helped convince the forensic and police experts that the skull was indeed that of Mengele. Concluding their study Lokash said:

> *There is no such thing as a 100% certainty in science, hence the investigators modification of their conclusion that they were reasonably certain that the remains were those of Mengele. "I don't think someone of age 69 would engineer anything like this" Lookash added. "The person was old, losing his teeth and had vascular problems."*

In general, the scientific community accepts that the grave in Brazil contained the remains of one of the most notorious criminals of the Second World War, a man who had escaped the dragnet of the Allies as they entered Germany.

The use of the photographic matching of skulls with known pictures of an individual was recently applied to study the cause of death of Mozart, the composer who died in December 1791 at the age of 35. Officially, his doctors diagnosed the death as caused by "heated miliary fever." Glausiusz, in a review of the new study of the possible causes of Mozart's death, tells us that "heated miliary fever" was 18th century medical jargon for "beats us!" Some people think Mozart was poisoned by his rival Antonio Salieri, others maintain he died of kidney disease, tuberculosis, rheumatoid arthritis, or of a streptococcal infection. Some people even believe that he was assassinated by Freemasons! Although it has commonly been assumed that Mozart's remains were lost because he was buried in a communal grave for paupers, a skull reputed to be Mozart's was actually unearthed ten years after his death by the grave digger who had interred him. The skull had a succession of owners until at the turn of the 20th century it ended up in the Salzburg Mozart Museum, with its authenticity still in dispute. This skull is shown is Figure 11.3(a). French anthropologists from the University of Provence examined the wear of the teeth in the skull and confirmed that it belonged to a man between 25 and 40 years of age. When researchers superimposed photographs of the skull on portraits of Mozart, they found that the features, notably the high cheek bones and the egg shaped forehead, matched

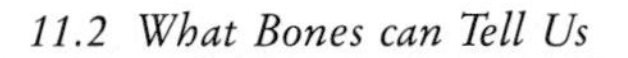

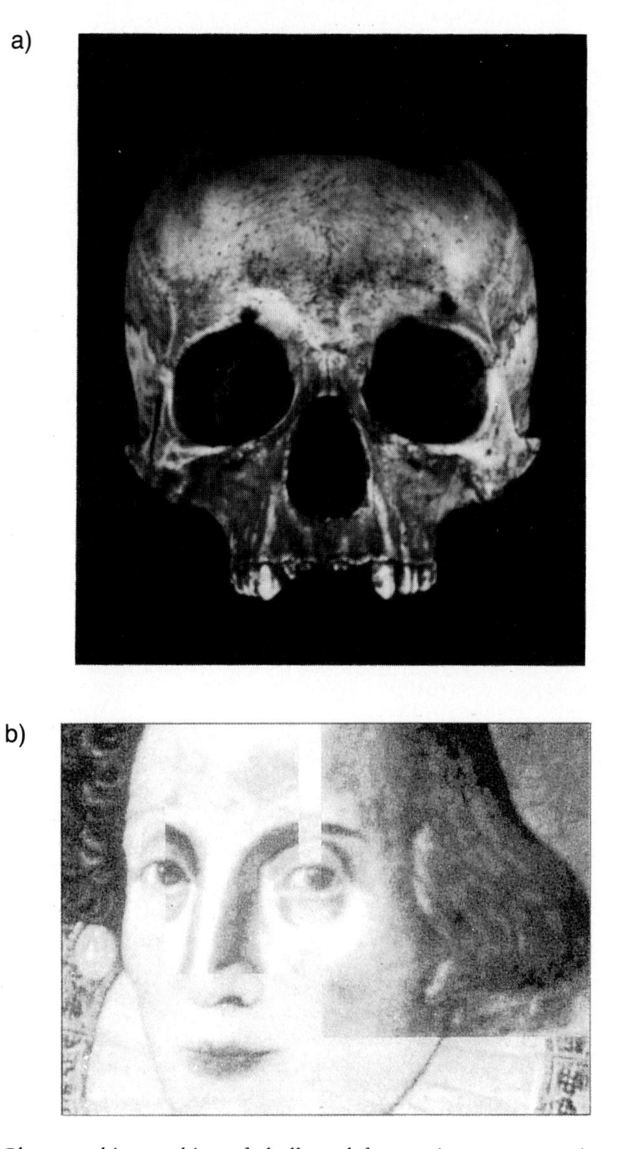

Figure 11.3. Photographic matching of skulls and faces using computers is part of the art of the forensic scientist. Studies in the area of historic personalities is suggesting some fascinating alternatives to historic ideas. a) Is this Mozart's skull? b) Was Shakespeare a pseudonym for Queen Elizabeth I of England? (Part (a) reproduced from *Discover*;[10] part (b) reproduced from *New Scientist*.[11])

perfectly. Recently the skull was examined by Miles Drake of Ohio State University, who describes himself as a neurologist and a frustrated musician. Drake was very intrigued by a crack in the skull's left temple. The fracture is only partly healed, indicating that it happened not long before Mozart's death, assuming that the skull was his. It is known that Mozart drank heavily so that the fracture may well have resulted from a drunken fall. The fall, says Drake, could have torn veins leading from the surface of the brain, allowing the blood to leak into the subdural space, which lies between the Dura (the brain's protective membrane) and the skull. As the blood accumulated and dried it would have put increasing pressure on the left hemisphere of Mozart's brain. That could have made him subject to mood swings and depression. It could also have impaired the ability to coordinate movements of the right side of his body and possibly his ability to compose music. Near the time of his death, his relatives recalled that Mozart could not dress himself and they had to help him, because his right hand was uncoordinated. (The left side of the brain controls the right hand side of the body and vice versa.) In today's medical language he would have been described as suffering from a **subdural hematoma** (hematoma means blood clot). According to Drake, Mozart could have survived the hematoma but he contracted an infection and a high fever that required the attention of a doctor. Drake claims that the probable treatment prescribed by the doctors at the time, known as bleeding, would have killed Mozart. The details of his conclusions and his views on the medical treatment can be found in reference 10.

The superimposing of images by computer to seek out the identity of historic figures is a subject that has been pursued by Lillian Schwartz, a consultant at AT&T Bell Laboratories.[11] Schwartz was intrigued by the fact that the face, which appeared in the 1623 edition of Shakespeare's plays, does not match the appearance of a statue that stands in Stratford on Avon. This statue was made in the same year that the manuscript of the play was published. The statue was commissioned by the Shakespeare family and is more likely to represent the real Shakespeare than the picture in the 1623 edition of the plays.

Over the years many different individuals have been suggested as being the real writer of the Shakespearean plays, among them is the Earl of Oxford, Edward de Vere, the contemporary writer Marlow, and Bacon (see discussion of style analysis in Chapter 10). Schwartz converted the image of Shakespeare from the book of these plays and compared it to pictures of Edward de Vere, but did not find a match. Another individual who has been suggested as the real author of Shakespeare plays is Elizabeth I, of Great Britain. It is said by some that she wrote the plays under a **pseudonym** (a word meaning a false name) since, as Queen, she did not want to draw attention to her literary activities. Lillian Schwartz found an exact match between the pictures of Queen Elizabeth and the 1623 picture of Shakespeare as shown in Figure 11.3(b). Schwartz suggests that either someone was having fun with

Queen Elizabeth's picture or wanted to leave to posterity a hidden message concerning the true authorship of Shakespeare. The interested reader will find details of the study by Schwartz in Reference 11, where Mona Lisa's real identity and that of the participants in the Last Supper is also considered.

In recent years forensic anthropologists have developed techniques for restoring the appearance of skulls.[7,12] The image was created from the probable structure of flesh on the skull. Anthropologists measure the thickness of flesh and other details of the physical appearance of people and from their records they reconstruct the appearance of a person by the procedure illustrated by the four pictures of Figure 11.4. First of all, pegs are attached to the skull as a guide for tissue thickness and then these pegs are connected with strips of clay. The modelling of the facial features is continued until the appearance of the fleshed out skull is reached, as shown in the final part of Figure 11.4. Ubelaker's book contains many pictures of reconstructed images and he discusses the successes of such pictures. James Taylor, Director of the Forensic Anthropology Team at Layman College tells us that in one case reconstruction was so perfect that a person's family identified him from photos of the restored face. Taylor also states that failures are commonplace, because there is an incredible number of factors that make one face different from another. He states that:

> *It is almost impossible to judge how the person's eyes looked or how deeply set they were in the orbits of the skull. We need a lot more information about the effect of muscle and tissue on facial bones. Right now we are computerizing our data. We are looking forward to the day when near perfect reconstruction will not just be a lucky accident.*

Ubelaker discusses some of the more recent developments in computer simulation of people's appearance in reference 2.

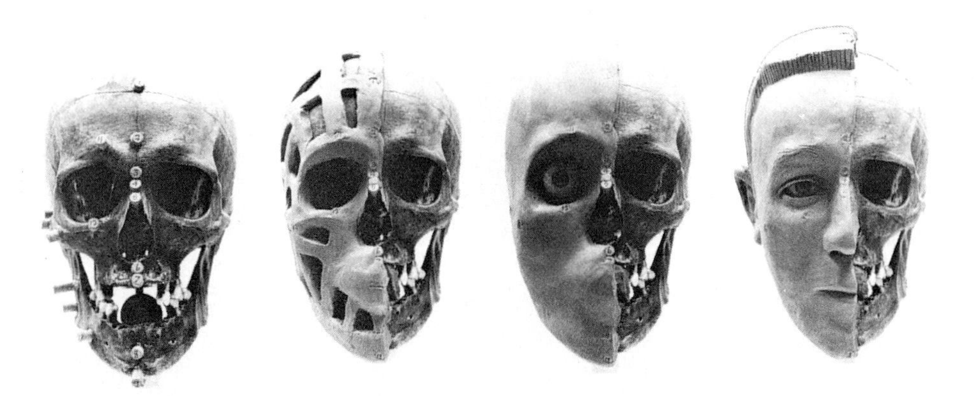

Figure 11.4. Forensic anthropologists have developed techniques to "flesh-out" a skull to recreate the appearance of a person from the structure of a skull. (Reproduced from *Science Digest*.[1])

Sykes describes a particularly successful reconstruction of a face built upon a skull. Thus, he tells us that In December 1989 the skeletal remains of a young female murder victim were discovered by workmen in the garden of a house in Cardiff, Wales. She had been dead for several years. With little more to go on than her hair and approximate age from dental records, a remarkable reconstruction lead to the identification of the girl. The facial reconstruction in this case was made by Richard Neave, a medical artist at Manchester University, England. Neave made a plaster cast of the skull and then built up the facial muscles using modelling clay. He completed his model within a day. Two days after pictures of the reconstructed head were released to the press, a Cardiff Social worker telephoned the police saying that she believed the victim to be Karen Price, who had run away from a children's home in July 1981. Dental records strongly supported the identification. In this particular case DNA studies, to be described in detail in the next chapter, confirmed without doubt the identity of the victim.

11.3 "Your Brother's Blood is Crying out to Me from the Ground"

The title of this section is a a quote from the book of Genesis, Chapter 4, Verse 10. In this chapter the crime of Cain, who killed his brother Abel, is told. When Cain tries to avoid being questioned about his brother's death, God is said to have replied with the sentence we have taken as the title of this section.

Often, the blood found at the scene of a crime, both by its constitution and its spread at the scene of the crime, cries out with information. Thus, in Toronto in 1994, a car went off the road and the male driver was injured. It was assumed that the passenger, a woman, was killed by the accident. However, further investigation showed that the blood of the woman was spread over parts of the car in a pattern that could not have been created by the road accident, and it was ultimately determined that the man had staged the road accident to cover up the fact that he had killed his wife by beating and strangling her.

In another bible story (Genesis, Chapter 37) Joseph was sold into slavery by his brothers and they took his coat of many colors and dipped it in animal blood.

> *the brothers killed a goat and dipped Joseph's robe in its blood. They took the robe to the father and said "we found this, does it belong to your son? He recognized it and said "yes, it is his, some wild animal has killed him."*

Today forensic specialists would be able to carry out a test to determine if the blood on the garment was animal or human. Most of us think of blood as being a rather simple liquid like tomato ketchup. In fact, biochemically, it is a very complex mixture of fluid and solids.[13] Two types of specialists involved in the study of blood are **hematologists**, (from the Greek word for blood), and serologists. An advertisement for a serologist reads:

The role of the **serologist** *involves identification and grouping of human blood and blood stains, investigation of disputed paternity cases and sexual offenses, visiting scenes of crime, attending court staff, etc. An honors degree in biology or a biological subject combined with at least five years experience in biological and serological investigations in a forensic science laboratory is required.*

The serologist studies serum, which is defined in a dictionary as watery portions of animal fluid remaining after coagulation of the blood. Daniel D. Garner, a forensic serologist, states that forensic serology involves the examination of physical evidence of a biological nature, and includes the specialities of immunohemotology, biochemistry, enzymology and human genetics. Obviously, in this book we cannot go into all of the details of the art of the forensic serologist. We can, however, touch on the highlights of such investigations and direct the reader to sources of information about the study of blood. As emphasized by Broad, first, one must determine if a red liquid left at the scene of the crime is actually blood and not tomato ketchup.[14] A suitable test is described by Broad in his book. Once it is known that a stain is indeed blood, then one must determine if it is animal or human blood. To do this one can carry out a test of the type illustrated in Figure 11.5. The first step in the test is to develop some antigens to human blood (see study of immunoassays in Chapter 8). To prepare this material human blood is injected into an animal such as a sheep or a goat. The animal responds by producing antibodies against the foreign antigens. These antibodies are then isolated from the animal. As described by Broad in the subsequent test:

A drop of liquid containing the nonhuman antibodies is placed in a small well made in the center of a thin layer of agar jelly (see Figure 11.5). Around this central well is a circle of six or eight wells of similar size. Into each outer well, except one, a specimen of blood of known origin (human, dog, cat, etc.) is placed. These are the controls. The blood of unknown origin is placed in the last well. Antibody and antigen molecules diffuse through the gel in all directions from each well. In between the wells they meet. If antibodies from the central well meet antigens from human blood, then complexes form. These complexes can be made to show up as light bands on a dark background. Antigens from nonhuman blood will not form complexes with the antibodies made in the animal blood.[14]

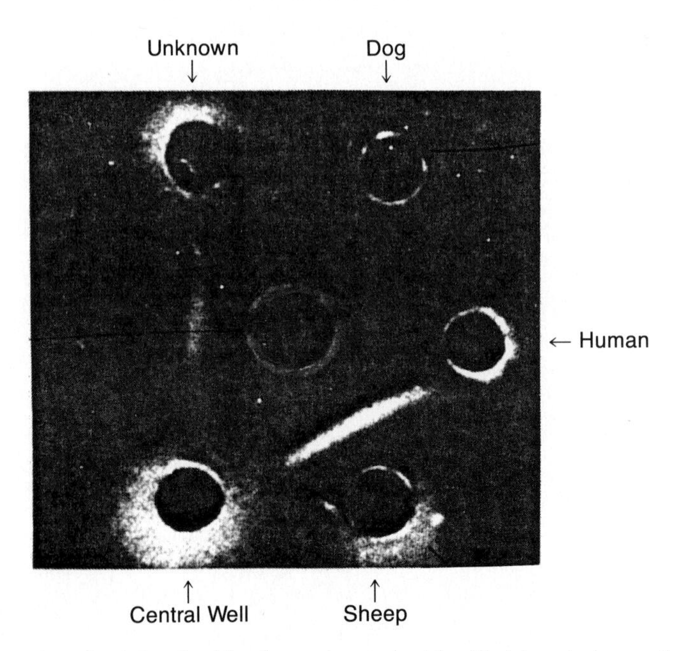

Figure 11.5. The species of origin of a blood sample can be identified by placing antibodies to human blood in the central well and control samples, along with the unknown sample, in the surrounding wells. A light band is formed between the central well and the human blood. A fainter band appears between the central well and the unknown, indicating that the unknown contains human blood.[14]

Broad points out that the process of diffusion in the jelly is quite slow so that often the diffusion of the various components of the test are accelerated by placing the agar jelly (also known as gel) in an electric field. This is similar to electrophoresis, discussed earlier. Electrophoresis techniques are also used to study the constitution of blood.[15]

If a forensic test of the type shown in Figure 11.5 had been available to Isaac, it would have proved that the coat had been dipped in goat's blood.

The way in which the configuration of blood patterns and splashes of blood can be used to solve crimes has been discussed by Sergeant M. Wolf, who is one of the experts on blood droplet patterns at the Royal Canadian Mounted Police. In one of his cases he was asked to investigate the death of a man who, it was suspected, had been killed by his younger brother. Wolf studied the blood stains at the death scene and was able to determine where the stabbing had taken place and that the brother had done the stabbing. He also concludes that the killing was probably an accident or an act of self defence. He told the local police that he believed bloody footprints indicated the younger brother had been fighting for his life when the crime was committed and that he had been trying to get away. Confronted with Wolf's interpretation the boy confessed that he had stabbed his brother after being attacked in bed. He was later charged and convicted of manslaughter. Wolf tells us that:

> *Blood stains are like still photographs. The pictures tell you what type of action took place at a crime scene.*

Analysts use fine instruments to study the size and shape of blood droplets. Size will reveal the type of attack, which caused the bleeding. Small mist-type stains reveal a high energy impact, usually a gun shot. From the shape of the drops, investigators can tell from which angle they hit an object, and thus, where the victim and the attacker were standing at the time of the assault. Patterns at different angles indicate a number of different blows.[16,17]

In Canada universities are beginning to offer courses for the police in forensic anthropology and blood stain analysis. Figure 11.6 shows Scott Fairgreive, Laurentian's forensic anthropologist, lecturing to police on the type of information to be deduced from various bones.[18]

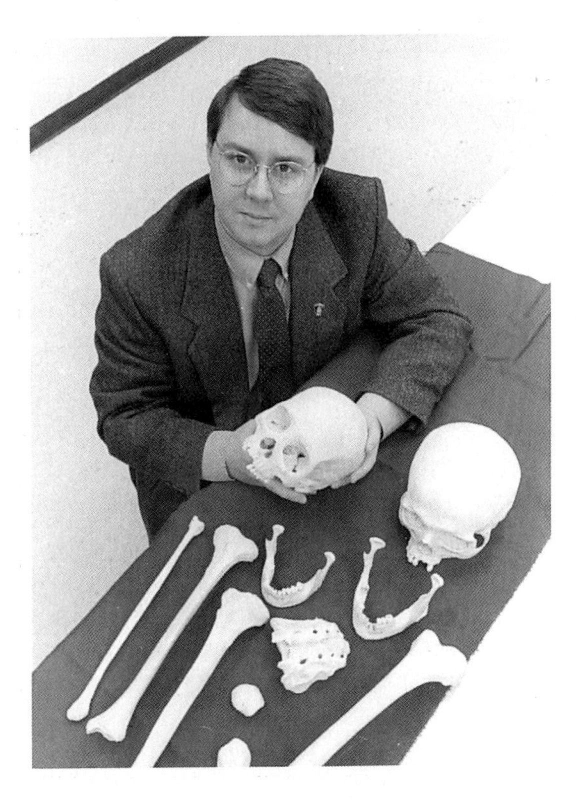

Figure 11.6. Dr. Scott Fairgrieve is a forensic anthropologist at Laurentian University. He is shown here giving a presentation on the use of forensic anthropology in criminal cases.

References

1. P. Tourancheau, L. Chauveau, "Corpses Can Have Talkative Teeth," *Readers Digest*, Canadian Edition, March 1994, 49.
2. D. Ubelaker, H. Scammell, *Bones, A Forensic Detective's Casebook*, HarperCollins Publishers, New York, 1992.
3. B. Knight, "Murder in the Laboratory," *New Scientist*, 25 December 1986/1 January 1987, 59.
4. B. Knight, "The Stiff Test That Crime Writers Fail," *New Scientist*, 29 May 1986, 38.
5. Z. Erzinclioglu, "Few Flies on Forensic Entomologists," *New Scientist*, 30 May 1985, 15.
6. I. Anderson, "Hominid Collarbone Exposed as Dolphin's Rib," *New Scientist*, 28 April 1983, 199.
7. L. Cherry, "Their Blood Cried Out for Vengence," *Science Digest*, May 1981, 60, 125.
8. T. Patel, "Teething Troubles Over Hitler's Body," *New Scientist*, 27 March 1993, 10.
9. C. Joyce, "How They Identified The Angel of Death," *New Scientist*, 4 July 1985, 19.
10. J. Clausiusz, "The Banal Death of a Genius," *Discover*, March 1994, 25.
11. R. Lewin, "Did Queen Bess have A Head For Shakespeare?" *New Scientist*, 16 November 1991, 15.
12. B. Sykes, "The Past Comes Alive," *Nature 352* (1991), 381.
13. L. Vronman, *Blood*, American Museum of Natural History, Garden City, NY, 1967.
14. J. Broad, *Science and Criminal Detection,* Macmillan, London, 1988.
15. B. W. Greenbaum, "Electrophoresis in Forensic Applications," *Industrial Research*, 15 November 1977, 13.
16. D. Grady, "Blood Will Tell," *Discover*, November 1980, 38.
17. "Written in Blood," (news story) *Sudbury Star*, 22 January 1987, 20.
18. For example, blood stain patterns were studied in a short course offered by Carleton University, Ottawa, Canada in 1982.

Chapter 12

Alphabet Soup and Genetic Fingerprinting

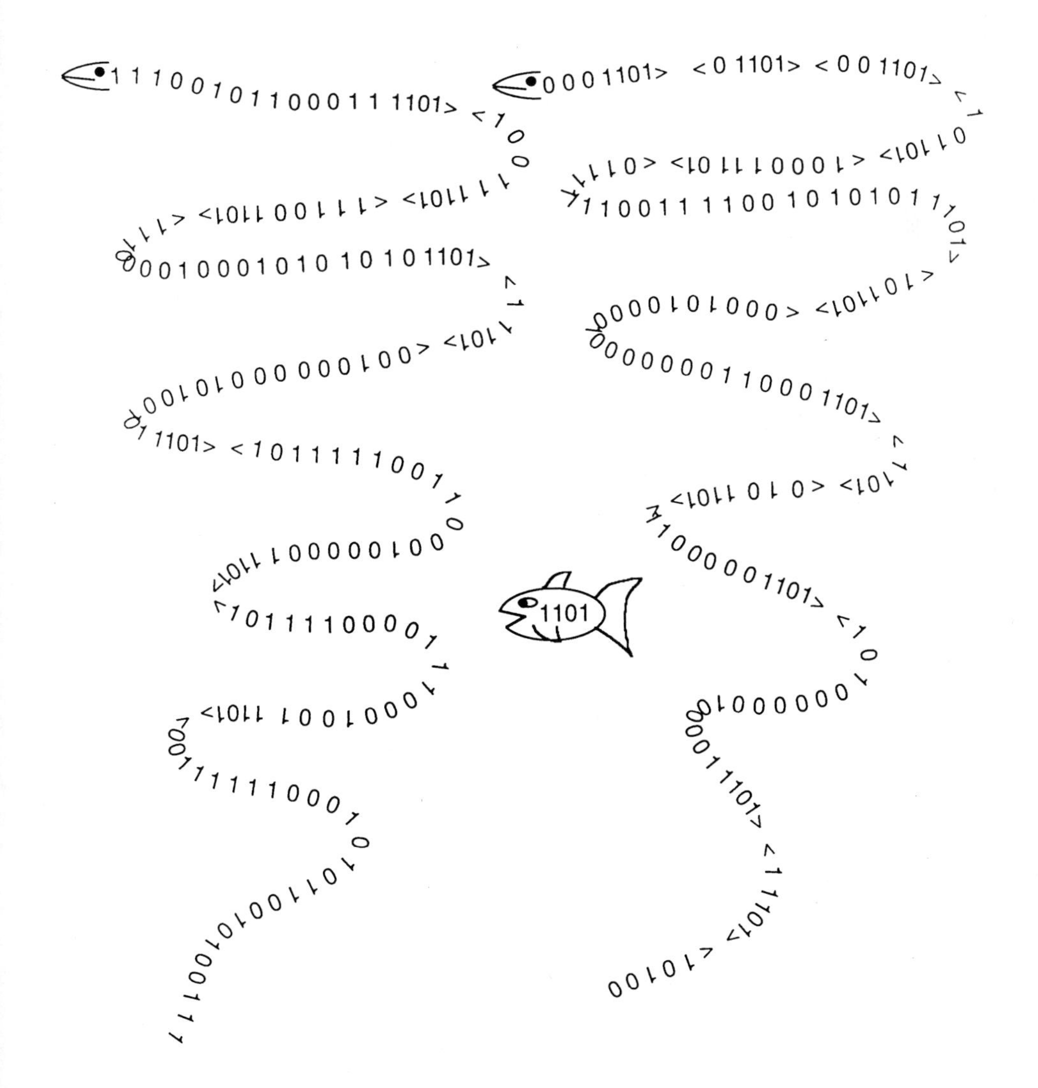

Chapter 12

Alphabet Soup and Genetic Fingerprinting

12.1 Coded Information for Building the Body

Probably the most dramatic development in forensic science in the 1980s was the development of a technique, which is known by the popular term genetic fingerprinting. **Genetic fingerprinting** involves studying the information for building the body, which is coded into cells (the basic building block of a body). Information is contained in the chromosomes in the nucleus of living cells and in other organelles, such as the mitochondria, which have their own DNA.

The location of the chromosomes and mitochondrion in a human cell is shown in Figure 12.1 (a). The appearance of a typical genetic fingerprint derived from a human cell, as used in legal evidence, is shown in Figure 12.1 (b). The human body is made up of trillions of cells and the names given to elements of the various structures assembled from these cells at different levels of complexity are shown in Figure 12.2. As can be seen from the last element of Figure 12.2, an important constituent of the human cells are chromosomes, which contain DNA. This is the substance that forms the basis of the genetic fingerprint. If one reads about the theory underlying genetic fingerprints, one soon finds oneself juggling clumps of letters such as DNA and PCR, and chemicals referred to by individual letters such as A, G, C, and T. In the process of exploring what is meant by genetic fingerprinting, one may start to think that one is swimming in alphabet soup, hence the title of this chapter.[1–3] We attempt to explain the meaning of the various clumps of letters to be found in the scientific literature on genetic fingerprinting. We also try to explain the chain of events that enables a scientist to generate a genetic fingerprint. In the last section we take a look at the controversy over the statistical theory used when interpreting the significance of the structure of genetic fingerprints presented in evidence. We will review the various areas of forensic science where genetic fingerprinting is revolutionizing methods of investigation and court procedures.

To begin our study of the biology behind genetic fingerprinting, we start with the basic structure of the cell and in particular with what is known as the nucleus of the cell (see Figure 12.1). The word *nucleus* in Latin simply meant a small nut. The idea is that the nucleus is at the center of the cell, like a nut inside its shell. Asimov tells us:

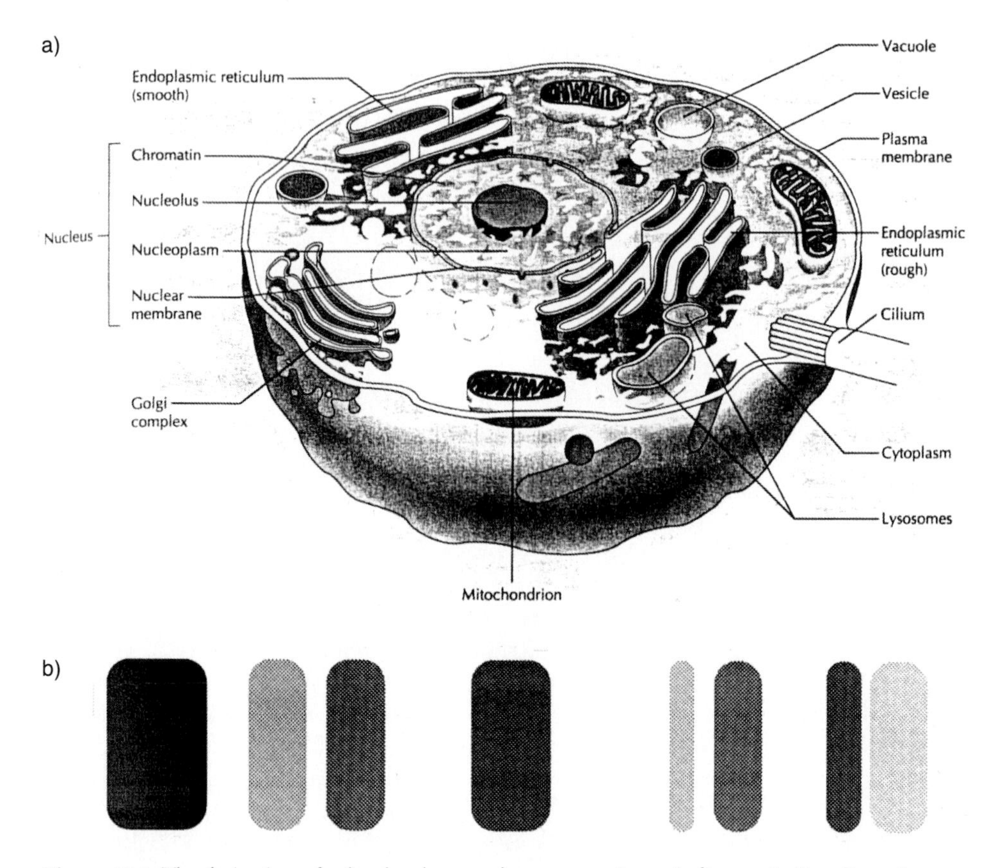

Figure 12.1. The derivation of what has become known as a "genetic fingerprint" or "genetic profile" from the genetic information in a human cell is a complicated process. a) Schematic representation of a typical human cell.[1] (Reproduced from M. J. Pelezer, E. C. S. Chan and N. R. Krieg, *Microbiology; Concepts and Applications*, © 1993, McGraw-Hill.) b) Stylized representation of the physical appearance of a genetic fingerprint derived from information built into the structure of the human cell.[2]

> *One of the first ways in which biologists tried to find out something about the inside of the human cell was to subject it to the action of various dyes. The different substances in the cell reacted differently and some objects showed up colored against a colorless background.*[4]

The Greek words for *colored* and *body* are "chroma" and "soma." Therefore, biologists call the colored bodies in the center of the cell **chromosomes**. It soon became apparent that chromosomes control the way in which a given cell grows into the complex system we call the body. Today, we say the chromosomes contain the information required to build a body. If we very carefully remove the nucleus of one cell and put a nucleus from a

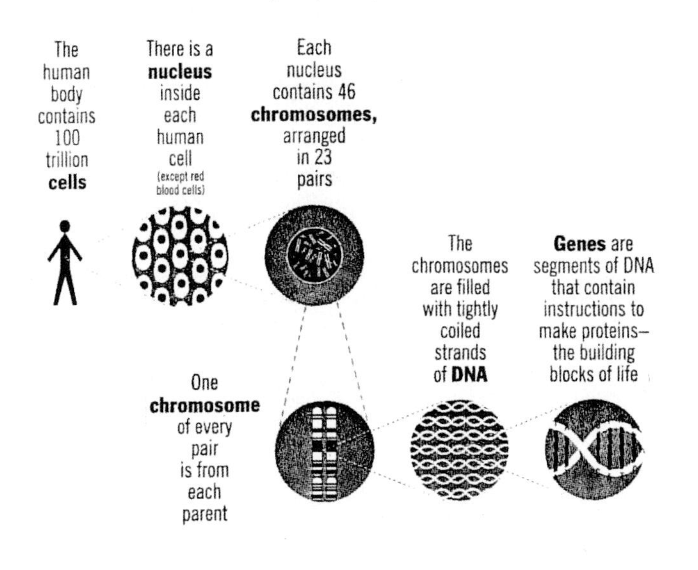

Figure 12.2. The details of the construction of the human body are amazingly complex.

different body into the cell, it will start to grow into the shape of the body of the donor cell, not into the shape of the original cell. This is the basis of what is known as **cloning**. The word *clone* in Greek means "twig" and refers to the growth of a body from the information coded in foreign chromosomes. The idea comes from the experience of a gardener. If the gardener takes a twig and grafts it onto the root stalk of another plant, then the tree that will grow on the root stalk is the type of tree represented by the twig, and not that of the root system.

As scientists probed further and further into the structure of the chromosomes, they discovered that they were made up of thread-like components, which became known as **genes**. Finally, it was found that the ultimate structure of a gene was the chemical known as **deoxyribonucleic acid**. This mouthful was shortened by working scientists to **DNA**. We now know that the DNA chains are made up from four different chemicals. One is called **adenine**, the second is **thymine**, the third is **cytosine** and the fourth **guanine**. The sequence of different combinations of these chemicals, represented by the letters A, T, C, and G, forms an information code in the thread of DNA used to direct the growth and type of a body. The study of DNA structures and similar aspects of biology are described as **molecular biology**, the study of living creatures at the molecular level. In an article on genetics, P. Dewitt states:

> *The thread of life (DNA) is deoxyribonucleic acid, the spiral staircase shaped molecule found in the nucleus of cells. Scientists have known since 1952 that DNA is the basic stuff of heredity. They know that human DNA looks like a biological computer program, some three billion items long that spells out the instructions for making proteins, the basic building blocks of life.*[3]

Further on in the article he explains:

> *the human set of genes is [like] an encyclopedia. It is divided into 23 chapters (chromosome pairs) each gene sentence is composed of three letter words which are in turn spelled by four molecular letters called nucleotides – adenine, cytosine, guanine and thymine. By scanning a database containing the complete sequence of letters, researchers can quickly end up at a particular gene's front door.*[3]

To explain the basic concept of how information is stored in the pattern of nucleotides and the basic procedure for making genetic fingerprint from stands of DNA, we will first look at a one-dimensional system for transmitting information. Radio messages are sent as a sequence of bleeps with silent gaps in between. Thus, Figure 12.3 (a) shows a sequence of bleeps, which form a message that – some suggested – should be broadcast into outer space in an attempt to communicate with extra-terrestrial beings (ET's) (The message is represented by the digit one separated by gaps, represented by zero). The inventor of this message, Drake, assures us that an extra-terrestrial intelligent mathematician would be able to recognize that, since the number of signals in the stream is a 1,271, the digits would have to arranged in a matrix of 31×41 as illustrated in Figure 12.3 (b). This message is packed with information, such as that humans reproduce sexually and have small children, that they live on a planet that has water, and that fish swim under the water. (If you would like to read a detailed discussion of the interpretation of the message, see reference 5). The important point for our discussion here is to see that a binary sequence contains enough information to create structures and patterns.

In the DNA structure of the human chromosome the information is more complicated than our string of 1's and 0's, since it uses four symbols and contains billions of instructions. (Some theologians suggest that all through life our experiences and memories are being transmitted in code into outer space to take up residence in some distant planet, where our existence may well continue after death; indeed, enough information could be transmitted this way to recreate our bodies somewhere on a planet in outer space.)

The human body is exposed to radiation from outer space and from radioactive substances on the earth. Energetic radiation, such as cosmic rays, ripping into the chromosome, can disrupt the DNA sequence. The body appears to have some ability to repair these damaged sequences, but if the body is running out of repair capacity or if the bombardment is severe, such

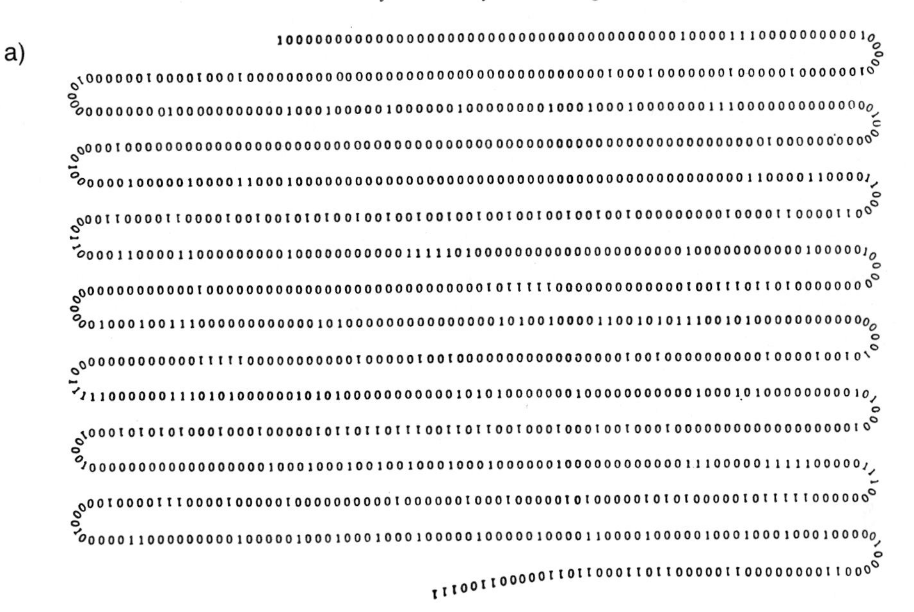

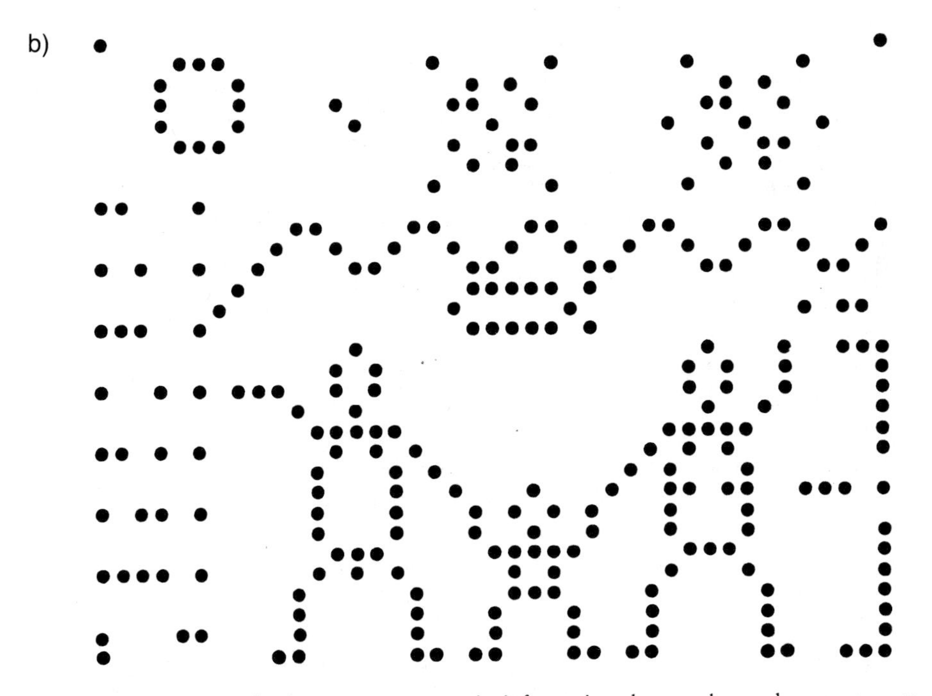

Figure 12.3. Binary codes in a sequence contain information that can be used to create patterns and structures. a) A binary sequence of bleeps, suggested as a possible broadcast into outer space. b) An intelligent extra-terrestrial being would presumably recognize the total number of bleeps and gaps, 1271, as the product of two numbers, 31 and 41, and could arrange the message into 31 lines of 41 dots and spaces. When this is achieved, the pattern above emerges.

as in a nuclear accident, the information in the cell goes haywire and produces abnormal growth in the body, which we call cancer. **Cancer** cells are actually immortal and proliferate in an uncontrolled manner because the building information contained in the cell has gone crazy. For example, it is thought that one of the reasons that asbestos fibers can create cancer is that the sharp ends of the fibers can penetrate into the nucleus of the cell, disrupt the genetic information, and initiate cancerous growth.

Imagine now that we are listening for messages from outer space and that we received the two messages shown in Figure 12.4 (a). How could we describe the difference between them? One relatively crude way of doing this is to segment the message as follows: Think of our long sequence of zeros and ones as a sea snake and imagine that a piranha (a vicious flesh-eating fish that lives in Brazil) swims alongside the snakes and that it is programmed to cut the snakes whenever it finds the pattern shown on the piranha After it had cut the sea snakes up into several portions, we would find that the segments of the message would look like the strips shown in Figure 12.4 (b). Although this seems to be a rather odd description of the two messages, it is obvious that it is a good way of looking at the differences between them. In the same way biologists have learned how to cut up the chains of DNA by using the equivalent of our piranha, molecules known as **restriction enzymes**. In the words of Lowrie and Wells:

Many bacteria make restriction enzymes to protect themselves from invading foreign DNA molecules such as viruses. The restriction enzymes attack the invading DNA by chopping it up. Each kind of enzyme recognizes a particular and different sequence of between four and six elements of the DNA strand and cuts them there. Many restriction enzymes have been purified from various species of bacteria and more than a hundred are commercially available.[2]

Although the analogy between snipping a binary message into segments and the snipping of DNA into stands is an oversimplification of the situation, the diagrams of Figures 12.3 and 12.4 help the reader to visualize the process of generating a genetic fingerprinting described in the next section.

12.2 Constructing the Genetic Fingerprint

In an excellent article, Lowrie and Wells set out the basic sequence of events in constructing genetic fingerprints.[2] A diagram based on their discussion is shown in Figure 12.5. Lowrie and Wells explain:

To make a genetic fingerprint you need a sample of any tissue that contains cells. Blood, saliva, hair roots or semen will do. Biologists extract the strings

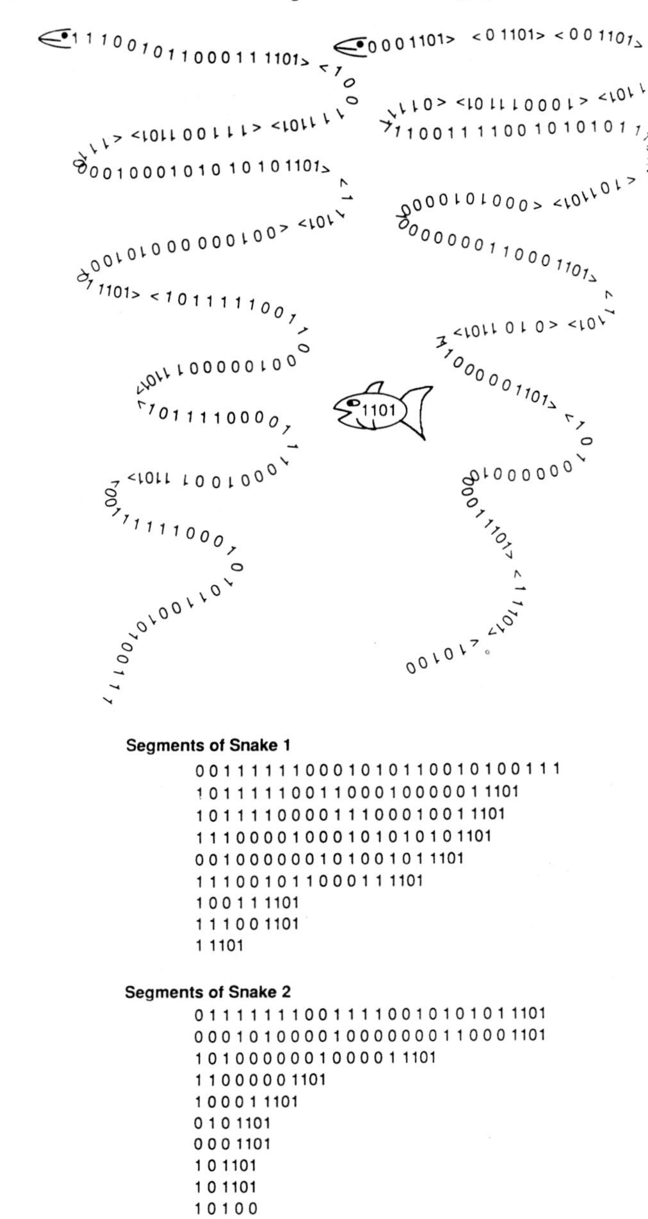

Figure 12.4. The idea of searching for a specific sequence to indicate where a chain should be broken is demonstrated here by a piranha which looks for the sequence 1101 – an appetizing site at which to bite a sea snake. a) Two messages represented as two sea snakes. When the piranha spots the sequence 1101 it bites the sea snake, thus subdividing it. b) When the segments created in (a) are ranked according to length, the difference between the two messages can be seen.

of DNA from the sample and cut them at specific points using so-called restriction enzymes. The fragments of DNA are placed on a gel and separated by running a current through the gel, a process known as electrophoresis. The pieces of DNA have a negative charge so that when a positively charged electrode is placed at the end of the gel, the charged fragments of DNA travel towards it through the gel. The shorter, lighter fragments travel more quickly through the gel while the longer, heavier fragments move more slowly. So this separates the fragments according to their size.[2]

The separation of DNA fragments is shown in Figure 12.5. The separated strands are invisible until they are developed using a combination of radioactive technology and x-ray film techniques. Biologists prepare special chemicals known as **DNA probes**, which can be made radioactive. When these probes are applied to the membrane containing the fractionated DNA, they bind to certain parts of the DNA. The excess unattached DNA probe chemicals are then washed away. The pattern of the DNA fragments is now made visible by placing an x-ray film over the membrane, the pattern appears as a visible set of bands on the film, as shown in the last part of Figure 12.5. It

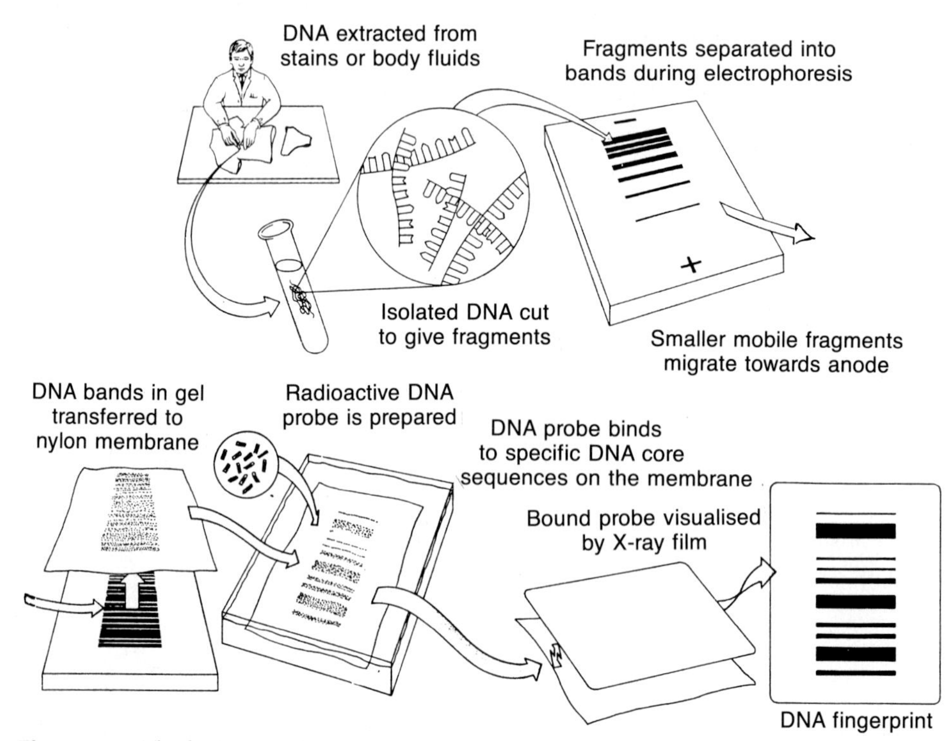

Figure 12.5. The basic procedure used to create a genetic fingerprint from segmented DNA as described by Lowrie and Wells.[2]

is not surprising that a DNA fingerprint produced like this has been called a "bar code for biologists." If we wished to compare genetic fingerprints from different sources, we must very carefully standardize all the stages shown in Figure 12.5. In other words, the electrophoretic gel plates be made in the same way and have the same physical properties.[2]

12.3 Using Genetic Fingerprints

Genetic fingerprinting was first used in forensic science in the late 1980s when Alex Jeffreys of the University of Leicester in England, who developed the technique, used it to identify a rapist and murderer. To check the identity of a rapist, the following steps are required. First a vaginal swab is taken from the victim; this contains cells from the victims body fluid and the rapist's semen. The subsequent fingerprint made from the swab will have the genetic information of the victim and that of the rapist in a combined pattern, as illustrated in Figure 12.6. Next, the victim's DNA fingerprint is made from a sample of fresh blood cells. Any bands from the original sample that are not present in the victim's sample can then be compared with the body fluids of any other suspects, as illustrated in Figure 12.6. In the simulated case shown in the figure, it is obvious that suspect 1 is the guilty person. Genetic fingerprints can be developed from very small amounts of material; it is estimated that currently a genetic fingerprint can be produced from as little as fifty microliters of blood, five microliters of semen or ten

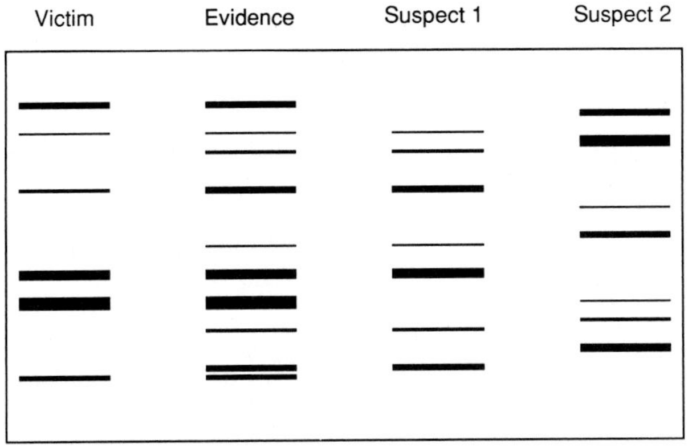

Figure 12.6. A sample, taken as evidence, which contains genetic material from both the victim and the culprit, can be compared with that of the victim and each suspect to determine who is the culprit. In this case, the evidence sample contains genetic material from the victim and suspect 1.

hair roots. Mouth swabs, fetal material and muscle tissue from dead bodies are also suitable sources of DNA.[2]

Of course, genetic fingerprinting can also be used to establish the innocence of the alleged rapist. Thus, in 1986 Lennie Callace was convicted of rape and sent to jail for 25 to 30 years. Subsequent examination of the body fluids kept in evidence cleared Callace and he was released from prison after six years.[6] However, in this case seems that it really should not have been necessary to wait for genetic fingerprinting, since the victim told the police that the rapist was five foot ten or taller, with reddish blonde Afro-style hair, a full beard, and a cross tattooed on his left hand. Callace was five foot eight, had straight blonde hair, a goatee and a tiny cross on his right hand.

Another major area where DNA typing has proved to be useful in establishing parentage. The human fetus acquires half of its DNA patterns from its mother and half from its father. Thus, DNA fingerprinting can be used to establish the identity of parents. DNA fingerprints from the mother, the child, and the two possible alleged fathers are taken. If the genetic fingerprint of the child is missing key elements from that of alleged father number two, it is highly probable that alleged father number one is the real parent of the child. Paternity matching by genetic fingerprinting has been used to establish the paternity of immigrant children claiming that they wished to come into a country to join their biological parents. Thus, in Britain the method was used to study 36 families wishing to enter Britain to join their relatives. The study showed that nearly 90 % of the cases were genuine. Previously, half of the applications had been rejected, due to lack of sufficient evidence to convince the immigration authorities that the people really were related.[2] In Argentina, genetic fingerprinting has been used to unite children with their biological parents or grandparents. Many of them were separated during a period of Argentinean history around 1977, when many people were kidnapped by the secret police.[2]

Paternity matching can also be used to establish the identity of a murder victim. In Chapter 11 we discussed the identification of Karen Price, a Welsh girl murdered in 1981, based on the reconstruction of her facial features from her skull by the forensic anthropologist. In the case, the final identification rested on an experiment involving a genetic fingerprint. Erika Hagelberg of the Radclift Hospital in Oxford, England, extracted a small amount of DNA from one of the bones of the body and sent it for analysis to Leicester University. As described by Lowrie and Wells:

> *the sample was heavily contaminated with microbiological DNA (stray DNA) from bugs and was decayed. But by using PCR the researchers amplified six areas of the DNA and by comparing these sequences with DNA from Karen Price's parents the evidence was strong that the body was indeed that of Karen Price.*

The letters **PCR** used in this quote stand for **polymerase chain reaction**. This is a technique, invented at the CETUS Corporation in the United States in 1985, for copying minute amounts of DNA many times over. This technique can be used to amplify the sample by making many, many copies of a tiny bit of DNA so that there is enough of the chemical to enable a genetic fingerprinting to be carried out.

A genetic fingerprint study, which has tremendous financial implications involves the rightful heir to the fortunes of the Russian Czar who was killed in 1918. During the revolution the communists claimed to have executed all of the family of the Russian Czar Nicholas II and his wife, Alexandra. They had five children, Olga, Maria, Tatyanna, Anastasia, and Alexei. In 1920 a young woman found wandering the streets of Berlin and claimed that she was Anastasia. She later became known as Anna Anderson and moved to Charlottesville, Virginia, where she died in 1968. In 1991 the Soviet government excavated a shallow pit in Yekaterinburg in Siberia. According to local rumors, this pit held the remains the Czar and his family. In 1992 Russian scientists tentatively identified the skeletons by superimposing computer images of their reconstructed skulls on old family photographs. They found nine skeletons in the grave and they concluded that they were the remains of Czar Nicholas, his wife Alexandra, their eldest children, (Olga, Marie and Tatyanna) the family doctor, and three servants. The bodies of the two youngest children Anastasia and Alexei were not found. Two scientists of the Forensic Science Service in England, Gill and Sullivan, extracted DNA from the bone cells of each skeleton by crushing the pieces of bone and partially dissolving the material. They were then able to use genetic fingerprinting to determine the sex of the individuals, and to establish that five of the bodies were from the same family.

So far in our discussions we have been talking about DNA from the cell nucleus. However, DNA is also to be found in the mitochondria, which is a small subunit of the cell referred to as an organelle. The **mitochondria** are involved in directing the production of energy by the cell. Mitocondrial DNA differs from that in the nucleus in that it is passed on only from mother to child, remaining unchanged for generations. Gill and Sullivan state that mitocondrial DNA only changes (mutates) once every 6,000 years. They compared mitocondrial DNA from the Russian bones at Yekaterinburg with material from a blood sample donated by Prince Philip, husband of England's Queen Elizabeth II. His maternal grandmother was Czarina Alexandra's sister. The material from the Czarina and the skeletons of the three daughters matched Prince Philip's mitochondrial DNA exactly. The mitocondrial DNA from the Czar's bones matched those of two descendants (who requested anonymity) of his maternal grandmother Louise, wife of Denmark's King Christian the IX. Gill stated "we are 98.5 % certain that this is Romanov DNA." (Romanov was the family name of the Czar of Russia.)

With regard to the missing skeletons of Alexei and Anastasia, the scientists state that since Alexei was a hemophiliac, he was unlikely to survive the shooting and bayonetting that did away with the rest of the family. (A hemophiliac is a person who's blood has difficulty clotting; the word hemophiliac means literally, "one who loves to bleed.") The British team are now trying to examine samples from the hair of two people who claimed to be Anastasia. If Anna Anderson of Charlottesville turns out to be the Anastasia of the royal family, the inheritance of large sums of money will finally be resolved.[7,8]

DNA fingerprinting has also been used to establish the identity of victims of an airline crash and soldiers buried in mass graves. Thus, in January 1992, an Airbus crash in France killed 87 people. 68 of the corpses were identified by matching dental records and blood tests. The remaining 19 were so badly burned that it was necessary to turn to DNA profiling. Comparing the results with those of the nearest relatives enabled the scientists to identify 16 more of the victims of the air crash. Commenting on this particular investigation, Patrice Mangelin, Director of the Institute of Forensic Medicine at the Louis Pasteur University in Strasburg, stated:

> *I don't think DNA fingerprinting will take over from conventional methods for the time being. The cost of fingerprinting after the Strasburg crash was $ 300,000 Canadian.*[9]

He also estimated that it would cost $ 100 to take and stock a person's DNA fingerprints for life. Mangelin pointed out that for this particular Airbus crash, it was very difficult to get information from conventional identification of the crew. If they had already had their DNA fingerprints on record, the crew could have been identified quickly. Mangelin stated that bodies could be tested to see whether a pilot had taken alcohol or medication before a flight and, therefore, it could be absolutely essential to know which of the badly burned bodies belonged to the pilot.[9]

At the time of writing this chapter, there was a growing movement to place DNA fingerprints of people convicted of serious offenses on permanent police files. As early as 1991, the state of Virginia had collected blood samples from 33,000 people and was adding 2,000 DNA fingerprints to their records every month. This aspect of genetic fingerprinting is currently a source of great controversy in legal circles.[10]

Another area of biological studies of interest to the legal profession is the determination of the genetic structure of **HIV** viruses (HIV is a virus which is the precursor to **AIDS** in humans). By 1992, there were already 200 lawsuits involving the transmission of HIV in the U. S. More than half of the states have specifically outlawed the sexual spread of HIV, thus, in a lawsuit it is not necessary to prove that the person intended to harm their partner. The problem of determining who infected whom with the HIV virus arises from the fact that HIV is itself constantly changing. In 1990 David Acer, a

Dentist in Florida, had allegedly infected several of his patients with HIV. Gerald Myers, Director of the HIV sequencing Data Base at Los Alamos National Laboratory in the United States, demonstrated beyond a reasonable doubt that the virus came from Acer.

In the Acer case, Myers and his colleagues recorded the genetic differences between the samples of virus taken from each person, who claimed to have been infected, and compared them with the variation in samples of virus in the local population. Scientists in the United States and Britain also showed, using viral genetic studies, that a HIV-positive surgeon in Baltimore did not infect a woman he treated. The woman said there were no other factors that put her at risk of AIDS, but it then emerged that she had received a blood transfusion at the time of the operation. In the case of the Florida dentist, sequences of DNA from five patients were virtually identical to those from the dentists strain. Scientists in the U. S. and at Edinburgh University in Scotland showed that the strains of HIV in donated blood were much more closely related to the HIV in the woman's blood than in the strains found in the infected surgeon.

Commenting on the techniques of collecting viral genetic data for forensic use, Edward Holmes, a scientist at the University of Edinburgh said:

> *The procedure is much more complex than other techniques, such as DNA fingerprinting. Scientists have to purify and clone samples of HIV from the individuals involved and then compare their sequences of viral DNA.*

One of the problems with HIV is that it changes rapidly, thus in an article on this type of test it is stated:

> *Someone who had been infected for more than about four years will be hard pressed to point blame at any single individual because the structure of their HIV strains would have diverged by mutation so much in that time.*

Because of the many pending suits on the transmission of the HIV-causing virus, this type of test is likely to receive a lot more publicity in coming years.[11]

DNA fingerprinting is not only of use in fighting crimes involving humans, but can also can be involved in detecting fraud, for instance in the breeding of race horses. In Australia trotters and pacers have to undergo DNA fingerprinting to verify their pedigree. Horses put out to stud will have their DNA fingerprinted, as will their offspring. The scheme is intended to protect buyers, because one horse can easily be confused with another, either deliberately or accidently. Foals can be swapped, mares may mate more than once, and it is important to know who actually sired the foal. Genetic fingerprinting of a horse, it is hoped, will also deter thieves, because it will be simple to check the horse's identity. It will also be useful to prevent swapping of horses before a race begins. The Australian Harness Racing Council expects to record genetic fingerprints from as many as 500 horses a month. The use of genetic fingerprinting in breeding lines is expected to be extended to Greyhounds, cats, cattle, and sheep.[12]

The genetic fingerprint of a plant was a major piece of evidence in a murder trial that took place in Arizona in 1993. Two seed pods from a Palo Verde tree were found in a pick-up truck belonging to Mark Bogen, who was accused of murdering a woman and then dumping her body at an abandoned factory. Bogen said that he spent time with the woman, but denied visiting the murder site. Proving that he had been to the factory became a key in the prosecution case. Quoting from the review of the murder by R. Mestel:

> *Investigators noticed that the site around the factory was dotted with blue Palo Verde trees, relatives of the pea. The trunk of one of the trees dipped low over the overgrown driveway and it had recently been scraped. A search of Bogen's truck turned up two Palo Verde pods. If Bogen's truck had knocked them off the tree as he drove into the factory the pods should be genetically identical to the ones on the tree.*[13]

Tim Helentjris, a plant geneticist at the University of Arizona, was asked to investigate whether or not the pods in the truck came from the tree in the driveway of the factory. As a first step in the investigation of the genetic structure, police provided Helentjris with 12 coded samples. Each one was ground up and DNA was extracted from them. Helentjris then used a technique to pinpoint key segments of DNA that differed from tree to tree. Every tree he found had a unique DNA fingerprint that could easily be distinguished from the others. Next, Helentjris was given a blind test of 18 tree samples from Bogen's old haunts and other sites around the city of Phoenix, chosen at random. None matched the two pods found in the truck, but a sample from the scraped tree matched them perfectly. The police also included another sample from the same tree at the murder site without telling Helentjris. Again he found that this matched the pods. When presenting his evidence Helentjris said:

> *In my professional opinion it was highly likely that the seed pods came from that tree and that it was highly unlikely that they came from any other tree by chance.* He also said that: *the judge wouldn't allow us to put numbers to that.*

The discussion as to how unique the evidence was is to be found in the review by Mestel.[13] On the use of genetic fingerprinting in general, Lowrie and Wells comment:

> *Although DNA techniques are now standard practice in molecular biology laboratories, great care must be taken in carrying out genetic fingerprinting tests. Forensic samples of DNA are rarely pure; typically blood stains on clothing or furniture fabric will be contaminated with all sorts of other material. In rape cases the vaginal or anal swab used to obtain a sample of*

the attackers semen also contains cell from the victim, so that the victim's DNA must also be analyzed alongside the sample from the alleged attacker.

DNA from fungi and bacteria are also invariably present and show up in the fingerprint. DNA decays rapidly, especially in warm or damp conditions. If the DNA has partially decayed, some of the sites attacked by a restriction enzymes may be lost and thus result in too few or too many DNA fragments. Any contaminants such as dyes from blue denim jeans can combine with the restriction enzymes, causing them to cut in the wrong places, again resulting in too few or too many pieces of DNA. When contaminated fragments of DNA are sorted in the electrophoresis gel, they may be mixed with ions that affect their charge, so that they travel different distances through the gel. Consequently, the bands are shifted to an unpredictable extent. Proteins from the environment can have the same effect, because they may become attached to the DNA fragments and weigh them down. If the DNA is contaminated, even DNA samples from the same person can give fingerprints where the ends of the barcode-type display do not align. However, their relative positions remain the same – this is known as **band shift**. In the probing stage at the end of the process (when the radioactivity material is added), the probes may combine with the contaminating DNA rather than with the sample, causing spurious bands, or the contaminant may interfere with the probe so that it does not bind as expected, leading to missing bands. In 1991 Lowrey and Wells stated that:

Despite these problems, genetic fingerprinting has been widely used in the British courts to secure convictions but responses to the technique in the U.S. courts have been more guarded. In some cases courts have rejected DNA evidence; the difficulty is to ensure that rigorous standards are always upheld. Blind trials have been carried out in the U.S.A. with different commercial laboratories running DNA tests on the same samples. In these trials some laboratories failed to identify mixed DNA as coming from two individuals or found false positives (declaring a match between samples when in fact they came from different individuals). European forensic laboratories are now trying to agree on standard techniques for producing genetic fingerprints in an attempt to overcome some of the criticisms raised against the technique in the United States. Standardizing the technique means agreeing on which restriction enzymes and probes to use as well as running a known set of controls for comparison and checking the probes are uncontaminated. Most experts agree that the technique still shows as much promise as ever, provided that proper precautions are taken to obtain uncontaminated samples and to avoid sloppy laboratory practice.[2]

Problems associated with the decay of DNA and contamination, for example, from the sweat of the laboratory technician, are examined in detail in an article by Roger Lewin discussing whether or not the basic idea of "Jurassic

Park" (the movie in which DNA from blood in a fossilized fly was used to clone dinosaurs) is possible.[14]

Probably the major confusion when juries are faced with genetic fingerprinting evidence is interpreting the statistics that indicate the probability that the genetic fingerprint is unique. This aspect of the problem will be discussed in detail in the next section.

12.4 Lies, Damn Lies and Statistics!

The title of this section is taken from a quote usually attributed to Desraile, a British Prime Minister in the late 19th century. It summarizes the attitude of many people when presented with statistics. Obviously, the public does not have a good grasp of probabilities or else they would never gamble. In forensic science, a big debate is in progress over how to interpret the probability that a genetic fingerprint is unique. The overall problem has recently been discussed in a review article by D. Pringle.[15] In this article Pringle reviews an appeal case in Great Britain, involving one Andrew Dean, who was convicted of rape in 1990. At the appeal, Peter Donnelly, a Professor of Statistics at a London University, pointed out that the forensic expert and the jury must answer slightly different questions. Thus, he points out that the forensic evidence answers the question:

What is the probability that the defendants DNA profile matches that of the crime sample, assuming that the defendant is innocent?

The jury on the other hand must answer a slightly different question:

What is the probability that the defendant is innocent, assuming that the DNA profiles of the defendant and the crime sample are the same?

At first sight there seems to be no significant difference between these two questions. But Donnelly explains the difference by considering the problem of playing poker with the Archbishop of Canterbury (a high church official in Great Britain). If the Archbishop were to deal a royal flush on the first hand one might suspect him of cheating, since the probability of dealing a royal flush with an honest card player is about l in 70,000. But if the judges were asked whether the Archbishop was honest, given that he had just dealt a royal flush, they would be likely to quote a probability greater than 1 in 70,000. Thus Donnelly points out that a very small answer to the first question above does not necessarily imply a very small answer to the second. In the card playing example the answer to the second question requires an assessment of prior belief in the honesty of the Archbishop. To a mathematician the odds of innocence are the ratio or the probability of innocence to the probability of guilt.

A further example of the way in which probable evidence must be interpreted is given by Pringle in his consideration of the hypothetical crime committed in Oxford, England, by an unidentified white man. The number of possible criminals could, at the upper limit, be Oxford's entire white male adult population of about 30,000. This implies odds of 30,000 to 1 in favor of the defendants innocence. If the probability of a random DNA match with the suspect were one in a million, then the odds of the man's innocence would be 33 to one (that is 30,000 multiplied by one in a million). It is the ratio of 33 to 1 that a jury should consider, not the ratio of one in a million. Donnelly stated that:

One of the biggest concerns with the use of probability in connection with DNA fingerprinting is that there can be misunderstandings by juries to the enormous disadvantage of defendants. Using the raw figure of one in a million instead of the proper 33 to 1 is known as the prosecutors fallacy.

(A *fallacy* is defined in the dictionary as an apparently genuine but really illogical argument. The word comes from a Latin word meaning "to deceive.") In the particular case of Andrew Dean's appeal, his conviction was quashed and a retrial ordered. (Quashed is a legal term meaning to annul, that is, cancel as if it had never happened, a decision that does not imply any sense of right or wrong.) Lord Chief Justice Taylor said that the decision was not to indicate that DNA profiling was unsafe, but that the trial had not proceeded according to good jurisprudence. (Jurisprudence is a word meaning the theory of law.)

It appears that some of the arguments between statisticians are misleading the public. Thus Balding, a statistician who has criticized the calculations used in presenting evidence with respect to the significance of DNA fingerprints, maintains:

In early 1993 the doubts over reliability of DNA evidence had been exaggerated.

Jeffreys, who developed the technique originally, told a conference on DNA fingerprinting that the court failures in Britain were a result of Britain's legal system rather than the technique itself. Balding said that if police scientists altered their statistical methods to take his criticisms into account (his argument involves the fact that people of the same race, even complete strangers, are distantly related) the odds of someone else having the same fingerprints as the accused would be reduced from one in ten million to one in one million. Lawyers believe the effect of such a change on the course of a case would be negligible, although Balding cautions that the effect may be greater in some cases than in others.

In a typical case in which DNA fingerprinting was challenged, a German judge ruled that criminal convictions cannot be made on the evidence of DNA fingerprints alone. Thus, in 1992, a 1990 conviction of a man for rape, in the state court of Hannover, was overturned. According to the prosecution at the original trial, the DNA fingerprint showed that there was a 99.98 % probability that the man committed the crime. However, if one takes into account the fact that the man was of Serbian descent, then the comparison with the DNA fingerprints of a random sample of 200 males from the Hannover area was not a proper match and the person's genetic fingerprint should have been compared with a comparable random sample from people of Serbian descent.[16]

The legal battle over the probabilities of DNA matching are likely to continue for some time with some very sophisticated mathematical arguments being bandied back and forth. We can, however, expect that as the dust settles, DNA fingerprinting, or genetic fingerprinting as it is popularly known, will turn out to be a very powerful weapon in the armory of the forensic scientist.

References

1. M. J. Pelezer, E. C. S. Chan, N. R. Krieg, *Microbiology: Concepts and Applications.*, McGraw Hill Inc., New York, 1993.
2. P. Lowrie, S. Wells, "Genetic Fingerprints," *New Scientist, Science Supplement,* 16 November 1991, 1.
3. P. Elmer-Dewitt, "The Genetic Revolution," *Time*, 17 January 1994, 40.
4. For a helpful discussion of the words used by biologists when describing the human body see I. Asimov, *Words of Science*, The New American Library, New York, 1969.
5. B. M. Oliver, "Radio Search for Distant Races," in R. Colbern (Ed.) *Modern Science and Technology*, Van Nostrand, Princeton, 1965, 563.
6. "Injustices Come Clean with DNA" (news story), *Sudbury Star*, 4 January 1993, B9.
7. J. Glausius, "Royal D Loops," *Discover*, January 1994, 90.
8. For a discussion of the fate of the Czar of Russian Nicholas II and his family and the supposed Anastasia see Gardener Assoc. (Ed.), "The Fate of a Family," *Great Mysteries of the Past*, Readers' Digest, Montreal, 1991, p. 90.
9. R. Keeler, "Uses for PER are Multiplying in Gene-Related Research," *R & D Magazine*, August 1991, 30.
10. D. Charles, "Convicts DNA Prints Added to U. S. Police Files," *New Scientist*, 21 September 1991, 19.
11. P. Brown, "Lawyers look to Genetics to Prove HIV Guilt," *New Scientist*, 11 July 1992, 5.
12. I. Anderson, "Genetic 'Hoofprints' for Harness Horses," *New Scientist*, 1 August 1992, 5.
13. R. Mestel, "Murder Trial Features Tree's Genetic Fingerprint," *New Scientist*, 29 May 1993, 6.
14. R. Lewin, "Fact, Fiction and Fossil DNA," *New Scientist*, 29 January 1994, 38.
15. D. Pringle, "Who's the DNA Fingerprinting Pointing at?," *New Scientist*, 29 January 1994, 51.
16. "Germans Wary of DNA Fingerprinting," *New Scientist*, 19 September 1992, 5.

Author Index

A
Asimov, I. 372
Ayto, J. 68

B
Barnsley, M. F. 158
Baynis-Cope, A. D. 101
Block, E. E. 43
Boese, R. A. 212
Bolt, R. H. 157
Brennan, R. P. 43
Broad, J. 9, 68, 101, 131, 180, 212, 352
Brunnelle, R. L. 330

C
Cantu, A. A. 330
Cartwright, M. 43
Chan, E. S. C. 372
Cheng, S. G. 68
Chisholm, J. J. Jr. 264
Colman, N. 9
Cooper, F. S. 157
Crone, J. C. 101
Crystal, D. 332

D
David, E. E. Jr. 157
Davies, G. 9
Delly, J. G. 332
Denes, P. B. 157
Diash, C. B. 43
Dudley, R. J. 9
Duffy, D. E. 43

E
Elmes 264

F
Faulds, H. 43
Fleming, S. 331
Forshufvud, S. 101
Fournier, J. M. 332
Freeman, S. K. 180
Frizzel, W. F. 43

G
Gilfillan, S. C. 264
Giovandi-Jukubczak, T. 101
Goldberg, B. 157
Goldwater, L. J. 263
Gordon, J. E. 132
Greenough, W. B. III 264

H
Hains, S. A. 43
Halliday, D. 43
Harper, M. 263
Hays, D. A. 330
Hebst, N. M. 332
Herschel, W. J. 43
Hirschhorn, N. 264
Hogben, L. 43

J
Johnson, R. I. 331
Joyce, C. 9

K
Kawakishi, S. 180
Kaye, B. H. 9, 43, 101, 131, 157, 263, 331, 332
Keenan, C. W. 330
Kleinfalter, D. C. 330
Kosler, W. G. 332
Krieg, N. R. 372

L
Laing, E. J. 68
Lew, J. S. 332
Liu, C. N. 332
LLoyd, J. B. F. 68
Locard, E. 101

M
Maddison, F. R. 101
Mandelbrot, B. B. 158
McCrone, L. B. 101
McCrone, W. C. 101, 331
Morimitsu, Y. 180
Morrison, I. K. 330

N
Neufeld, P. J. 9
Nickell, J. 331
Nickolls, L. C. 131,212,330
Noakes, G. R. 43

O
Oliver, B. M. 372
Orear, J. 43

P
Painter, J. D. 101
Palenik, S. 101, 332
Paterson, J. L. 9
Pelzar, M. J. 372
Penn, W. A. 43
Perkins, R. M. 101
Pickett, J. M. 157
Polsten, C. J. 9

Q
Quinn, D. B. 101

R
Resnick, R. 43

S
Scammell, H. 264, 332, 352
Scharfe, M. 330
Shaw, D. F. 43, 158
Shen, S. P. 68
Simpson 264
Smaldon, K. W. 9
Smale, D. 101
Smith, L. S. 330
Smolinkske, S. C. 263
Smyth, F. 43, 68, 131, 157
Song, H. O. 68
Spencer, F. 331
Spoerke, G. Jr. 263
Starkey, M. L. 212
Stevens, K. N. 157
Stover, E. 9
Sykes, B. 352
Szles, D. M. 158

T

Timbrell, J. A. 263
Tipler, P. A. 101, 330
Tucker, A. 263
Tyler, P. A. 43

U

Ubelaker, D. 264, 332, 352

V

Vienot, J. C. 332
Vredenbregt, J. 332
Vronman, L. 352

W

Wallis, H. 101
Walls, H. J. 101, 132, 212, 330
Waugh, J. B. 157
Wellinhofer, P. 331
Wheeler, J. P. 43
Willis, C. I. 331
Wilson, C. 9
Wood, J. H. 330

Y

Yuan, B. 68

Z

Zheng, D. C. 68
Zhu, Y. 68

Subject Index

A
Absinthe 230
absinthism 230
accelerant, detection of 166, 179
accelerometer 318
acinetobacter calciacatieus 30
acoustic spectrogram 137
adenine 357
adversarial system of justice 5
aerosol, viable 236
agaricus campestris 235
age
  determined from bone 343
  determined from teeth 344
aging forged items 309
AIDS 366
alcohol
  blood level 184
  detection of impurities in 189
  ethyl 183
Alice in Wonderland 236
alkaline 163
alkaloids 194, 207
alveoli 187
amalgam, dental 243
amanita 236
amber 299
amino acids 15, 23
ammunition, round of 105
amphibole asbestos 252
anabolic steroids 207
anachronism 314
anaerobic 232
anaerobic bacteria 232
Anastasia 365
anatase 100
Anderson, Anna 365
Ångström 113
animal feed, accidental contamination 249
anode 113
antibody 202
antigen 202
aphrodisiac 223
apiezon oil 173
apple pips, cyanide in 227
archaeological hazards 237
archaeopteryx 311
Archimedes 268
Archimedes principle 268
argon-ion laser 23
Aronson, J. 229, 263
arsenic 94
  as a pigment 95
  as an aphrodisiac 223
  content in bronze 225
  detection in hair 94
  effects of 94
  immunity to 222
  in copper alloys 225
  poisoning
    industrial hazard 224
    symptoms of 224
  release of by mold 95
  use of in flypaper 223
  use of to promote growth 222
  use of to reduce fever 222
arsenic trioxide 221
asbestos 251
  and smoking, synergism 259
  as industrial hazard 252
  chrysotile 251
  crocidolite 251
  fibres, cell damage 259
  link to cancer 360
  types of 252

athletic performance enhancing
drugs 207
atom, description of 30
atomic absorption (AA) 111, 113
author verification
by diffraction 320
by style analysis 323
autopsy 91
autoradiography 30
autoradiography, technique 31
autumn crocus 227

B

backscattered electron imaging
(BEI) 111
baking soda, odor absorption 163
ballistic pen 318
ballistic signature 318
ballistics 122
ballistics, first use 122
band shift 369
barbecue syndrome 231, 232
prevention of 232
barrister 6
beat frequency 148
beating speeders 148
bendectin 261
Bicycle Bandit 317
biological toxin 220
biopsy 92
Birmingham Six 121
Bisbing, R. E. 83, 86
bit mapping 154
bite marks, matching 341
bitter almonds 226
black box 150
black light 20
bladder cancer 254
blind comparison of DNA 368
blood
alcohol content 184
drinks per time 186
vs. content in breath 187
vs. urine alcohol content 186
determining the origin of 349
serum 349
stains
as evidence 348
on the Shroud of Turin 305
pattern of 350
bloodhound 161
robotic 171
blue asbestos 251
body armor 127 ff
efficiency of 128
body cooling, factors affecting 326
Bonaparte, Napoleon 94
bone
age of 338
determination of age from 343
boot polish, analysis of 55
bootlegger 189
borenol 211
botulism 232
braille markings on postage stamps
298
Brazil, mercury poisoning in 243
breath
alcohol analysis 187
alcohol content 187
breech markings 124
Brunelle and Cantu 277, 330
Bulger, J. 150
bullet
deformation on impact 128
high speed photograph 103, 108
speed of 107
bullet cytology 120
bullet hole, powder residue in 111
bulletproof jacket 127 ff

C

caffeine 209
calcite 66
calibre 124
camera, infrared, operational range
143
camera, surveillance 151

camphor 211
cancellation of stamps 298
cancer 360
  bladder 254
  genetic link to 255
  how asbestos may initiate 360
  lung, increasing incidence of 256
  nasal 254
cannabis 200
capillary action 279
capillary tubes 279
carbon dating 306
carbon monoxide 238
Cargille Laboratories 81, 82
Carol, Lewis, magic mushrooms 236
carrier gas 116
Cartesian geometry 224
cashmere, fraud involving 316
castor bean bush 216
cathode 113
cheque protection 290
chlorophyll 235
Christie, John 339
chromatogram 116
chromatography 115
  detecting explosives by 116
  gas, *see* gas chromatography
  liquid *see* liquid chromatography
  origin of 115
  thin layer, in ink analysis 277
chromosomes 356
chrysotile asbestos 251
cilia 163
cloning 357
cocaine, standard samples 204
coded messages 358
codeine 193
coherent light 22
colchincine 227
colloid science 29
color copying 284
comparison microscope 123
Conan Doyle, Sir Arthur 71, 301
contamination
  of DNA samples 368
  of test equipment 121
conviction, wrongful 325
cooking oil poisoning 221
cooling of a body, factors affecting 326
Cooper, J. A. 74, 101, 111, 131
counterfeit cheques from prison! 291
counterfeit currency, xerographic 287
counterfeiting
  by color copying 287
  prevention of 290
coup poudres 234
cracks, uniqueness of 58
cricket bat, Piltdown man 303
criminalistics 1
criminology 1
crocidolite asbestos 251
crystal glass 79
curare 216
currency
  American 289
  Australian 289
  British 289
  Canadian 288
  diffraction grating on 290
  variable color tag on 287
  watermark on 289
curse of the pharaoh 237
cuts, tears, in a fabric 87
cyanide 226
cytology 120
cytosine 357
Czar of Russia 365

D

data falsified 326
death, time of, estimation 335
degradation of DNA samples 368
denatured 221
dental offices, mercury use in 243

dental records, identification from 343
dental x-rays for identification 341
dentist, forensic 339
deoxyribonucleic acid (DNA) 357
depression hysteresis 50
designer clothing, forgery of 315
diastema 343
diatoms 338
diesel soot, bladder cancer 254
diffraction
  by a sieve 39
  by fingerprints 37
  grating 36
  of light 36
  on currency 290
  patterns 35
    of writing 318
diffusion of an odor 168
digitalis 229
dispersal pattern from a shotgun 125
distance from a target
  effect of 124 ff
  shotgun pattern 125
DNA
  blind comparison of 368
  increasing quantities 365
  mitochondrial 365
  probes 362
  profiling (fingerprinting) 360, 364
  samples
    contamination of 368
    degradation of 368
documents
  altered, detection of 271
  authenticity 320
dog *see* sniffer dog
doping, athletic 207
Doppler effect 147
Doppler gun 147
Doppler, C. J. 147
dose-response curve 217
Drake, Sir Francis 308
Dravnieks, Dr. A. 33, 161, 177, 180
Dreyfus case 320
dropsies 229
drug
  analysis
    by gas chromatography 194
    by liquid chromatography 194
  definition of 191
  identification
    by IR absorption spectroscopy 198, 200
    by x-ray diffraction 198
  identification of a dealer 194
  olfactronic signature of 191
  testing
    by gas chromatography 208
    by immunoassay 203
    standard samples 203
  use, evidence in hair 202
dry ink (xerography) 282
dust, establishing jurisdiction 8, 83
dynamite, detection of odor of 179

E
EDAX analysis of a Stradivarius 97
EDXRF 75
  analysis of glass fragments 82
  application of 111
effective dose, determination of 217
ejector mechanism marks 124
electronic surveillance 150
electrophoresis 362
electrostatic
  charge on a carpet 48
  deposition analysis (ESDA) 67, 121
  development of a footprint 48 ff
  fingerprint development 18
electrostatics 16
Elmes and Simpson 253, 259, 264
eluant 115

Emsley and Pallister 219, 221, 263
energy dispersive x-rays spectroscopy (EDAX) 97
energy dispersive x-ray fluorescence (EDXRF) 75, 82
entomologist, forensic 336
envelope, saliva on seal 274
epidemiology 224
eraser 63
  composition of 66
  effect of 63
ergot poisoning 205, 207
ESDA 67
ethyl alcohol 184
etymology 268
Euclidean geometry 152
Evans, T., style analysis 324
evidence
  presentation of 5
  uniqueness of 4
Excedrin, tampering with 226
excipient 200
execution, cost of 325
expert witness 260 ff
  acceptance of 7
explosives
  components of 107, 116, 118
  detection of 116, 173, 175, 179
  olfactronic signature of 179
  testing for contact with 121
extender oils, fluorescence of 56

F

fabric, cuts and tears in 87
facial reconstruction from a skull 347
Fairgrieve, Dr. S. 351
falsified data 326
Famous People Players 20
Faraday, Michael, mercury poisoning 241
Faulds, Dr. H. 13, 43
fiber identification 86, 90
filament, headlight, as evidence 73
fingerprint
  as a diffraction grating 36
  as evidence 34
  composition of 15
  development
    by electrostatics 18
    on oil paintings 30
    with gold sol 29
    with iodine vapor 28
    with metal vapor 26
  diffraction by 37
  electronic recognition 35
  features of 17, 35
  odor analysis of 33
  spatial filtering of 40
  Tolansky's method 26
fingerprinting
  DNA, method 360
  DNA, uses of 364
  first use of 14
  genetic 355
firing distance, effect of 124 ff
firing pin marks 124
flak jacket 127 ff
flame photometry 113
flies, life cycle of 337
fluidized bed 173
  odor collection 171
fluoram 23
fluorescent 20
fluorine, in teeth 340
fly agaric 236
flypaper, arsenic used in 223
food products, improperly labeled 316
footprint, identification
  by electrostatics 48 ff
  by infrared absorption 53
  by interference holography 50 ff
  by laser luminescence 53
  uniqueness of 55
forensic
  anthropologist 260
  linguistics 324

ondontologist 335
science, definition 1
forged artifacts, detection of 307
forged metal items, detection of 309
forgery
aging of forged items 309
errors in 295, 314
of art 312
of postage stamps 292, 296
of pottery, detection by x-ray 311
fossil flies 299
Fournier, J.M. 320, 332
foxglove 229
and Van Gogh 229
source of digitalis 229
tea, effects of 229
fractal dimension 152
fractal geometry 152
fractal transform 154
image compression 154
franking of stamps 298
Franklin expedition 249
Franklin, Sir John 249
fraudulent insurance claims 317
frequency spectrum 137
Frye procedure 260
Frye rule
overturn of 261
views on 261
fungi 92
fungicide 92

G
Galton, Sir F. 14
gamma rays 93
gas chromatograph 189
gas chromatograph/mass spectrometer (GC/MS) 191
gas chromatography (GC) 116, 173, 176
analysis of alcohol by 189
detecting impurities 189
drug analysis by 194
use in athletic testing 208
gasoline, olfactronic signature of 179
GC/MS testing of horses 211
genes 357
genetic fingerprinting (profiling) 355
genetic predisposition to cancer 255
genetic profiling 355
Germani, M. S. 111, 131
Gibson, D., dental assistant 243
glass
analysis by EDXRF 82
fragments, analysis of 78, 82, 83
insulation fibres 86
glazes, pottery, lead leaching through 244
glory lily 227
gold mining, use of mercury 243
gold sol 29
fingerprint development 29
gold wire, production of 310
Gordon, J. E. 128, 132
Grant, J., forgery detection 269
Griess test 121
concentration of caustic soda 122
false positive 122
method and results 121
Gross, H. 72
ground beef, barbecue syndrome 231
guanine 357
Guildford Four 120
gunshot residue 109
detection of 116, 119
time after firing 117
Gutenberg Bible, analysis of 282

H
Haciliar artifacts 307
hair
analysis by NAA 92

as evidence 92, 93
chemical analysis of 92
detection of arsenic 94
evidence of drug use in 202
NAA to show poisoning 222
structure of 91
half-life of an isotope 306
hamburger disease 231
hard x-rays 75
harmonics 137
of a wave 137
hashish 200
headlight filament as evidence 73
Hedges, R. 306, 331
height, determination of from a leg bone 343
hematologist 349
Henry System 14
Henry, Sir Edward 14
heroin
library of chromatograms for 198
olfactronic signature of 191 ff
Herschel, Sir W. 14, 43
Hertz, H. 148
high energy photons 93
high explosive 105
Hitler, Adolf 341
HIV 366
hologram, postage stamp 299
holography, interference 50
honey, botulism in young children 233
horse racing, monitoring of 210
horses, DNA matching 367
hysteresis 50

I

image
compression, fractal transform 154
enhancement 151
intensifier 146
recognition 154
immunity to arsenic 222
immunoassay 202
impressions of writing, detection 67
incandescence 21
incoherent light 22
infrared absorption spectroscopy
detection of impurities in alcohol 189
detection of tire prints 60
drug identification by 198, 200
infrared absorption, development of footprints by 53
infrared camera
lens material 146
operational wavelengths 143
infrared imaging 143
infrared radiation 20
absorption of 144
carbon dioxide 144
water vapor 144
infrared transmission spectra of paint films 77
ink analysis, forged documents 280
ink
historical use of 276
identification of 277
structure by chromatography 115
tagging for identification 282
inorganic compounds 239
inquisitorial system of justice 5
instant verification ink 276
insurance, fraudulent claims 317
interference fringes 36
interference holography 50
interference holography, development of footprints by 50
iodine vapor for fingerprint development 28
ion 175
iron oxide 77
isotopes 31, 92
iterated Systems, image recognition 156

J
Jack the Ripper 223
Jekyll and Hyde 217
jequirity plant 217
jibberisch 240
Jurassic Park 299
jurisdiction established by dust 8, 83
justice
  adversarial system 5
  inquisitorial system 5

K
Kennedy assassination 109
keratin 91
Kevlar 127
King's Lynn 156
knife, cuts made by 87

L
Laing, E.J. 53, 68
lameness in metal workers 225
lands 105
laser 21
laser luminescence
  development of 18, 23
  development of footprints by 53
  examples of 25
  sources of 23
latent fingerprints 15
leaching of lead in pottery 245
lead
  as color in pottery 244, 245
  as sweetener 245
  detection in hair 93
  industrial hazard 246
  link to sterility 245
  paint pigment 247
  pipes, water contamination by 247
  poisoning from canned food 250
  solder used in canning 250
lead oxide 77
leak detection, pipeline 169
lens, infrared 146
lethal dose
  determination of 217
  of various poisons 220
lexicographer 7
light filament as evidence 73
light
  black 20
  refraction of 78
  wave nature of 36
  wavelength of 19
liquid air 82
liquid chromatography (LC) 114
Locard's principle 71
Locard, Dr. E. 71, 101
logarithmic scale 217
long-lived isotopes 92
Loveland and Williams 194, 202, 212
low explosive 105
low light imaging 143
LSD 205
  olfactronic signature of 191
luminescence 21
lung cancer, increasing incidence of 256

M
mad hatters 240
maggots 337
Maguire Seven 121
maple syrup, botulism in young children 233
marijuana
  olfactronic signature of 191
  standard samples 204
Markov, G. 215
Marymount, M. 222
masking agents 209
masking odor 170
mass spectrometer 175
mass spectrometry of wines 317
Maybrick, J. 223
McCrone Associates 82, 101, 84, 86, 98

McCrone, W. 305, 331
medulla 91
Mengele, J. 341
Mercurochrome 53
mercury
  in dental offices 243
  in paint 242
  methyl 239
  poisoning 239
    in Brazil 243
    symptoms of 240
  pollution 239
  use in gold mining 243
  vapor in the laboratory 241
messages, coded 358
metal
  artifacts, detecting forgeries 309
  vapor fingerprint development 26
  workers, lameness in 225
methyl alcohol 184
methyl mercury 239
Ménière's disease 229
micrometer 19
micron 20
Minamata 239
mistletoe 217
mitochondria 365
mitochondrial DNA 365
mold, release of arsenic by 95
molecular biology 357
monochromatic light 22
moonshine 184
morphine 193
Mozart
  cause of death 346
  identification of remains 344
multichromatic light 22
mushrooms 235
Mussolini, Benito, diaries of 269
mutation of a virus 367
mycology 207
mycotoxins 207

N
NAA
  analysis of hair 92
  evidence of poisoning 222
  of Drake plaque 309
  uses of 93
nanometer 19
Narcisse, C. 234
nasal cancer 254
nemesis 328
neural network 154
  applications of 154
  facial recognition 154
neutron activation analysis (NAA) 92
Newton Sir Isaac, mercury poisoning 241
Nicholl, S., Excedrin tampering 226
Nickols, L.C. 183, 212
night vision 143
Ninhydrin test 28
nitrocellulose 29
non-carbon copies 274
North-West passage, search for 249
nosocomiasl disease 236
nylon 127

O
odor
  absorption by baking soda 163
  accelerant, detection of 179
  analysis of a fingerprint 33
  collection, methods of 171, 173
  diffusion of 167
  masking 170
  movement of 167
  persistence of 165
  sensation of 163
  signature 176
oil painting, fingerprints on 30
olfactronic signature 176
  of drugs 191
  of explosives 179

of gasoline 179
variations in 178
ondontologist, forensic 339
opium 192
optical computing 39
optical spatial filtering 40
optical variable device (OVD) 290
optically variable ink (stamp) 297
organic compounds 239
osteons 339
effect of age 343
OVD *see* optically variable device 290

P
paint
age from pigment size 315
chip
analysis by EDXRF 76
analysis by pyrolysis 74
as evidence 72
films
analysis by infrared transmission spectra 77
mercury as a slimicide 242
pigment
lead compounds as 247
common 77
x-ray diffraction 315
structure of 73
palaeography 320
Palenik, S. 71, 101, 82, 85, 86, 317
paper
collection of odors from 171
content 270
manufacture of 270
micrographs of 172, 272
opacity of 80
recycled 271
surface of 271
top/bottom differences 271
watermark 270
paraffin test 109
parallel processing 42
Particle Atlas 82, 84
paternity, establishing 364
pathologist 325
patina, production of artificial 310
PCR *see* polymerase chain reaction
peach pits, cyanide in 227
peer review 260
penetration pressure 128
Penn and Duffy 37
perforation, postage stamp 294, 296
perylene 63
phonemes 141
phosphorescence 21
phosphors 21
photo radar 149
photo recognition 154
photographic fraud 315
photographic matching to a skull 344, 346
photon 93
photon activation analysis (PAA) 92
picture enhancement 151
pigments, common 77
Piltdown man 301
plagiarism 329
plant debris as evidence 85
plants, DNA matching 368
plaster of Paris 48
plumbing, lead, contamination of water 247
poison identification 219
Polarproof 316
pollen from the Shroud of Turin 305
polymerase chain reaction 365
pore patterns 25
postage stamp
Braille markings 298
cancellation/franking 298
detection of forgery 294
forgery 292, 295
holograms on 299

identification of rare 295
origin of 292
perforations 294, 296
preventing forgery 296
value of 292, 295
washing 298
potassium cyanide 226
pottery
detection of forgery by x-ray 311
forged, Mexican 308
pozzolana 97
prism 19
Priss, J. 4
probability of innocence 370
probability scale 217
probenecid 209
prosecutors fallacy 371
proton induced x-ray emission (PIXE) 282
pseudonym 346
pseudoephedrine 209
psychemedics 204
pumice 66
pyrolysis-gas chromatography 74
pyrolysis, analysis of a paint chip 74

Q

quartz 79

R

radar 147
photo 149
speed detection 147
radio immune assay (RIA) 203
radio-carbon dating 306
of a bone 338
principles of 306
radioactivity 31
radiochemistry 30
Rasberry, S. 203, 212
recording tachometer 150
recycled paper, problems with 271
red herring 170
refraction of light 78
refractive index 78
determination of 80 ff
effect of 78 ff, 106
residue, gunshot 109
restriction enzymes 360
revolver 124
ricin 216, 219
rifle 105
rifling 105
rigor mortis 336
rock salt 20
rolled fingerprints 15
round of ammunition 105, 124
roxarsone 222
rubber debris, identification 63
Russia, Czar of 365

S

saccharin 234
Salem, Witches of 205
sawdust, link to nasal cancer 254
scanning electron microscope (SEM) 73
Scheeles' green 95
scientific fraud 326
Scoose, F. 121
Scotland Yard 8
scratches, uniqueness of 58
seasnake 360
sebaceous glands 15
secondary electron imaging (SEI) 111
serologist 349
serpentine asbestos 252
serum, blood 349
sex, determined from bones 343
Shakespeare, William 346
Shirley Institute 48
shock wave, speed of sound 106
short-life radioactive material 92
shotgun, dispersal pattern of 125

Shroud of Turin 303
 blood stains on 305
 description of 303
 history of 305
 pollen from 305
 radio-carbon dating 306
sieve
 damaged apertures 39
 diffraction by 39
 odor 176
 olfactronic 176, 178
 structure of 318
 verification 318, 320
silica 79
silicon dioxide 79
skull
 photographic matching 344, 346
 reconstruction of face 347
sleuth 161
slimicide, mercury 242
Sloan, U. S. vs., ink analysis 280
smoking
 asbestos, synergism 259
 link to lung cancer 256
Snow, S., Excedrin, tampering victim 226
sniffer dogs 169
 fatigue 170
 for detecting explosives 169
 robotic 171
soft x-rays 75
solder, lead, in canning 250
solicitor 6
sound waves 135, 137
spasm signature 318
spatial filtering 40
spectrogram, speech 135, 138
spectrometer, mass 175
spectrum 19
speech
 recognition 141
 spectrogram 135
 template 141
spider silk
 collection of 130
 properties of 131
Spoerke and Smolinkske 228, 263
stamp, postage see postage stamp
standard samples for drug testing 203
statistical presentation of evidence 370
sterility due to lead exposure 245
steroids, anabolic 207, 208
Stevenson, R.L. 217
stochastic wear signatures 56
Stradivari 96
Stradivarius, EDAX analysis of 97
striations 58
 on a bullet 123
 on a bullet casing 124
style analysis, author verification by 323
subdural hematoma 346
super poisons 212, 216
surveillance
 acceptance of 156
 effect of 156
 electronic 150
 operator fatigue 157
Swinehart and Gore 200, 212
synchronous fluorescence 63
synergism 258
syphilis 228

T

tachometer, recording 150
tag, variable color, on currency 287
TAGA 175
teeth
 determining age from 344
 fluorine in 340
 gap (diastema) 343
 identification of a victim 339
 lifestyle indication 341
 matching bite marks 341

Teflon 128, 129
Tekscent, leak detection using 169
templates, speech 141
teratogenic chemical 234
teratologist 234
tetrodotoxin 233
thalidomide 235
The Old Bailey 292
therapeutic dosage window 216
thermoluminescence testing 307
thin layer paper chromatography (TLPC) 115
thin layer chromatography in ink analysis 277
thymine 357
Timbrell, J. A. 215, 234, 263
time of death, estimation 335
tire prints
  identification by fluorescence 56
  detection by infrared light 60
  enhancement of 60
tire, wear of 58
titanium dioxide 66, 77
toadstool 235
Tolansky, fingerprint development 26
toner 67, 282
toxic dose, determination of 217
toxicology 215
toxin 215
trace atmospheric gas analyzer (TAGA) 175
transmission electron micrographs (TEM) 73
transmutation 31
tribology 18
truffle 236
trimethoxy arsenic 95
Tylenol, tampering with 226

U
Ubelaker and Scammell 325, 332, 336, 347, 352
Ubelaker, Dr. Douglas 260, 264
ultraviolet radiation 20
ultraviolet, tire print fluorescence 56, 61
uniqueness
  of breakage patterns 58
  of evidence 4
  of footprints 55
  of scratches 58
  of wear patterns 58
urea 15
urine alcohol content 186

V
Van Gogh 228
  absinthism 230
  digitalis poisoning 229
  link to Foxglove 229
  venereal disease 228
Van Meegeren, fake Vermeers 313
venereal disease 228
verotoxigenic coli 231
viable aerosol 236
Vinland map 98
Visuprint System 28
voice
  analysis, effect of speaker 138
  lineup 142
volatile 163
Voodoo 234
VTEC 231
  elimination of 232
  factors affecting 231
  source and spread of 231
  symptoms of 232
Vucetich System 14
Vucetich, J. 14

W
wadding, forensic value of 124
wallpaper, pigment used in 95, 96
Walls, H. J. 125, 132, 183, 187, 212, 274, 330
washing of postage stamps 298
water contamination by lead pipes 247

watermarks
  on currency 289
  on paper 270
Waters Associates 116, 131, 194, 212
wave nature of light 36
wavelength of light 19
wear patterns, uniqueness of 58
whistle blowing 327
white arsenic 221
white asbestos 251
whiting 66
Williams, W., evidence used in case 86
Witches of Salem 205
witness, expert 260 ff
word frequency, author verification by 323
World Trade Center 274
writing impressions, detection 67
wrongful conviction 325

X

x-ray diffraction
  identification of drugs by 198
  of paint pigments 315
x-ray fluorescence 76
x-rays
  dental, for identification 341
  detection of forgery 311
xerography 282
  and counterfeiting 287
  color 284

Ż

zombie 233